Multiphysics in Porous Materials

Zhen (Leo) Liu

Multiphysics in Porous Materials

Springer

Zhen (Leo) Liu
Michigan Technological University
Houghton, MI, USA

ISBN 978-3-030-06573-7 ISBN 978-3-319-93028-2 (eBook)
https://doi.org/10.1007/978-3-319-93028-2

Softcover re-print of the Hardcover 1st edition 2018

Printed on acid-free paper

This Springer imprint is published by the registered company Springer International Publishing AG part of Springer Nature.
The registered company address is: Gewerbestrasse 11, 6330 Cham, Switzerland

I am dedicating this book to my wife, Ye (Sarah), and my son, Brent. Their understanding and support enabled this work

The support from the United States National Science Foundation (NSF Grants 1562522 and 0700524) is acknowledged. The support helped me step in and stay active in multiphysics and research on porous materials

Preface

This book started from a course that was first offered in the spring of 2015 at Michigan Technological University. This course was possibly the first one on multiphysics in porous materials worldwide and even likely among the first in the general area of multiphysics. Back at that time, I could not find an appropriate textbook for multiphysics, let alone one with a focus on porous materials. Therefore, I had to develop all the lecture notes and computer labs from scratch. This book was created based on this educational endeavor. Many of the book materials have been tested multiple times via both teaching and learning in that course. In addition, the book has been enriched with materials from several completed and active research projects including cutting-edge studies supported by the US National Science Foundation. In short, the book represents a very implementable and practical learning guide for multiphysics and a path from the basics of multiphysics to advanced research topics.

This book is intended to be a general reference for multiphysics. By now, more than a dozen of books with a title including "multiphysics" can be found. However, they are mostly conference proceedings, introductions to the development of specific multiphysics applications, or summaries on multiphysics in specific areas. A relevant fact is that multiphysics researchers usually tend to focus on the multiphysics applications in their own areas, no matter in the general multiphysics context or in porous materials. There is no effective reference for people to systematically learn about multiphysics, especially to grab a big picture and the basics. This book is presented to fill this gap.

This book is also designed to be a comprehensive introduction to the area of multiphysics in porous materials. Porous materials are an ideal breeding ground of multiphysics due to their multiphase and multicomponent nature. As a result, porous materials are related to most common types of multiphysical phenomena. For example, the coexistence of thermal, hydrological, and mechanical processes is a major interest in geotechnical engineering, hydrogeology, soil science, and petroleum engineering. Particle transport and chemical reactions together with heat transfer and water movement have been extensively studied in environmental engineering, agricultural engineering, and forestry. Electromagnetics, vibration, and acoustics are the primary interest of geophysicists, earthquake engineers, mechanical engineers, and exploratory geologists. Therefore, this book will be a handy tool to researchers and practitioners in these areas involving porous materials, no matter they are new to multiphysics or they want to step into a multiphysics topic out of their home disciplines.

This book is organized in a way to serve readers with different backgrounds and with different levels of knowledge accumulation in multiphysics. For the purpose, this book provides materials that would be essential to obtain a general understanding of the concepts, theories, and implementations of multiphysics. It is believed that multiphysics is an "art" of mathematics, physics, applications, and numerical analysis. Accordingly, the book is organized into five parts: introduction, mathematics, monolithic physics, multiphysics, and numerical analysis. In Part I, a general introduction to the history, state, essence of multiphysics, and its applications in porous materials is offered. The definition, scope, and classification of multiphysics will be given based on the discussions on the existing studies. In Part II, necessary mathematical backgrounds such as tensor and fields, tensor analysis, partial differential equations (PDE), and numerical solution of PDEs are provided. Eight most representative monolithic physical fields are introduced in Part III, each of which is presented as a chapter. Then in Part IV, typical multiphysical processes are introduced. The introduction is made based on the mathematics in Part II and the monolithic physics in Part III. Each multiphysical topic is introduced in the order of background, theory, implementation, and applications. The theory will be discussed in a way that mathematics and physics are integrated. Implementations are demonstrated with a simple but representative example, which is solved using a PDE solver. In Part V, a brief tour will be given to show how to implement a multiphysics model with three typical numerical simulation methods, i.e., finite difference method, finite volume method, and finite element method, using self-developed computer code.

The use of physical fields extends over most science and engineering disciplines, while a discipline or sub-discipline usually primarily deals with one or a few physical fields. This fact makes multiphysics an extremely interdisciplinary topic and poses challenges in enumerating the fields. The eight most common monolithic physical fields introduced in this book are heat transfer (thermo-), pore water movement (hydro-), concentration field (concentro or diffuso/convecto/advecto), stress and strain analysis (mechano-), dynamics (dyno-), chemical reactions (chemo- or chemico-), electrostatics (electro-), and magnetostatics (magneto-). For multiphysics, the following nine most representative topics in porous materials will be discussed: thermomechanics, hydromechanics (static poroelasticity), thermohydromechanics, electrokinetics, electromagnetics, fluid dynamics, hydrodynomechanics (e.g., fluid-structure interaction), thermoelectromagnetics (including Joule heating, dielectric heating, and induction heating), and electromagnetomechanics (e.g., piezoelectricity).

It is my hope that this book can significantly flatten and shorten the learning curve of multiphysics. Usually, it takes many years to pick up the basic concepts, preparation knowledge, and software skills before we can excel in or even get a big picture of multiphysics. Such a multiyear knowledge preparation includes math courses such as partial differential equations and tensor analysis; specialty courses such as mechanics, heat transfer, and electromagnetics; and numerical analysis courses such as finite difference method, finite volume method, and finite element method. As a result, the learning curve for multiphysics is both steep and long. Commercial software such as COMSOL made a great effort at lowering the bar for implementing multiphysics and expediting the process for research and design with multiphysics. However, more advanced use especially that requires an in-depth understanding of multiphysics and bottom-level theoretical and numerical operations still calls for a framework and summary of multiphysics. Based on my 10 years of multiphysics implementations and even more time in knowledge preparation, I want to share the opinion that not all the knowledge in the above courses/subjects are equally needed for understanding and practicing multiphysics. Moreover, it is possible to organize the most relevant and significant knowledge into a book by which people can obtain a smooth and pleasant ride to the kingdom of multiphysics especially for the applications in porous materials. The book is presented and will be continuously improved to show and prove this opinion.

A continuously updated website, multiphysics.us, is available to complement this book. This website provides updated technical details, applications from communities, and networking information for multiphysics and its applications in porous materials.

Houghton, MI, USA Zhen (Leo) Liu

Contents

Part II Mathematical Background

Part III Monolithic Physics

Part I
Introduction

Chapter 1
History and Future

Introduction

The chapter starts the introduction with the ubiquitousness of multiphysics. Examples are given to demonstrate the wide existence of multiphysical phenomena in both naturally occurring processes and man-made applications. These examples traced back to ancient activities of human beings, showing that multiphysics and their applications are neither new nor rare. We will also show that the understanding of such phenomena via computers, mostly in terms of multiphysics simulation, has also been around for decades. Brief introductions will also be made to reveal the relationships between multiphysics and the trends of interdisciplinary research, industrial collaboration, computer engineering, software engineering, rapidly increasing complexity of new scientific problems, and engineering design and policy-making. These will help readers understand the clear uprising trend of multiphysics and foresee its future developments.

Multiphysics Is There

We live in a multiphysics world. Real-life processes are inherently multiphysics. From atoms to galaxies and from amino acids to living organisms, naturally built systems involve various types of physical phenomena operating at different spatial and temporal scales. Loads and deformations on solids, complex flows, fluid-structure interactions, plasma and chemical processes, and thermomechanical and electromagnetic systems are just a few representative examples in fundamental and applied sciences (Krzhizhanovskaya 2007). Take a cell phone in Fig. 1.1, for example – the antenna receives electromagnetic waves, the touchscreen or buttons handle the interaction between mechanical and electric components, and the battery involves chemical reactions and electric currents [Multiphysics Cyclopedia].

Z. Liu, *Multiphysics in Porous Materials*,
https://doi.org/10.1007/978-3-319-93028-2_1

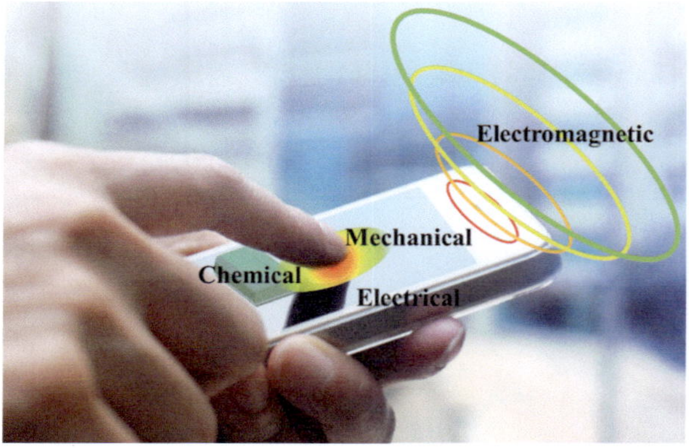

Fig. 1.1 An example of multiphysics

Therefore, multiphysics is neither a research concept far from daily life nor a theory or technique developed recently.

We have been dealing with this multiphysics world even before the civilization of human beings. For example, a nature fire from lightning is a typical electrostatic, thermal, and chemical phenomenon. Fire starting with a hand drill involves both mechanical and thermal processes. Flood protection and irrigation required to understand the interaction between hydrodynamics and groundwater movement. Construction and earthwork called for the understanding of a coupled mechanical and hydrologic process. The modern industrialization even stemmed from multiphysical processes such as steam engines, which is an art of thermal, mechanical, and dynamics. What would be more surprising is not that the modern lives are full of multiphysics, but instead, we have been taking advantage of most of these multiphysical processes in such an adept way.

Surprisingly, while multiphysics is part of nature and history of lives, the study of multiphysics is not new either. We can describe what happens in the world using sets of physical laws. Originally, due to the limited understanding of complex physical processes and computing resources, physical effects were observed and described in isolation. For example, traditional courses in engineering disciplines were usually built upon the understanding of one type of physical process. For example, various courses in civil engineering were established on the basis of mechanics while many in electrical engineering were on electrostatics. However, there are still many multiphysical phenomena, which are very fundamental and "multiphysics," such as electromagnetics and elastodynamics. These multiphysics topics were studied and matured much earlier than other multiphysical phenomena, even before the advent and prevalence of the word "multiphysics." Since the 1940s, we have been using computers to understand these physical phenomena. Thus, regardless of the trademark "multiphysics," simulations that incorporate multiple physical phenomena are as old as simulations themselves (Keyes et al. 2013).

Despite the studies which are by nature multiphysics but were not traditionally counted, multiphysics research has been on the playground for a couple of decades. On one hand, many physical problems of interest these days are complex in nature and consist of separate physical processes that each contributes to the problem as a whole. Also, some problems which were historically investigated as an isolated physical process are now needed to be considered as a holistic process. This fact requires researchers to consider multiple physical processes when constructing numerical models (Groen et al. 2014). On the other hand, the uprising computing power in virtue of the innovations in computer engineering and science has been lending us the power to deal with more complicated problems, which may be insurmountable in the old days due to complicated couplings between fields. The higher speed of computing unit, higher storage, and especially the popularization of computer clusters enable the consideration of the whole process in a holistic way, giving rise to various multiphysics techniques. Multiphysics studies under the name of multiphysics have been conducted and widespread at least for two decades.

In fact, multiphysics has been developed beyond a research concept. For example, in industrial design and product testing, engineers used to solve one physical phenomenon at a time, such as structural integrity, and then import the results into another system to solve for the aerodynamic behavior. Multiphysics simulation tools now allow them to simulate and analyze both these behaviors and many others at the same time. Similar to those physical processes discussed in research papers, in reality, product performance usually depends on several physical phenomena interacting with each other: multiphysics. To acquire dependable and realistic simulation results that allow better design decisions, coupled simulations with many physical effects involved become necessary and essential. Also, in light of the advance and promotion of multiphysics computer-aided design tools such as COMSOL, multiphysics has become an accepted consensus, trend, and even standard practice in many industries. With a multiphysics capable simulation tool, designers and engineers now can better capture the important aspects of the behavior of the product.

Multiphysics Is Coming

Multiphysics has rapidly developed into a research and application area across many science and engineering disciplines. There is a clear trend that more and more challenging problems we are faced with do not involve physical processes covered by a single traditional discipline in science or engineering disciplines. This trend requires us to extend our analysis capacity to solve more complicated and more multidisciplinary problems, advance the front edge of engineering and applied sciences by exploring these problems, contribute back to the software development and computing techniques, and establish and explore innovative but urgent application areas.

Modern academic communities are confronted with problems of rapidly increasing complexity, which straddle across the traditional disciplinary boundaries between physics, chemistry, material science, and biology. To respond to this challenge, many disciplines have spontaneously initiated efforts at reaching out to other disciplines to target at emerging issues which appear as a traditional topic in those disciplines while also extend beyond the scope. At the same time, computational science, which is more focused on techniques rather than problems, has been receiving a steadfast development. Innovative modeling techniques have been investigated for the purpose of offering an optimal handling of the information transfer which connects different aspects, components, scales, and levels involved in the quantitative description of those multidisciplinary phenomena. This trend is entailing the seamless coupling between different mathematical representations of distinct physical natures across a wide spectrum of scales from geologic scales and scale of lab specimen to atomistic trajectories.

Multiphysics has also become a frontier in industrial applications. Simulation programs have been evolving into a tool familiar to many engineers in design, product development, and quality control. This change is reshaping traditional creation processes: from design blueprints to prototypes to testing to validation to production, which was once simple, to extremely complex ones involving engineers from several disciplines. During these creation processes, engineers now must think in areas outside of their training, even with the assistance of the simulation tools. As commented by Courtney (2015), it is more and more necessary for the modern engineers to know and grasp the concept of what is known deep inside the engineering world as "multiphysics." Take automotive industry, for example; traditionally, there are different people focusing on the structures, fluids, electromagnets, and the other individual aspects separately. Now, the intersection of aspects, which may represent two physics topics and once was a gray area, can be the essential link in the life cycle of the product. Such needs will enable multiphysics applications to attract a much wider engineering user base and allow more designers to come together. As commented by Nilsson (Thilmany 2010), "Design engineers are running more and more multiphysics simulations every day because they need to add reality into their models."

Multiphysics will continue to benefit from the progress in computer engineering and science. The advent of commodity computing has been leading to profound changes in numerical analysis of continuum mechanics problems. In the meanwhile, the computing power, in terms of both the maximum computing flop rate and energy efficiency, has been increasing exponentially as shown in Fig. 1.2. Besides, teraflop rate computing facilities, which were available only to a few until recently, are becoming commonplace in an industrial setting (Jasak 2006). This just reflects the fact that the range and complexity of problems in numerical analysis are expanding, leading to increasing interest in coupled problems which was computationally prohibitive. Besides, spectacular advances in computer performance and emerging technologies of parallel distributed grid computing are offering tools to break the barriers and bring simulation to a higher level of detail and accuracy. In the meanwhile, the advances in hardware and computing techniques also call for new efficient numerical algorithms specific to the field where coupling different models or scales within one simulation is essential (Krzhizhanovskaya 2007).

The developments in multiphysics have been in turn contributing to the software engineering. The need and purpose of multiphysics software are to bring different strands of engineering under one umbrella. However, this is not an easy and straightforward task as it sounds. This is because research on complicated multiphysical phenomena or computer-aided design can include issues as diverse

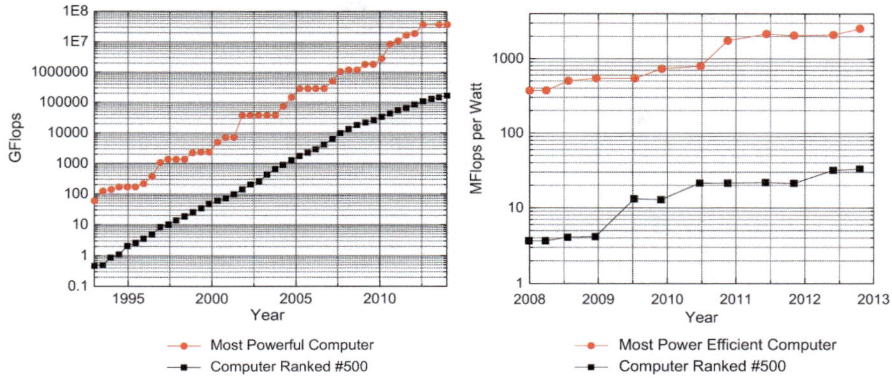

Fig. 1.2 Development of computing (**a**) power and (**b**) efficiency. (After top500.org)

as predictions of structural strength, electromagnetic behavior, and fluid flow. Fortunately, the needs of the academic and industrial communities have generated considerable momentum in the software market to transfer or extend from traditional software packages with separate physics modules to tools with multiphysics genes which can promote better communication and interaction during the project life cycle. A general rule is that the more multiphysics phenomena, the more accurate the final model will be, though the cost of building and solving more complicated model usually discourage such effort. This would lead to a compromise between accuracy and effort. Despite the progress, multiphysics is still a field in need of integration (Thilmany 2010). An easy way is to "stitch" different packages together using file imports, which is still common today, and continuous effort is being made to render it smoother. More recent techniques aim at building up all the applications on the same platform all the way from the bottom.

Multiphysics is not only reformatting traditional applications but also giving rise to new applications in scientific prediction, engineering design, and policy-making. In the automotive industry, multiphysics can be employed for traditional testing such as forces of rotating wheels, environmental stresses, and shock and vibration, under the conditions of different geometries, materials, and load scenarios. In the meanwhile, new application areas are under development in virtue of multiphysics. For example, multiphysics is a significant impetus for the research and application of cyber-physical systems, which combine electronic control with physical problems. It is challenging but essential to understand the multi-domain problems underlying these systems, particularly for those humanoid robots equipped with multiple systems with more complex electrical and mechanical elements. For example, engineer Joel Gibbard's Open Hand Project (Web 2013) explored the idea of how an amputee can create his or her prosthetic hand using a 3D printer. This project involves the design using National Instruments LabVIEW software and data acquisition hardware and the measurement of the electrical activity of muscles at rest and during contraction to understand how the nerves send electrical signals.

Chapter 2
What Is Multiphysics

Introduction

As introduced, multiphysics has been studied since the beginning of numerical simulation and at least two decades under the name of "multiphysics." However, there has been no consensus on the definition of multiphysics. First of all, this word might be a little misleading or controversial if used in areas such as physics or in a context when the concept is not well known. Second, even in numerical simulation, the word could still refer to significantly different things.

In this chapter, the possible meanings of multiphysics will be discussed based on four definitions that can be found in the literature. Based on that, a clear definition of multiphysics will be reached to lay down the scope of this book. With this definition, more details will be shown about how to move from monolithic physics to multiphysics. The purpose is to bridge this common knowledge gap to help some readers who are familiar with traditional monolithic physics simulations move toward multiphysics. Detailed discussions will be made regarding the couplings and methods for developing multiphysics models. Finally, common types of multiphysics, as well as the essence and scope of "(physical) fields," will be discussed to outline a big picture for multiphysics. This chapter aims to present the first comprehensive discussion on "what is multiphysics" to help readers understand and compromise the widely existing discrepancies in the definition, scope, and use of terminology in multiphysics.

Definitions and Scopes

Definitions of multiphysics can be found in some literature which can be categorized according to the broadness of the scope. The typical definitions are listed and explained as follows in the sequence of scope broadness, i.e., from broad to narrow.

Z. Liu, *Multiphysics in Porous Materials*,
https://doi.org/10.1007/978-3-319-93028-2_2

Definition 1

Multiphysics treats simulations that involve multiple physical models or multiple simultaneous physical phenomena [Wikipedia]. While the use of "physical phenomena" semantically rephrases "multiphysics," the inclusion of "multiple physical models" makes this definition a very broad and general concept. For example, combining chemical kinetics and fluid mechanics is a multiphysical process according to "physical phenomena" in the definition. Likewise, combining finite elements with molecular dynamics is then also a type of multiphysics as it includes "multiple physical models." This definition is a little bit self-contradictory as the implication of physical models may include that of physical phenomena. In fact, the definition of physical models is possibly more appropriate for multimodel. The multimodel concept may include multiscale, multirate, multilevel, and multilevel problems (Keyes et al. 2013). In this definition, multiphysics typically involves the solution of coupled systems of partial differential equations.

Definition 2

In a relatively narrow sense, multiphysics includes (1) coupled physical phenomena in computer simulation and (2) the study of multiple interacting physical properties [COMSOL]. This definition narrows the scope of Definition 1 to physical phenomena and the materials associated with them. Thus a coupling between either physical phenomena of different types or materials properties associated with these different physical phenomena is an essential characteristic. The use of physical phenomena implies that multiple physical processes, each of which should be a distinct physical field, such as thermal field, can be described by a partial differential equation (s) based on underlying physical mechanisms such as conduction. The use of "physical properties" in the definition makes it more closely related to applications because in many phenomena of practical interest, such as piezo resistance, these physical properties instead of the coupled process are the major interest. However, it is worthwhile to mention that the physical properties should be more than two properties and the properties are associated with different fields. Otherwise, density changes due to variations in water content in porous materials can be accounted as a multiphysical phenomenon.

Definition 3

Semantically, a multiphysics system consists of more than one component governed by its own principle(s) for evolution or equilibrium, typically conservation or constitutive laws (Krzhizhanovskaya 2007; Groen et al. 2014). This definition is very close to Definition 3 except for that it does not emphasize physical properties.

Definition 4

In a more strict way, multiphysics can be defined as those processes including closely coupled interactions among separate continuum physics phenomena (Cross & Slone 2005). In this definition, two-way exchange of information between physical fields, which could involve implicit convergence within a time step, is the essential feature. Accordingly, those processes only involve using data generated by one code as input into another is not qualified as multiphysics but, instead, as multidisciplinary.

Definition in This Book

The book defines multiphysics in the following way. First, the above definitions all point out several features and leave out some others. Among these features, multiple physical processes which are governed by their own laws of evolution or equilibrium are necessary. It seems that this is a feature directly reflecting the name and characteristics of most multiphysical processes. Second, the multimodel definition is abandoned to differentiate this topic from multiscale and multilevel problems. Third, couplings are not adopted in the definition as those are more an implementation technique than a fundamental characteristic.

Besides, a few neglected details need to be further discussed. First, "physics" in "multiphysics" means a physical field, which is a physical state variable varying with respect to space and/or time according to physical laws for its evolution or equilibrium. Thus, a field is either a time-dependent spatial distribution of a state variable in a transient problem or a spatial distribution of a state variable in an equilibrium problem. The word physical process/phenomenon is not that clear and closely related to mathematics. Second, the simultaneous occurrence of these physical fields seems to be more a rule than an exception. Though sequential occurrence is also possible, it is far less common and leads to the direct coupling of physical processes involved. Therefore, simultaneous occurrence will be emphasized to leave out this possibility. Third, all the definitions implicitly imply a deterministic process where stochastic factors do not seem to be present. However, if "physical field" is used, then it will be unnecessary to emphasize this feature. Fourth, the above definitions were made from the perspective of processes, phenomena, or system. All of these make multiphysics equal to the problem under investigation. But in a more common context, we also frequently use multiphysics to refer to the studies of and/or knowledge pool for these problems. Finally, a consensus on the above definitions is that multiphysics is about computer simulations. However, the implementation of multiphysics is not necessary via computer simulations and is also relatively independent of the methods of discretization and solution. Finally, it is believed that a major pool of knowledge is about the establishment of the mathematical model based on physical laws, which will also be the focus of this book. Thus, computer simulations will not be emphasized in the definition.

Based on the above discussion, in this book, multiphysics is defined as the coupled processes or systems involving more than one simultaneously occurring physical fields and also the studies of and knowledge about these processes and systems. For example, objects moving due to Newton's second law will be viewed as an example of multiphysics as long as velocity will be calculated at every point of the domain, that is, to obtain a field. However, a model of a star cluster that resolves Newtonian gravitational interactions or molecular dynamics, which is built upon Newton's second law, is not multiphysics according to this definition. From a mathematical perspective, systems consisting of partial differential equations (PDEs) of different types (e.g., elliptic-parabolic, elliptic-hyperbolic, or parabolic-hyperbolic) may be classified as multiphysics because each of the classical PDE archetypes possibly represents a different physical phenomenon (Keyes et al. 2013).

Multiphysics problems vary widely in their nature, both in the physics involved and in the manner of coupling. To further clarify the scope, we will first classify multiphysics into two major categories based on the way in which different physical processes are coupled. In the first category, the coupling occurs in the bulk through source terms or constitutive relations that are active in the overlapping domains of the individual components. In the second category, the coupling occurs over an idealized interface or a narrow buffer zone through boundary conditions that transmit fluxes, pressures, or displacements (Dennis et al. 2015). Typical examples of bulk-coupled multiphysics systems are radiation with hydrodynamics in astrophysics (radiation hydrodynamics), electricity and magnetism with hydrodynamics in plasma physics (magnetohydrodynamics), and chemical reaction with transport in combustion or subsurface flows (reactive transport). Typical examples of interface-coupled multiphysics systems are ocean-atmosphere dynamics in geophysics, fluid-structure dynamics in aeroelasticity, and core-edge coupling in tokamaks.

Multiscale is frequently mentioned with multiphysics. This is possibly because multiscale and multiphysics are different sides of a physical process. Multiphysics, which lies in fundamental physical laws, comes into being as a result of interactions originating from atomic scales and below and upscaling level by level to the scale of interest. Here we use the term multiscale modeling to refer to both multiscale modeling and multiscale simulation of physical problems, and the term multiscale application to refer to the program used to do the modeling. Groen et al. (2014) differentiated multiphysics and multiscale using the concept of submodel. Multiscale and multiphysics are therefore two distinct concepts, but they are common in that both of them consist of a number of submodels which are combined or coupled. Therefore, both multiphysics and multiscale face the same challenge of coupling these submodels such that the overall model is both accurate enough to be scientifically relevant and reproducible and efficient enough to be executed conveniently by modern computational resources (Groen et al. 2014).

From Monolithic Physics to Multiphysics: Couplings

Though multiphysics is not necessarily obtained from monolithic physics, an intuitive and widely accepted idea is to reach multiphysics by means of monolithic physics. For such an idea, the first step to establish a general multiphysics platform is to provide the required models in a "single-physics" mode (Jasak 2006). This is possibly related to the historical development of numerical simulation. Monolithic physics simulations are the traditional commonplace and still predominant, and accordingly, numerous commercial software programs exist. It is thus intuitive and possibly more reasonable to attempt to establish multiphysics frameworks based on the current knowledge and tools for monolithic physics.

The relationship between monolithic physics and multiphysics can be treated in two ways. A common and straightforward way is to approach multiphysics bottom-up as the assembly of individual physical fields. As illustrated in Fig. 2.1, a bottom-up approach, taking the thermo-hydro-mechanical models for example, is established based on the understanding of thermal, hydraulic/hydrologic, and mechanical processes and necessary couplings to stitch them together. The other way is to view the multiphysical process from a complementary perspective: that problems are intrinsically coupled, while a monolithic physics application is merely an idealization made in asymptotic limits (Keyes et al. 2013). This approach is exemplified by the non-isothermal poroelasticity, which can be simplified into isothermal poroelasticity and thermomechanics for porous materials. The basic physical fields used in the above example for the two methods are actually similar, though the terminology is different.

The strength of couplings between the physical fields can be used to determine which way to take. The strength of a coupling here is measured by its influence on the overall process. For example, in most cases, a flow of pore water within a porous material could change the porous skeleton, which in turn changes the flow regime. If this flow occurs in stiff rocks, the effect of this coupling to the flow or the fluid flow-rock deformation process is likely to be small. When this small effect is not of concern, a feasible way to consider this problem is to establish a model based on pore water flow and deformation of solids, for which monolithic models are available. Then the major effort to be made is to find a way to formulate this deformation and, if

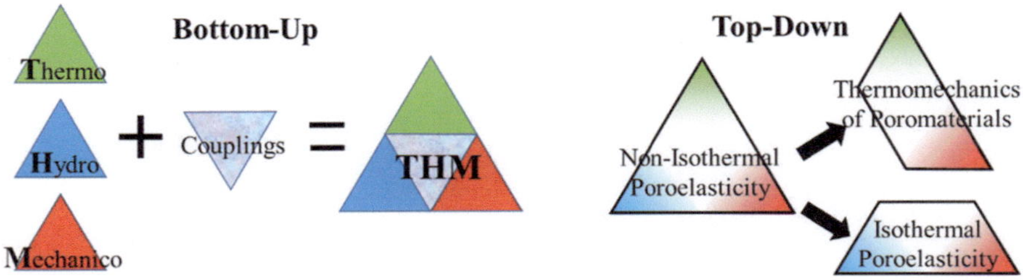

Fig. 2.1 Two ways for developing multiphysics model

needed, changes in permeability, both of which are caused by the flow. However, if this process occurs in a soft clay, then there will be a process called consolidation, which can take place over years. The couplings between water flow and solid deformation are very strong, and a bottom-up coupling may not be able to provide the desired accuracy. This is the reason why van Terzaghi's theory of consolidation and Biot's theory of poroelasticity came into play. In this book, the introduction to multiphysics will be made in a bottom-up way. That is, monolithic physics will be introduced first to prepare the ingredients for cooking multiphysics. However, whenever possible, multiphysical processes will be described from a complementary perspective.

Couplings between different physical fields can be classified as one-way and two-way couplings according to the way in which two fields interact and as explicit and implicit according to solution techniques. The definitions of one-way and two-way coupling are pretty straightforward. A one-way coupling indicates that, for a specific type of interaction, one field affects the other field but not in the opposite way, while for two-way coupling, two fields will influence each other via the same or relevant physical mechanism(s). Further examinations of coupled-field analysis reveal that two types of coupling generally exist for multiphysics problems including implicit (directly) coupling and explicit (sequential) coupling. In an implicitly (directly) coupled system as shown in Fig. 2.2, an algebraic matrix system of equations based on all of the relevant physics is assembled and then solved. One drawback of direct-coupled systems is that finding a solution can be costly in terms of required processing power and computer memory. Alternatively, an explicitly (sequentially) coupled system adopts a segregated solver, where the solution to the first set of governing equations is passed to the second set of governing equations, which is solved and then passed to a third set of equations, etc. This segregated process is repeated until a final solution is obtained. Many commercial multiphysics finite element analysis packages now are able to automatically choose an appropriate default coupling depending on the physics involved or let the user customize these settings, as needed, based on solution time, available computing power, and numerical convergence of the problem at hand (Dede et al. 2014).

The two classification methods can then be combined to categorize couplings based on the level of coupling. The first category is the one-way explicit coupling,

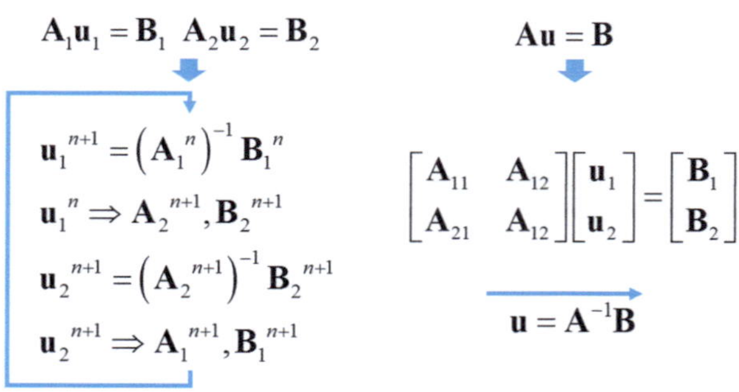

Fig. 2.2 Explicit (sequential, left) and implicit (direct, right) couplings

which is usually shorted as one-way interaction. This is the simplest way to couple two fields. One-way implicit coupling has rarely been mentioned possibly because the incorporation of a one-way coupling implicitly into the stiffness matrix may not make a significant difference. The next level is the explicit two-way coupling, which is typical when multiple software or discretization methods are involved. In this coupling, simulations run side-by-side and exchange coupling information during the run. This approach formally operates in Picard iterations in computational fluid mechanics and fails even in modestly interacting problems. The level with even stronger coupling is implicit two-way coupling. This level provides closer model-to-model coupling that involves matrix-level interaction, where two physical models and model-to-model coupling terms are discretized separately and combined into a single linear system before the solution. This is an implicit variant of the above, with additional stability provided by the linear solver. Here, we can see the benefit of shared matrix and solver modules between multiple discretization schemes. The highest level of coupling is the equation-level coupling, which is however only available for special cases. This type of coupling can be achieved based on the fact that continuum mechanics models originate from mass, momentum, and energy balance models. Thus, a fluid-structure system may be considered as a single continuum governed by a single PDE, with different constitutive relations in various regions. Representing the strongest coupling, this level of interaction, in fact, represents a coupled physical model and may require special numerical techniques. Such techniques are particularly suitable for strong coupling or cases involving phase change and transition regions, e.g., melting and solidification.

Some facts have not been explicitly discussed in the literature but may deserve our attention in the future. The first fact is that most existing couplings were understood in the context of binary field systems. It is thus hypothesized that physical fields interact in a pair-wise way. Therefore, the possibility that three or more physical fields work together with multichannel entangled couplings is totally ruled out. It is believed to be necessary not only to mention this possibility here but also to point out that the couplings introduced in this book are pair-wise couplings if not mentioned otherwise. The second fact is that it is usually unclear regarding how to count couplings. One way is to define the interaction(s) between two different physical fields as "a" coupling. The other way is to define a specific interaction as a coupling, which is identified according to nature of this interaction including its direction and physical mechanism. This book adopts the latter way. Again, take the interactions between water flow and contaminant transport, for example – advection and dispersion will be treated as two different couplings because their underlying mechanisms are different. Similarly, the water flow due to particle movement in electrophoresis and particle movement due to water flow are believed to be two couplings because they have different directions. However, when coupling is used as an uncountable noun, coupling also means any or the overall interaction between two fields.

Establishing multiphysics based on monolithic physics is not a "one plus one equals two" game (Keyes et al. 2013). The job of coupling individual simulations may introduce limitations on stability, accuracy, or robustness that are severer than the limitations imposed by the individual components. Furthermore, the data motion and data structure conversions between independent simulations for each component may be more costly in latency and electrical power than those of the individually tuned components. As a result, "one plus one" may cost significantly more than "two" and may be less amenable to scalable execution (Keyes et al. 2013). Therefore, though the involved single physics, e.g., fluid flow, structural mechanics, electromagnetic fields, and acoustics, are well-understood, coupling them into one multiphysics simulation is not trivial.

The "plus" in this one-plus-one procedure is the main difficulty due to several reasons. First, discretizing the whole set of equations for a multiphysics problem is prone to ill-conditioned system matrices that are hard to solve with sufficient accuracy. Also, the implementation of a new code for every possible (and required) combination of single-physics phenomena would be an immense effort. Thirdly, reuse of existing single-physics codes by just gluing them together requires a lot of numerics such as data mapping between non-matching grids and numerical iteration schemes to regain the solution of the fully coupled system and technical solutions for code-to-code communication. Fourth, the high accuracy of multiphysics models can only be exploited with a very high resolution of the underlying computational grids, necessitating the use of massively parallel supercomputers.

Perhaps one of the greatest challenges for multiphysics simulation is to understand all of the relevant physics involved and to set up a geometrically accurate model with appropriate loads and boundary conditions (Dede et al. 2014). The purpose of this book is to provide such a framework to multiphysics especially those in porous materials. There is currently no such a comprehensive framework. Therefore, whenever a designer would like to conduct a multiphysics analysis for which he or she does not have such background, it is intended that this book can be either a comprehensive introduction or a quick reference. However, we do not consider multiphysics applications from algorithmic and architectural perspectives, where "algorithmic" includes both mathematical analysis and computational complexity and "architectural" includes both software and hardware environments. In fact, many multiphysics applications can degenerate into an algebraic paradigm via linearization in which individual components, i.e., monolithic physics, are represented by diagonal blocks and the multiphysics coupling between them, as off-diagonal blocks (Keyes et al. 2013).

Physical Fields and Types of Multiphysics

The part "physics" in "multiphysics" denotes "physical field." There, multiphysics means the coexistence of multiple physical fields in a process or a system. In physics, a field is a physical quantity that has a value for each point in space and time. For example, on a weather map, a vector at each point of the map is assigned to represent the surface wind velocity, including both speed and direction of the movement of air at that point. In another classic example, an electric field can be viewed as a "condition in space" emanating from an electric charge and extending throughout the whole space. If a test electric charge is placed in this electric field, the particle will accelerate due to the force associated with the "condition in space." Possibly attributed to the latter example, physicists tend to think of the notion of a field as the cause of a force.

The origin of "field" is also related to force. In the eighteenth century, a new concept was proposed to simplify all these gravitational forces. This quantity, namely, the gravitational field, was believed to exist at each point in space, whereby the total gravitational force can be felt by an object with a unit mass at that point. This concept just provided another way to interpret the force rather than change the physics. With the field concept, two treatments will be equivalent: one calculates all the gravitational forces on an object individually and then adds them together, and one first adds all the contributions together as a gravitational field and then applies it to an object (Weinberg 1977). Then in the nineteenth century, the concept of a field received substantive developments due to the development of the theory of electromagnetism. In the early stages, Newton-style laws were employed to express the forces between pairs of electric charges or electric currents. Later the field concept became much more predominant for expressing these laws in terms of electric and magnetic fields. It is believed that Michael Faraday was the first to coin the term "field" in 1849.

The scope of field may be wider than what is needed in multiphysics. In this book, the fields we are dealing with are constrained to classical fields. Therefore, the quantum field will not be considered. Actually, the discussion of the quantum field in the context of multiphysics is not only interesting but also necessary. First, the quantum field has seldom been discussed in the multiphysics literature. This has been at least a convention, though it will not necessarily be in this way forever as the scope of multiphysics is expanding. The quantum field will not be counted or discussed in the book. However, it is necessary to point out that the physical fields we are dealing with occur in virtue of quantum mechanics. This is because common physical processes occur due to the basic forces. All of these basic forces except gravity, i.e., strong interaction, electromagnetic force, and weak interaction, can find their origins in the quantum field. As a result, multiphysics will be constrained above the scale of atoms, though it is not necessarily a solely macroscopic theory. For example, hydrodynamics can span over scales from geologic scale down to Kolmogorov microscale, and chemical field can reach down to the molecular scale. Another presumption is that the fields in multiphysics are non-relativistic. Descriptions of physical fields were given before the advent of relativity theory and then revised in light of this theory. Consequently, classical field theories are usually categorized as non-relativistic and relativistic. Relativistic field theory can explain gravity. But we do not need to worry too much about the origins of the basic forces because we just deal with the non-relativistic fields and have physical laws above the quantum level.

The discussion in the above paragraph essentially determines the mathematical languages to be used in multiphysics. A field can be classified as a scalar field, a vector field, a spinor field, or a tensor field according to the type of physical quantity they represent: a scalar, a vector, a spinor, or a tensor. A field has a unique tensorial character at every point where it is defined. That is, a field can only exclusively be a scalar field or a vector field. Take the Newtonian gravitational field, for example – it requires three numbers to specify the components along the three spatial dimensions at any point, leading to a vector field. However, within each category (scalar, vector, tensor), a field can be either a classical field or a quantum field, depending on whether it is characterized by numbers or quantum operators, respectively. As we do not deal with the quantum mechanics here, so we just to use the normal operators in calculus analysis.

The use of physical fields extends all over most science and engineering disciplines. However, a discipline or subdiscipline only primarily deals with one or a few physical fields. This fact makes multiphysics an extremely interdisciplinary topic and also poses challenges in enumerating the fields. However, the following fields can be identified by overviewing the core courses in the disciplines and publications on multiphysics: heat transfer (thermo-), pore water movement (hydro-), concentration field (concentro or diffuso/convecto/advecto), stress and strain (mechano-), dynamics (dyno-), chemical reactions (chemo- or chemico-), electrostatics (electro-), and magnetostatics (magneto-). In multiphysics, a multiphysical process is usually titled using compound words such as "thermo-hydro-mechanical." These words now appear as the birthmarks or tattoos of multiphysics. The prefixes, roots, or suffixes in the parentheses after each physical field are the ingredients for cooking multiphysics. Each physical field is called a monolithic physical field, uniphysics, or single physics.

Combinations of the above monolithic physics can lead to 247 possible types of multiphysics. However, the world of multiphysics is not only a game of combination in statistics. But instead, we coin a name and investigate a type of multiphysical phenomenon under this name based on the observations in nature and sciences and the needs from practice. Based on a review of multiphysics with an emphasis in porous materials, the most representative multiphysical processes are believed to be the following:

- Thermomechanics
- Hydromechanics
- Thermohydromechanics
- Electrokinetics
- Electromagnetics
- Elastodynamics
- Fluid dynamics
- Hydrodynomechanics
- Thermoelectricity
- Thermoelectromagnetics
- Electromagnetomechanics

Another point deserving attention is that the orders of the roots in the name. Generally, it would be desirable to put the major process, cause or process of primary interest in the front. However, there is no widely accepted rule. For a process such as

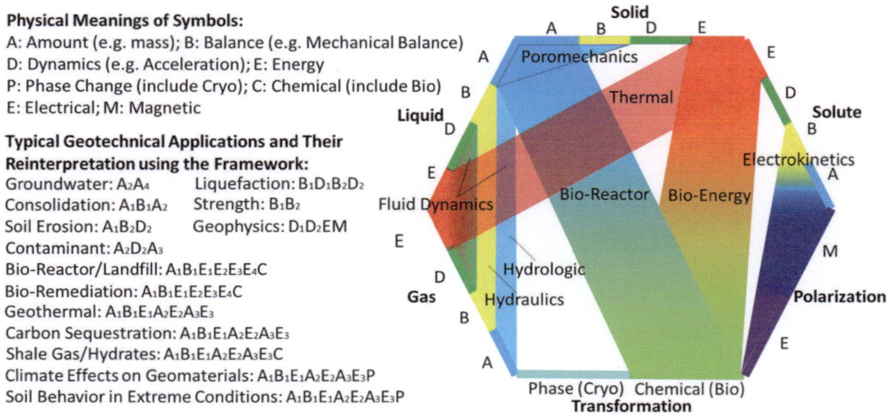

Fig. 2.3 Holistic multiphysics framework

poroelasticity, it may even be a difficult job to judge which process is dominant, i.e., water movement and solid skeleton deformation.

Figure 2.3 shows an example of a comprehensive multiphysics framework which was developed based on the compositions and physical mechanisms in porous materials. Any multiphysical process can be reinterpreted using four fundamental mechanisms in the real phases: mass conservation, balance (static), oscillation (dynamic), and energy conservation, and mechanisms in two "latent" phases: polarization (electric and magnetic) and transformation (phase change and chemical reactions). Typical multiphysical phenomena are combinations of different components (right). Also, classic and highly active research topics in geotechnical engineering can be reinterpreted using the framework (left). As we move from classic topics to upcoming ones, the level of difficulty raises in the context of multiphysics.

There is more than one way to categorize and name multiphysics. For example, the multiphysics software COMSOL categorizes multiphysical phenomena based on applications. The most common multiphysics applications were divided into four major categories: electrical, mechanical, fluid, and chemical. Electrical applications include Joule heating, induction heating, microwave heating, piezoelectric effect, piezoresistive effect, and electromechanical effects. The mechanical section consists of thermal expansion, thermal stress, and acoustic-structure interaction. The fluid section contains Navier-Stokes equation, Boussinesq approximation, non-isothermal flow, fluid-structure interaction, poroelasticity, and squeezed and sliding films. The chemical section comprises convection and diffusion. The advantage of this classification is that it is conducted based on applications, so it uses terms closer to applications. But the disadvantages are also very clear. The adopted terms do not effectively reveal the underlying mechanisms and cannot really reflect the real hierarchy and evolution of multiphysics and their relationships to monolithic physics. On the contrary, a classification purely based on the combinations of monolithic physics is not able to reflect the applications in the real world. For example, many combinations of monolithic physics may not have any existing applications at all. The classification system used in this book is a compromise between theories and applications. So you can see it still keeps the birthmarks or tattoos but, at the same time, adopts terms that are widely accepted but actually do not reflect the involved monolithic physics, such as hydrodynamics and poroelasticity.

Chapter 3
How to Do Multiphysics

Introduction

The implementation of multiphysics includes several essential sequential components: identifying a multiphysical process/system, developing a mathematical description of this process/system, discretizing this mathematical model into an algebraic system, and finally, solving this algebraic equation system and postprocessing the data to obtain information of interest. The abstraction of a multiphysical problem from a complex phenomenon and the description of such a problem are very critical to the success of the multiphysics analysis. This requires to identify the system to be analyzed, including geometry, materials, and dominant mechanisms, which will be interpreted using mathematical languages (function, tensor, differential equation) as computational domain, boundary conditions, auxiliary equations, and governing equations. Mostly, discretization, solution, and postprocessing are carried out using computers. Therefore, the above procedure is very similar to those in general numerical simulation based on partial differential equations.

This chapter is intended to give out a general but definite answer to the question: "How to do multiphysics?" First, the abovementioned essential sequential components of the implementation of multiphysics will be introduced. Emphasis will be placed on both the abstraction of a multiphysical problem from a complex phenomenon with physical and mathematical languages. Finally, the discretization, solution, and postprocessing using computers will be explained. The understanding underneath the above introduction is that multiphysics can be viewed as an art of physics, mathematics, applications, and numerical analysis.

Z. Liu, *Multiphysics in Porous Materials*,
https://doi.org/10.1007/978-3-319-93028-2_3

Mathematical Model

A mathematical model boils down to a bunch of equations. According to the nature and intended role, the equations can be divided into three categories. Couplings between fields can be achieved in each category. As shown in Fig. 3.1, the mathematical model which is defined in a continuous domain will be discretized into an algebraic model defined on a meshed domain for solution.

The first category is governing equations. A governing equation describes the major physical mechanisms and process without further revealing the change and nonlinearity of the material properties. For example, in a heat transfer problem, the governing equation could describe a process in which the thermal energy (represented using temperature or enthalpy) at an infinitesimal point or a representative element volume is changed due to energy transferred from surrounding points via conduction, advection, radiation, and internal heat sources or any combinations of these four heat transfer mechanisms as the following equation:

$$\underbrace{\frac{\partial u}{\partial t}}_{\text{Accumulation}} \underbrace{+\nabla\cdot(u\mathbf{v})}_{\text{Advection}} \underbrace{-\nabla\cdot(K\nabla u)}_{\text{Diffusion (Conduction)}} \underbrace{-\nabla\cdot(D\nabla u)}_{\text{Dispersion}} = \underbrace{Q}_{\text{Source}} \tag{3.1}$$

According to the definition of multiphysics in this book, one physical field of one component in one region only has one governing equation. Therefore, the mathematical model for any multiphysical process should contain at least two equations. Likewise, a mathematical model for a physical process containing two or more governing equations indicates multiphysics according to this criterion. Couplings can be embedded into governing equations. For example, in the thermal field, heat advection as a separate term describes the coupling from the fluid movement to the thermal field.

A governing equation is mostly a partial differential equation, that is, a differential equation with two or more independent variables. There could be two or more spatial coordinates (x, y, ...) in a 2D or 3D equilibrium problem or one temporal coordinate (t) with one or more spatial coordinates (x,...) in any time-dependent

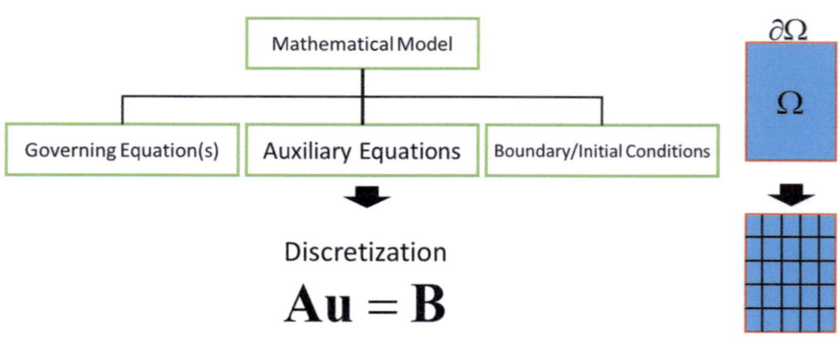

Fig. 3.1 Mathematical model and its discretization

(transient) problems. The dependent variable, i.e., the state variable of the field such as temperature in the thermal field, is a function of these independent variables. However, there are two exceptions when a governing equation is an ordinary differential equation instead of a partial differential equation. This first exception is 1D equilibrium problems, in which a spatial coordinate is the only independent variable. The other is the chemical field. Technically speaking, chemical reactions are not a part of physics, but in multiphysics, it is a convention to count it as a physical field. This treatment also brings difficulties in the establishment of the mathematical model because a chemical field is usually described using an ordinary equation. However, it can be viewed as a "field" mathematically if we associated every point or REV in the computational domain with such an ordinary equation. Every ordinary differential equation describes a chemical reaction at the point it is associated with and is independent of the others. However, connections can be made via other fields such as mass transport.

The second category of equations is auxiliary equations. These auxiliary equations describe (1) the relationships between state variables and the relationships between material properties and (2) other state variables or/and material properties. These two types of auxiliary relationships represent two different ways that coupling can occur or can be implemented. Two typical examples are shown in Fig. 3.2. The first type exists because there are many cases in which there are more state variables than the number of governing equations. As the number of independent variables and number of governing equations should be equal to ensure a unique solution, only some of the state variables can be selected as dependent variables. Then the unselected state variables then can be related to the selected ones via constitutive relationships. This treatment is quite common in both monolithic physics and multiphysics, such as stress and strain in the mechanical field, suction and water content in unsaturated flow, and pressure, density, and temperature in compressible fluid mechanics. The second type is the dependence of a material property on state variables or/and other material properties. For example, it is not difficult to imagine that the heat capacity of a porous material would be a function of water content and

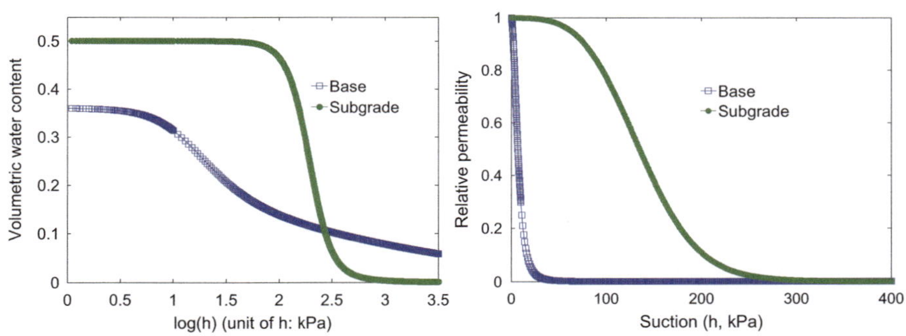

Fig. 3.2 Relationship between state variables (SWCC between water content and suction) and between state variable and material properties (relative permeability and suction) in different porous materials (base material and subgrade soil under pavement). (After Liu and Yu 2011)

heat capacity of the constituents of the porous material. In reality, these two types can possibly be mixed up, leading to even more complicated auxiliary relationships. The auxiliary relationship is also a material property, such as the stress-strain relationship, which defines the stiffness of an elastic object. It is not unusual that the stiffness is a function of other state variables such as temperature and material properties such as moduli of the constituents.

The third category of equations is the boundary and initial conditions. One PDE or a PDE system may have an infinite number of solutions. This is very similar to the solution to ordinary differential equations. When inadequate constraints are provided, that is, lack of enough boundary/initial conditions, we can only obtain general solutions. These general solutions contain arbitrary constants and represent an infinite number of possibilities. However, boundary and initial conditions as additional constraints can turn the general solutions into a particular solution. Boundary conditions correspond to spatial coordinates, whereas initial conditions correspond to the temporal coordinate. Therefore, no initial condition is needed or allowed in an equilibrium problem. The existence of the boundary and initial conditions also has practical implications. This is because, in numerical simulations, it is infeasible to simulate an infinitely large domain or a process lasting forever. In fact, the discretization size, algorithm, available computing resources, and the nonlinearity of the problem determine the upper limit of the geometrical size and duration of the process that can be simulated.

Multiphysics is an art of mathematics, physics, and applications. Just as shown in Fig. 3.3, math, physics, and application provide three independent angles to perceive multiphysics. As a result, a complete description of a mathematical problem needs input from all of the three aspects. In detail, the mathematical model should be established based on the real physical conditions. However, there are distinct differences between the real physical processes and the conceptual mathematical models. These differences result in differences in the sequences and languages adopted in each and the differences between the real physical process and the solution to the mathematical model.

Assuming there is a physical process/system to be simulated, from the perspective of multiphysics, the first thing to think about is "whether this is a multiphysical

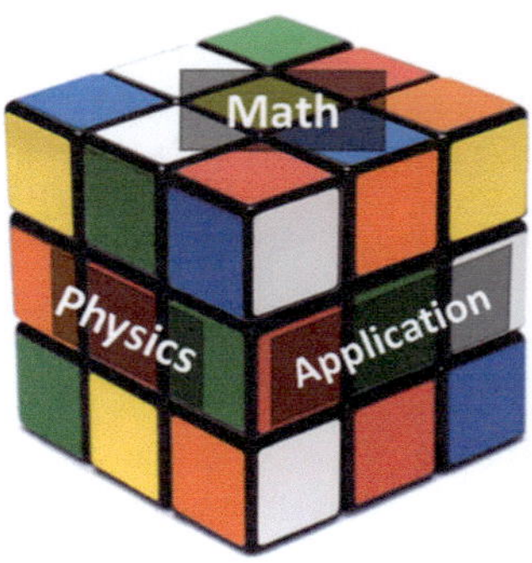

Fig. 3.3 Three aspects of multiphysics

process, if it is, then how many different physical fields are involved?" In most cases, the answer is not that clear as multiphysics is more a rule than an exception, and we can possibly find the trace of most physical processes if we zoom in enough. So the question then turns into which physical fields are the primary ones. The answers can also change with the expectation of the accuracy and effort for the analysis. Then the question that needs to be considered is what object will be simulated. In physics, this question includes what are the objects/areas of interested, what are the interactions between these objects/areas, and how these objects interact with the remainder of the world. After being translated into mathematical languages, the questions turn into how many regions in the computational domain and what are the boundary conditions on boundaries between regions and on the external boundaries of the whole domain. Boundary conditions need to be defined for each physical field, so do initial conditions. For multiregion problems, different physical fields and axillary relationships may also be assigned to different regions.

In most commercial numerical analysis software, a user interface is usually provided to help users to build up mathematical models. To effectively use this user interface, we still need to understand the physics and determine the physical fields and physical objects to be simulated. Once we have them in mind, it is usually fairly easy to choose the physics, to build up the computational domain and to set up boundary and initial conditions. The computational domain could be a line in 1D, an area in 2D, or a volume in 3D. It can also turn out to be very complicated 2D or 3D geometries with complex inclusive regions. Many programs offer a powerful graphical user interface to help generate these geometries or provide an interface to import geometries from other computer-aided design software. But repairing and defeaturing of the geometry are often required to fix the parts of the geometry that are not "watertight" and remove slender surfaces or merge unnecessary small edges.

Discretization

Discretization is the step right following the establishment of a mathematical model. This is because the mathematical model is a bunch of equations and usually we do not solve these equations directly using computers. It is possible to solve some mathematical models by hand or by computer programs such as Mathematica or Maple using symbolic computation. But for most mathematical models, it is possibly easier to use numerical computation. Numerical computation requires an algebraic equation system such as

$$\mathbf{AX}=\mathbf{B}, \tag{3.2}$$

where $\mathbf{A}$ is the stiffness matrix, $\mathbf{X}$ is the matrix of unknowns, and $\mathbf{B}$ is the force matrix.

Discretization is the step to transfer a mathematical model into a numerical model or an algebraic equation system as shown above. In mathematics, discretization concerns the process of transferring continuous functions, models, and equations into discrete counterparts. This process is usually carried out as a step toward making them suitable for numerical evaluation and implementation on digital computers.

There are many options for discretization, among which finite difference method, finite volume method, and finite element method are the most popular ones. These numerical methods employ significantly different ways for discretization. The finite difference method approximates the differential equations with difference equations, in which finite differences approximate the derivatives. The finite volume method employs volume integrals in a partial differential equation, in which a divergence term is converted to a surface integral using the divergence theorem. The finite difference method and finite volume method are similar to each other, and both work on values calculated at discrete places on a meshed geometry. The finite element method also works on a meshed geometry, but it connects many simple element equations over many small subdomains, named finite elements, to approximate a more complex equation over a larger domain. The method uses variational methods from the calculus of variations to solve the problem by minimizing an associated error function. In summary, these methods involve discretization of both the computational domain and the equation system.

The discretization of the computational domain is called meshing in numerical analysis, which is considered to be one of the most difficult tasks of preprocessing in traditional numerical finite volume and finite element analysis. The difficulties involve both the effort to obtain good quality grids/meshes automatically and also mesh settings complying with the nature of the physical problem. For example, in wave propagation problems, such as elastic waves in structural mechanics and electromagnetic waves in radio frequency analysis, the size of the largest element has to be substantially smaller than the wavelength to ensure a solution without losing significant physical details. In fluid flow, boundary layer meshes are necessary to simulate boundary layers, for which the cell Reynolds number may determine the element size in the bulk of the fluid. In the modern numerical analysis, adaptive meshing has become a widely accepted technique, in which an initial mesh may be automatically altered during the solution process for the purpose of reducing the error or changing the geometry.

Solution and Postprocessing

Compared to meshing, selecting and setting the solvers, and obtaining a solution to the equations, constituting the numerical model within a reasonable computational time is possibly an even more difficult task. The difficulties have different sources.

The first source is the size of the discretized model, which can be very large in terms of algebraic equations. The degrees of freedom of numerical simulation can easily reach several millions. Also, nonlinear, parametric, eigenvalue, and time-dependent problems are usually transferred into a system of linear equations and solved using iterative methods. Therefore, the challenge in numerical simulation ends up at the solution of a large linear system of algebraic equations. However, it is usually not easy to solve large linear systems. General purpose solvers are available yet usually not the optimal choice considering the high computational cost due to lack of optimization for realistic problems. For example, direct solvers based on lower-upper factorization methods or general purpose iterative methods usually approach the solution without a clear consideration of the structure of the system. Moreover, the structure of a multiphysics problem may be difficult to identify or even does not exist. One way to help address this problem starts from the structure of the included monolithic physics. For iterative methods, algebraic (Wesseling and Oosterlee 2001) or geometric multigrid methods (Ruge and Stuben 1987) can also be helpful.

The second source is the nonlinearity of the numerical models. Multiphysics problems usually involve extreme nonlinearity due to the strongly coupled governing equations, complex nonlinear material properties in terms of auxiliary equations, and more complicated initial and boundary conditions. The nonlinearity will lead to complex morphology in the structure of the solution. This is analogous to a terrestrial region full of complex topographic features such as mountains and canyons. In iterative methods, the search of the solution is like looking for a point in the region, usually the highest or the lowest point. The complex topographic features, or nonlinearity in this case, will tremendously increase the difficulty of a successful search. Even without an iterative method, the numerical solution also requires the use of approximation functions to approach the true solution. Whenever the solution changes fast over a small region or a small amount of time, the effort for an acceptable approximation will multiply. To solve this problem, different simplifications or relaxations of the problem are usually adopted. For example, less significant nonlinearities may be removed to obtain a linear problem that can be solved easily. This type of solver is the segregated and continuation solver (Heil et al. 2008).

Difficulties in solution can also come from the numerical model or numerical schemes. For time-dependent problems, the stability of the numerical scheme needs to be considered. Inappropriate settings of time steps and mesh size can lead to instability, which can make the numerical solution deviate from the true solution or even fail to converge. Sometimes, extremely small time steps or mesh sizes are needed in order to obtain a reasonable solution, leading to a high requirement of computational resources. The solution to such problems is much more difficult than for a more benign and well-behaved numerical model. For such a condition, it is usually helpful to investigate the numerical model and the schemes for discretization and their relationship to the structure of the problem and the true solution. Tuning solver settings may help identify the problem but may not be the most effective cure.

Postprocessing is the last link of the numerical solution. It usually involves visualizing and investigating 2D/3D plots of the modeled fields and variation of quantities with respect to time or spatial coordinates. Old-fashioned packages require that the plots and derived values to be analyzed had to be defined before the solution process. Omission of some key definitions means that the solution has to be carried out again from the beginning. The definitions of these needed quantities and/or derived values are, therefore, deemed as part of preprocessing. More recent software usually tends to allow any derived quantities and derived values to be calculated after the solution process. Error estimation is also an essential part of the numerical solution. It is usually helpful to try the numerical model with different mesh sizes in order to estimate the convergence of the numerical solution. Besides, sensitivity analysis, which involves the solution of the same model with different input, such as material properties, initial conditions, and boundary conditions, tests the influence of some parameters and the overall performance of the numerical model. This can help understand the structure of the problem, hidden bugs in the code, and errors in the model.

Chapter 4
Multiphysics in Porous Materials

Introduction

This chapter is intended to explain the strong connections between multiphysics and porous materials by overviewing the history and applications of multiphysics in porous materials. We will start with the relationship between multiphysics and porous materials and then explain the reason for the existence of monolithic physics as the mainstream in traditional porous studies and the transition of these studies to multiphysics for pressing and upcoming issues. Finally, we will explain the relationships between the major multiphysics topics and common porous material studies by listing and elaborating well-known existing applications of multiphysics in porous materials.

History

Porous materials are susceptible to multiphysical phenomena due to their nature as a composition material consisting of a porous skeleton, pore fluids, and solutes. This is because this distinct nature provides favorable conditions for energy transfer, water migration, geomechanical responses, contaminant transport, and electrokinetic phenomena. Many porous materials have complex constituents in both solids and liquids, imperfections at different scales, and amorphous structures. For example, soils as a natural product could contain hundreds of different minerals, structural deficiencies from nanoscale partial substitutions to macroscale ink holes, and amorphous structure as a particulate material. This makes any process in it hardly a monolithically physical one. Almost most common multiphysics problems can find their trace or applications in porous materials. That is also the reason why it is necessary to cover the whole multiphysics when introducing the multiphysics in porous materials.

Z. Liu, *Multiphysics in Porous Materials*,
https://doi.org/10.1007/978-3-319-93028-2_4

Porous materials are so closely related to multiphysics; no wonder multiphysics has its footprints in the whole history of porous material studies. A major discipline dealing with porous material, i.e., geotechnical engineering, was believed to be developed into an independent discipline when von Terzaghi published his work on the one-dimensional consolidation theory in the 1920s. This theory, however, is about the coupled process between pore water flow and deformation of the solid skeleton. The later rapid development of geotechnical engineering in the 1980s was triggered by the earthquake investigations and geoenvironmental applications, which boil down to considerations of more physical processes such as dynamics, contaminant transport, and chemical reactions. In other disciplines related to porous materials, the story is similar. For example, petroleum engineering featured the multiphase and multicomponent flow, the influence of many chemicals and magnetic waves on the flow, and also the explorations of the reserve with various electrical, seismic, and magnetic approaches.

As physical processes in porous materials are, more or less, coupled in nature, it is very difficult to separate one from the others. However, a single physical phenomenon can be analyzed individually without significantly affecting the analysis results when the couplings between different physical fields are weak enough. This is, in fact, the mainstream in the traditional studies of porous materials, though multiphysics has always accompanied. Multiphysics was considered only in the cases in which the couplings between some physical processes are too strong. Typical examples include consolidation and porous materials under frost action. However, multiphysics is turning to be more and more of a necessity as we are faced with more and more challenging problems such as climate effects and more and more stringent requirements on the accuracy of the analysis. Examples of such applications include shale gas, carbon sequestration, geothermal energy, climate effect on porous materials, the behavior of porous materials under extreme conditions, soil erosion, extraterrestrial soils, synthetic of new materials, exploration of traditionally unexplored areas, design and fabrication of actuators and sensors, biomechanics, and microfluidics. More concerns have been shifted from individual physical phenomena to coupled multiphysical processes due to emerging needs in porous material-related disciplines and advances in multiphysics research.

Figure 4.1 schematically shows the structure of a typical thermo-hydromechanical process in porous materials. Many multiphysical phenomena are triggered by a change in the thermal field. This change serves as an excitation, similar to an external force in the mechanical field, and gives rise to changes to the whole system in the form of a multiphysical process. For example, thermal changes can alter the compressibility and shear strength (mechanical), induce water migration, and change hydraulic properties (hydraulic). These are the typical couplings from the thermal field to mechanical and hydraulic fields, respectively. These changes in mechanical and hydraulic fields will in return exert "reacting forces" to the thermal field. For example, heat transfer can be promoted by energy convected by migratory water, and thermal properties such as heat capacity and thermal conductivity can be changed due to the variations in the structure and composition of the material. The "reacting forces" represent the couplings from the other two physical processes to

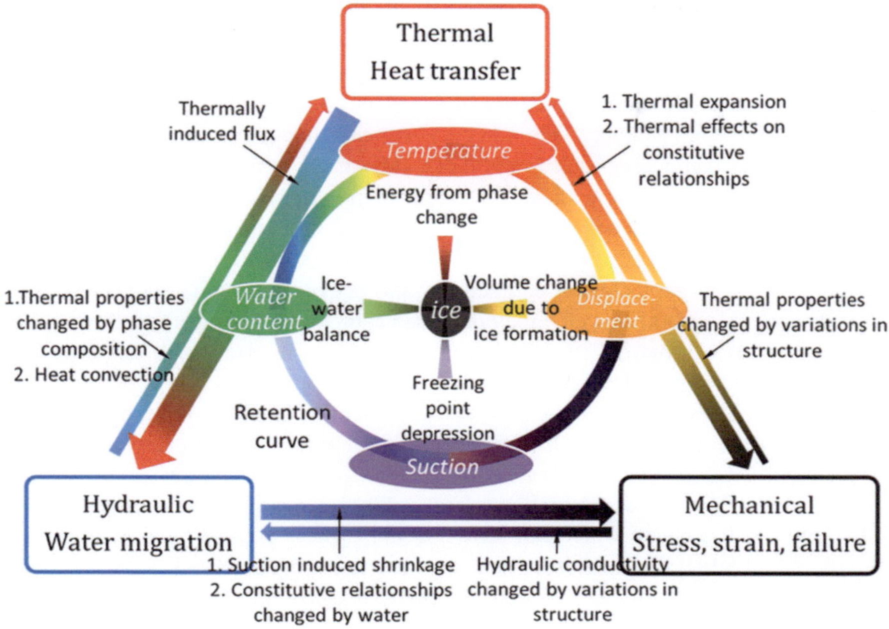

Fig. 4.1 Couplings between physical fields in soils

the thermal field. It is worthwhile to point out that the couplings from the thermal field to other fields are usually strong couplings. Among these couplings, the thermally induced water flux is especially significant and thus of special interest. This process of water transfer due to temperature gradients is critical to many multiphysical processes in porous materials. When the phase change of water (freezing or thawing) gets involved, the multiphysical phenomenon could be more complicated, and the significance of the thermally induced water flux can be seriously exaggerated. One typical example is frost heave, in which water is sucked from deeper positions to the advancing frost front due to temperature changes on the ground surface.

Applications

Thermomechanics is the coupling between the thermal field and mechanical field. Typical applications can be dived into two categories. The first category includes temperature-induced mechanical responses, that is, the thermomechanometry of porous materials and its consequent physical processes. Thermomechanometry is the measurement of a change of a dimension or a mechanical property of a sample subjected to a temperature regime. It is also responsible for the mechanical behavior of various materials under non-isothermal conditions. Typical examples include thermo-cracking of pavement, energy piles, microstructural changes of porous materials such as thermal barriers or heat exchangers, and effect of climate change on geomaterials. The second category involves an opposite process: heat generated by mechanical movements. This category attracts much less interest but is still responsible for some very basic phenomena such as frictional heat.

Hydromechanics is more of an extension of poroelasticity. Poroelasticity considers the coupled processes between the pore water movement and deformation of porous materials. A typical application is the soil consolidation, which is a fundamental process in geotechnical engineering and petroleum engineering. Hydromechanics is more general as it also includes nonelastic material properties. Several hot applications are primarily related to multiphysical processes such as shale gas recovery, carbon sequestration, and hydraulic fracturing.

Thermohydromechanics is possibly the most widely existing multiphysical phenomenon in porous materials. It is more general than thermomechanics and hydromechanics because (1) non-isothermal conditions or heat transfer is usually unavoidable, (2) temperature gradients can cause water movement in porous materials, and (3) migratory water can advect heat. Typical applications include frost heave, gas hydrate dissociation, cement hydration, geothermal reservoir, and enhanced oil recovery using hot water flushing. It is also common that more fields can be added. For example, chemical reactions and particle transport can be incorporated so that phenomena such as a bioreactor, sanitary landfills, ground remediation, ground improvement using chemical or biologic methods, and erosion can be included. If the electric field is included, it then can be used to consider fuel cells.

Electrokinetic phenomena are a family of several different effects that occur in heterogeneous fluids, in porous bodies filled with fluid, or in a fast flow over a flat surface. According to Lyklema [ref], the complete family of electrokinetic phenomena includes electrophoresis (motion of particles under influence of electric field), electroosmosis (motion of liquid in porous body under influence of electric field), diffusiophoresis (motion of particles under influence of a chemical potential gradient), capillary osmosis (motion of liquid in porous body under influence of the chemical potential gradient), sedimentation potential (electric field generated by sedimenting colloid particles), streaming potential/current (either electric potential or current generated by fluid moving through porous body or relative to flat surface), colloid vibration current (electric current generated by particles moving in fluid under influence of ultrasound), and electric sonic amplitude (ultrasound generated by colloidal particles in oscillating electric field). Typical applications are ground remediation, electrokinetic chromatography, and electrokinetics in microdevices for biotechnology.

Electromagnetics has a very long history and serves as the background of electrical engineering. In porous materials, its significance is primarily attributable to the application of various electromagnetic technologies in porous materials, such as sensors based on electromagnetics (ground penetration radar, time domain reflectometry, and capacitive-type water content sensor), microwave and induction heating of porous materials for agricultural and petroleum applications, and novel materials treating using electromagnetics.

Elastodynamics is the study of elastic waves and involves linear elasticity with variation in time. An elastic wave is a type of mechanical wave that propagates in elastic or viscoelastic materials. The elasticity of the material provides the restoring force of the wave. When they occur in the Earth as the result of an earthquake or other disturbance, elastic waves are usually called seismic waves. Therefore this multiphysical phenomenon is one of the primary mechanisms underlying earthquakes. It also determines the applications of seismic waves in exploratory geology and the applications of any acoustic methods in porous materials.

Fluid dynamics deals with fluid flow in motion. It usually becomes a concern in porous materials in two ways. The first is that the flow of surface water interacts with porous materials. This is very common in soil erosion, bridge scour, and piping. The other way is obvious when pore water moves at very high velocity, such as those microfluidics applications (e.g., inkjet print heads, micro-propulsion) and wicking in porous materials. This is also of concern whenever pore water flow is connected to a large body of water such as the interaction between groundwater and surface water and natural convection affected by groundwater.

Hydrodynomechanics refers to the involvement of both hydrodynamics and mechanical responses. In fact, this word has seldom been used in the literature. However, an equivalent term, fluid-structure interaction (FSI) is used to represent the interaction of some movable or deformable structure with an internal or surrounding fluid flow. Fluid-structure interactions can be either stable or oscillatory. In oscillatory interactions, the strain induced in the solid structure causes the solid to move such that the source of strain is reduced. Sometimes the structure returns to its former state so that the process can be repeated. This multiphysical process will possibly attract more and more attention as we move from lands to seas and oceans.

Thermoelectricity has two major categories of applications: thermoelectric effect and Joule heating. The thermoelectric effect is the direct conversion of temperature differences to electric voltage and vice versa. A thermoelectric device creates a voltage when there are different temperatures on two sides. Conversely, when a voltage is applied to it, it creates a temperature difference. At the atomic scale, an applied temperature gradient causes charge carriers in the material to diffuse from the hot side to the cold side or vice versa. This is actually how thermocouples work in porous materials. Joule heating provides a theoretical basis for the measurement of thermal properties of porous materials. It is also of critical significance to many electrical devices consisting of porous materials.

Thermoelectromagnetics includes microwave heating and induction heating. They are of concern whenever electromagnetic waves with high energy exist in porous materials. This would especially be an issue as the world is filled with more and more types of microwaves. An extremely significant issue is the influence of electromagnetic waves on biomaterials, which are another big category of porous materials. In addition, this term also refers to the magnetism in non-isothermal conditions and newly found phenomena such as thermoelectromagnetic convection in liquid steel.

Magnetomechanics, electromechanics, and electromagnetomechanics are closely related. These three types of multiphysical phenomena represent the interaction between magnetostatics, electrostatics, and electromagnetics with mechanical responses. These three mechanisms are the fundament of microelectromechanical systems. They also provide a theoretical basis for the development of most electronic sensors and smart materials and systems. Porous materials are both the raw materials and the areas of application of these sensors and systems.

Chapter 5
How to Use This Book

Introduction

This chapter provides an overview of the organization of the book and suggestions to readers with different backgrounds and interests regarding how to use the book. We will first summarize the existing literature on multiphysics, the normal ways to learn about multiphysics, and especially, the lack of a general book on multiphysics. In this way, the users can better understand the purpose and organization of this book. The book is organized to reflect the understanding of multiphysics as a fusion of mathematics, physics, applications, and numerical analysis, which led to the major parts of the book: introduction, mathematics, monolithic physics, multiphysics, and numerical implementation. This book is designed for people with diversified backgrounds, from college-level students, engineers, and numerical simulation practitioners to experts in specific multiphysics applications, as a textbook, a reference book, or a technical book. Accordingly, detailed suggestions will be provided for the use of the book for different purposes.

Organization of the Book

This book is intended to be a general reference for multiphysics. By now, a dozen of books can be found whose title includes multiphysics. However, they are mostly conference proceedings, introductions to the development of a specific multiphysics application, or summary on multiphysics in a specific area. A relevant fact is that multiphysics researchers usually tend to focus on the multiphysics applications in their own areas, no matter in a general multiphysics context or in porous materials. There is no such an effective reference when people would like to know about multiphysics, especially to get a comprehensive understanding and a big picture of this topic of increasing interest. While an overall understanding is usually very

Z. Liu, *Multiphysics in Porous Materials*,
https://doi.org/10.1007/978-3-319-93028-2_5

difficult, a common way to get familiar with the topic is to check papers and online materials in the area related to the intended multiphysics applications. Therefore, knowledge is disseminated in a very ineffective way due to the lack of a comprehensive multiphysics book. Some commercial software provides possibly the most comprehensive introductions as the developers have a much broader interest than researchers specialized in any single area. However, such interest tends to lean more toward industrial needs and possibly lags a little behind the front edge in many research areas. In addition, the documentation is also not that comprehensive as most of it is designed for users rather than developers or researchers.

This book is designed to be a comprehensive introduction to the area of multiphysics in porous materials. As introduced, porous materials are an ideal breeding ground of multiphysics due to their multiphase, multicomponent, and possibly amorphous structures. Their ubiquitous existence also makes the multiphysics in them closely related to human lives. It is also necessary to point out that porous materials are related to most multiphysical phenomena. For example, the coexistence of thermal, hydrological, and mechanical processes is often of the most interest in geotechnical engineering, hydrogeology, hydrological engineering, soil science, and petroleum engineering as well as many other energy application domains. Particle transport and chemical reactions together with heat transfer and water movement have been extensively studied in environmental engineering, agricultural engineering, and forestry. Electromagnetics, vibration, and acoustics are the primary interest of geophysicists, earthquake engineers, structural engineers, and exploratory geologists. Therefore, this book will be a handy tool to researchers and practitioners in these areas of porous materials, no matter they are new to multiphysics or they want to step into a multiphysics topic out of their area of specialty or an interdisciplinary one. More emphasis is placed on porous materials to reach the depth of research for many applications.

This book is organized in a way to serve readers with different backgrounds and with different levels of knowledge accumulation in multiphysics. For the purpose, this book provides materials that, the author believes, would be essential to obtain a general understanding of the concepts, theories, and implementations of multiphysics. It is believed that multiphysics is a fusion of mathematics, physics, applications, and numerical analysis. So the book is roughly organized in this way. First, a general introduction to the history, state of the art, the idea of multiphysics, and its applications in porous materials are offered in the first part. Right after this part, necessary mathematical backgrounds such as tensor and fields, tensor analysis, partial differential equations (PDEs), and numerical solutions of PDEs are provided in the second part. For physics and applications, a bottom-up way is taken for the introduction. The most representative monolithic physics is introduced in Part III, for which each monolithic physical phenomenon is presented as a chapter. Then in Part IV, typical multiphysical processes are introduced. The introduction is made based on the mathematics in Part II and the monolithic physics in Part III. Each multiphysical topic is introduced in the order of background, theory, implementation, and applications. The theory will be discussed in a way that mathematics and physics are integrated. Implementations are demonstrated with a simplified but typical example, which is solved using a PDE solver. In Part V, a brief introduction will be made to show how to implement a multiphysics model using three typical numerical simulation methods: finite difference method, finite volume method, and finite element method.

The focus of the book is "physics," which is currently discussed a little, rather than mathematics and numerical methods, which have been more thoroughly discussed in many other monographs. Therefore, Part II for mathematics and Part V for numerical methods will only provide the most fundamental and necessary components. For example, the mathematics will be restrained to Cartesian coordinates which are mostly encountered in applications. In numerical methods, accuracy and error analysis are only briefly or even not discussed, and only the simplest schemes and element types will be introduced. For the introduction to these numerical methods, there are several ways. But in this book, only the simplest and most straightforward way in the author's opinion is taken. The purpose is to provide a broad but focused picture with minimal effort.

Suggestions to Readers

As indicated, this book is prepared for readers with different backgrounds. Here are a few suggestions to potential users:

1. The necessary background for this book is college-level mathematics on calculus and linear algebra and a little understanding of computer simulations. If some of these are missing, it is suggested that the readers take some time to read or review these parts. For calculus and linear algebra, most textbooks should be able to fill the gaps. For computer simulations, many introductory materials such as Wikipedia may help fill the vacancy.
2. Given that you have the above knowledge but do not know much of multiphysics, then possibly reading the books sequentially is a good choice. If you would like to know all about multiphysics, then try to read the chapters one by one. However, this is possibly unnecessary for most people. This is because multiphysics is a very interdisciplinary area and it has not been developed into an independent discipline. Thus most people from a specific area are interested in some applications that are closely related to their area. Cracking a book with equal effort at all the chapters may turn out to be unnecessary.
3. If you are a researcher from a specific area, you possibly know the general idea of multiphysics and have the background in mathematics. Then you possibly would like to know more about a specific application or several applications. Then it is advised that you quickly go through Part I if necessary, skip Part II if possible, and then directly jump into those chapters for monolithic physics and multiphysics related to your applications. The introduction to these specific topics is made in the premise of striking a balance between generality, depth, ease of understanding, and constraints of length. As a result, there may be a distance from

the presented materials and your topic or the front edge of the research. However, the distance should be able to be shortened with a limited amount of work on recent publications specified for the topic.

4. If you are an engineer, who is interested in some applications. Depending on whether you know the concept of multiphysics and the required physics, you may decide whether and how you should read Part I and Part II. For applications, it is suggested that you start from the example provided at the end of the chapters corresponding to the application that you are interested in after reading the theories. You should be able to consider more complicated or specified cases by further integrating and/or expanding the code.
5. If you are a numerical analysis practitioner or a software developer, you may want to go directly to chapters in Part III and Part IV. The theories in these chapters are ready for implementation. However, more and accurate axillary relationships may be needed if down-to-earth applications or computer programs are targeted at.
6. If you are an expert in multiphysics, you possibly agree that it would be enjoyable to have such a reference in the playground, though the book is still far from adequate due to the author's limited vision and capacity. The author had been waiting for such a book for several years and, at the same time, could not stop reviewing his own notes. He finally decided to make one himself and share the knowledge with anyone who may have the same needs. You possibly once had the same feeling or thought. So it would be a great encouragement if you can pay a smile and, if possible, leave your comments. The authors strongly desire to improve this book with you.
7. All in all, the author would like to prompt the spirit of sharing. Some materials are excluded from the book, such as problems for learners, due to a variety of concerns. But please feel free to contact the author for these materials. Also, please do not hesitate to contact the author if you have anything to share or anything you would like to put into the next version of the book. Any feedback on the grammatical, editorial, technical aspects of the book will be sincerely welcomed.

Part II
Mathematical Background

Chapter 6
Tensor and Field

Introduction

As introduced in Part I, the physics that we are dealing with in multiphysics is classical physical fields that can be described using mathematical languages such as tensors and partial differential equations. Part II of this book is intended to present such necessary mathematical languages to enable later introductions to monolithic physics and multiphysics.

In this chapter for tensors and fields, we will first explain scalars, vectors, and tensors from both the physical and mathematical aspects. The explanation of the tensor will start with the definition and understanding of this concept with the assistance of daily examples and end with the operations and changes of tensors. Next, we will discuss the physical field including its definition; relevant concepts such as coordinates, domain, and grid; and the materialization and storage of the physical field in the computer.

From Scalar, Vector to Tensor

A scalar represents a physical quantity that can be completely described using a real number. As shown in Fig. 6.1, the expressions of its component are independent on the choice of the coordinate system. Examples include temperature, mass, density, potential, and so on. In contrast, a vector can be used to describe a physical quantity that has both a direction and a length. The expression of its components is dependent on the choice of the coordinate system. Examples include displacement, velocity, force, heat flow, and so on. Tensors are a generalization of scalars and vectors: a scalar is a zeroth-order tensor, while a vector is a first -order tensor. While orders of tensors can vary from 0 to infinite, the orders of the tensors that we encounter in multiphysics usually do not exceed 4. The tensors that we mostly see are

Z. Liu, *Multiphysics in Porous Materials*,
https://doi.org/10.1007/978-3-319-93028-2_6

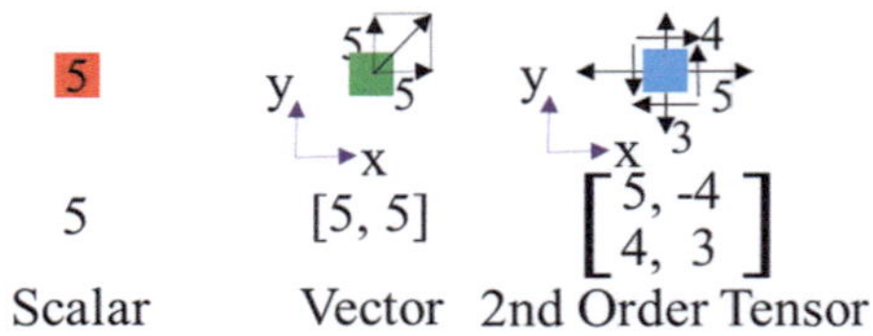

Fig. 6.1 Understanding and structure of scalar, vector, and second-order tensor

second-order tensors that define an operation for transforming a vector to another vector. A second-order tensor can be visualized as a two-dimensional array (matrix), while tensors of the third or higher order are difficult to visualize. In a narrow sense, tensors only refer to those with an order of two or higher.

The concept of field or physical field is frequently used with scalars, vectors, and tensors. In general, a scalar field associates a scalar value to every point in a space (and time if it is involved). The scalar at a point is a real number quantifying the state or property at that point. Examples of scalar fields in applications include the temperature distribution throughout a space and the concentration of a solute in a liquid. These fields are the subject of scalar field theory. A vector field is an assignment of a vector to each point in a subset of a space. A vector field in a plane, i.e., 2D, can be visualized as a collection of arrows with a given magnitude and direction assigned at every point in the plane. Vector fields are often used to describe the dynamics of a fluid moving throughout a space under the influence of some forces such as the magnetic or gravitational force. The magnitude and direction of the vector quantity may vary from point to point.

The operations and changes of tensors including scalars and vectors are usually of interest as they the mathematical languages to describe the operations on and the physical mechanisms underlying a multiphysical process. The operations and changes of scalars are studied in the context of arithmetic and calculus, which are usually introduced in fundamental and advanced mathematics, respectively, and will not be discussed in this book. In the following chapters of Part II, the operations and changes of vectors will first be introduced sequentially as vector algebra and vector calculus. Tensor analysis including its algebra and calculus will be briefly introduced later as an extension to vector analysis. We will stay with the fundamental but most useful contents to multiphysics for the sake of simplicity and clarity. As a result, concepts such as covariance and contravariance, metric coefficients, and curvilinear coordinate systems, which are more advanced and less common in applications of multiphysics, will be excluded. A useful notation, index notation, which is different from the vector notation used in the first few sections of this chapter, will also be introduced. In addition, knowledge of partial differential equations needed for multiphysics studies will be presented. Finally, tools for the numerical solution of PDEs will be introduced.

Physical Field

Similar to tensors, the introduction to physical fields should also include both physical and mathematical aspects. The physical implications of physical fields have been discussed in Part I. The general physical meaning of a physical field is the values of a state variable associated with spatial coordinates, which can change over time in a transient problem or be stable in an equilibrium problem. The state variable can be a scalar, a vector, or higher-order tensors. The spatial coordinates can be a scalar (1D), a vector with two (2D) or three numbers (3D) at any point within the computational domain. As the computational domain is usually discretized (meshed), the physical fields can be visualized as the *associated* presentation of state variables and its corresponding spatial coordinates. The spatial coordinates are determined by means of the discretization methods, which are usually defined at the grid points, vertexes of elements, centroids of volume surfaces, or even points of Gauss integration. Usually, grid points are used for FDM and FVM, and vertexes are used in FEM. The result of a discretization is a finite number of points/elements and consequently a collection of a finite number of coordinates.

Both the state variables and associated coordinates can be depicted using arrays or matrices. However, the definitions of arrays and matrices could be different depending on the context. Both terms are commonly used to refer to two-dimensional datasets. But different conventions are adopted by people from different areas. For example, people working with images tend to call two-dimensional datasets arrays, while mathematicians tend to use matrices. While these two terms are interchangeable, the different conventions may still lead to confusion.

It is helpful to understand the relationship between the mathematical definitions of matrix, data structure, and computer storage. Take a two-dimensional dataset with dimensions m and n, for example – this dataset is stored in a one-dimensional set of memory addresses in a computer. It is the indices used by the software to access the data in memory that make the data "two-dimensional." Further understanding of the process of mapping multidimensional data into one-dimensional computer memory needs more in-depth and specified materials. Strictly, a matrix can be defined as a rectangular array: a two-dimensional array of numbers, symbols, or expressions, arranged in rows and columns, which is treated in certain prescribed ways. One such way is to state the order of the matrix. For example, the order of the matrix below is 2×3 because there are two rows and three columns. The individual items in a matrix are called its elements or entries.

$$\begin{bmatrix} 5 & 2 & 1 \\ 13 & -14 & 0 \end{bmatrix}.$$

If two matrices have the same size or order, in other words, have the same number of rows and the same number of columns, then the two matrices can be added or

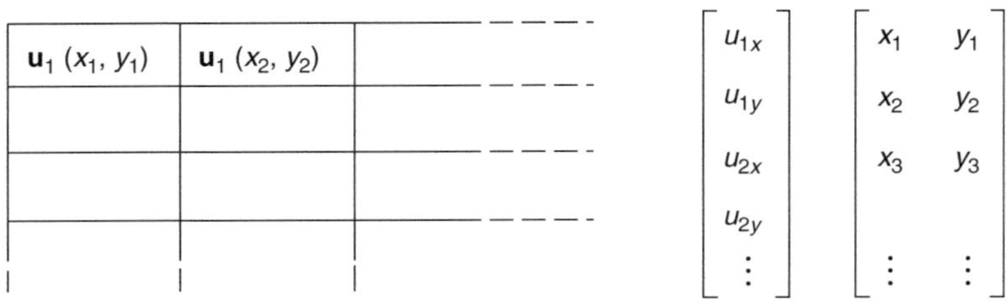

Fig. 6.2 Storage of matrices in multiphysics

subtracted element by element. However, it would be better to use arrays for cases where there are more than two dimensions.

The description of a physical field is straightforward but could vary in different programs. First, the dimension of arrays is related to the order of vectors. For example, in a 2D water flow problem, the state variable, velocity, is a vector containing two elements. Therefore, the enumeration of the values of state variables at all points, say *N*, could be *N* by 2 or 2 by *N* matrix (or array) as shown in Fig. 6.2. However, in many cases, a 2 *N* by 1 or a 1 by 2 *N* matrix is used instead. Accordingly, the coordinates at every point also involve two numbers. Hence, the enumeration of the coordinates can be described in the same way. If the same type of arrangement is used for the variables and coordinates, then they can be *associated* in an easy way. But if they are described in distinct ways, rules to correlate them need to be defined so that computer programs can establish the mapping from one to the other. It is less common but also possible that variable values and coordinates can be combined and indexed based on the point number. All in all, the word *associated* in the definition of the physical field is very loosely defined.

Chapter 7
Tensor Analysis

Introduction

The chapter presents the basic mathematical knowledge for tensors, including both vectors and higher-order tensors. Tensor analysis is frequently called tensor calculus. But it actually includes both algebra, which involves the basic operations, and calculus, which involves more advanced differential and calculus. This chapter first gives out the notation for the description of tensors of different orders using different symbols. Considering that vectors are a special case of tensors and are the ones that we are more familiar with, we will start from the introduction to the algebra and calculus of vectors. Extensions to higher-order tensors will be made based on that. This extension covers both differential and integration of tensors. Important identities in vector calculus will also be given out to facilitate the introductions to monolithic physics and multiphysics theories in later chapters.

Notation

A notation defines the way to describe a mathematical term using symbols. Thus we first need to address this most fundamental element of the mathematical language for tensors. Using vectors as an example, we here define three notations that will be used throughout this book and in most multiphysics literature: tensor notation, full notation, and array notation. One more notation, i.e., index notation, will be introduced in the last chapter of this part of the book. The tensor notation uses distinct symbols for different types of tensors. In this book, a scalar is written in italics, e.g., α; a vector is usually represented by a bold English letter, e.g., $\mathbf{u}$; and a tensor can be a bold Greek letter or a bold capital English letter, e.g., $\boldsymbol{\sigma}$ and $\mathbf{A}$. In handwriting, it is recommended to use italics, italics with an arrow, and italics with a bar, i.e., u, $\vec{u}$, and $\bar{u}$, for scalars, vectors, and higher-order tensors, respectively.

Z. Liu, *Multiphysics in Porous Materials*,
https://doi.org/10.1007/978-3-319-93028-2_7

We can conveniently write down physical quantities with different natures and physical mechanisms/processes involving these quantities using the tensor notation. This is in fact what we usually do. However, such practice requires the users to have a good understanding of the meanings of these symbols and their operation. Unfortunately, we cannot directly read the meanings from the symbols without training. To better illustrate what tensors and their operations stand for, we can employ the full notation and the array notation.

In the full notation, a scalar is a real number while a vector is the sum of the products of a number and a basis vector. Considering a three-dimensional Euclidean space, then there will be an arbitrary basis ($\mathbf{e}_1$,$\mathbf{e}_2$,$\mathbf{e}_2$) associated with the three dimensions. In general, the basis vectors are unnecessarily unit vectors nor mutually orthogonal. However, they must be linearly independent. Then a vector **u** can be expressed as

$$\mathbf{u} = u_1\mathbf{e}_1 + u_2\mathbf{e}_2 + u_3\mathbf{e}_3. \tag{7.1}$$

The Einstein summation convention is usually used with the full notation to achieve notational brevity. A term with an index variable appearing twice implies summation of that term over all the values of the index. Accordingly, the above vector can be written as

$$\mathbf{u} = u_i\mathbf{e}_i. \tag{7.2}$$

In a three Cartesian coordinate system, i will enumerate from the index of the first axis, 1, to that of the third axis, 3.

The array notation is also helpful for understanding the meaning of tensors and their operations, especially for visualization. For example, the above vector can be represented using an array such as

$$\mathbf{u} = [u_1, u_2, u_3] \tag{7.3}$$

or

$$\mathbf{u} = \begin{bmatrix} u_1 \\ u_2 \\ u_3 \end{bmatrix}. \tag{7.4}$$

Either a row or a column array works. However, operations with the tensor need to be defined accordingly to ensure consistency.

Vector Algebra

The basic algebraic operations, i.e., non-differential, in vector calculus are referred to as vector algebra. These operations are defined for a vector space and globally applied to a vector field associated with space. Basic algebraic operations consist of:

1. Vector addition: addition of two vector fields, yielding a vector field: $\mathbf{u} + \mathbf{v}$
2. Scalar multiplication: multiplication of a scalar field α and a vector field $\mathbf{u}$, yielding a vector field: $\alpha\mathbf{u}$
3. Dot product: multiplication of two vector fields, yielding a scalar field: $\mathbf{u} \cdot \mathbf{v}$
4. Cross product: multiplication of two vector fields, yielding a vector field: $\mathbf{u} \times \mathbf{v}$
5. Scalar triple product: multiplication of three vectors with two different multiplication operators, yielding a scalar field: $\mathbf{u} \cdot (\mathbf{v} \times \mathbf{w})$
6. Vector triple product: multiplication of three vectors with two cross product operators, yielding a vector field: $\mathbf{u} \times (\mathbf{v} \times \mathbf{w})$

Vector Additions and Scalar Multiplication

Vector addition can be understood via the resultant of two forces. When two vectors are placed head to tail, their sum is the diagonal pointing from the free tail to the free head. As shown in Fig. 7.1, the subtraction operation is similar. Therefore, the operation can be represented using the full notation and array notation in the following way:

$$\mathbf{u} + \mathbf{v} = u_i\mathbf{e}_i + v_i\mathbf{e}_i = (u_i + v_i)\mathbf{e}_i, \tag{7.5}$$

$$\mathbf{u} + \mathbf{v} = [u_1 + v_1, u_2 + v_2, u_3 + v_3]. \tag{7.6}$$

Basic laws of algebra, i.e., associative, commutative, and distributive laws, are valid for scalar multiplication and vector addition whenever applies. The above equations show that the vector addition is the addition of corresponding components. Since the additions of the corresponding components obey the commutative law, the commutative law holds for vector addition:

$$\mathbf{u} + \mathbf{v} = \mathbf{v} + \mathbf{u}. \tag{7.7}$$

Similarly, we can easily obtain the associative law

$$(\mathbf{u} + \mathbf{v}) + \mathbf{w} = \mathbf{u} + (\mathbf{v} + \mathbf{w}) \tag{7.8}$$

is valid. As a result, the parentheses in the above equation can be omitted without any ambiguities. The distributive law does not apply to the vector addition.

In scalar multiplication, a scalar will be multiplied with each component of the vector. That is, $\alpha\mathbf{u}$ or $\mathbf{u}\alpha$ is defined as a vector which has a length of $|\alpha||\mathbf{u}|$ and a direction identical to that of $\mathbf{u}$ when α is positive and opposite to that of $\mathbf{u}$ when α is

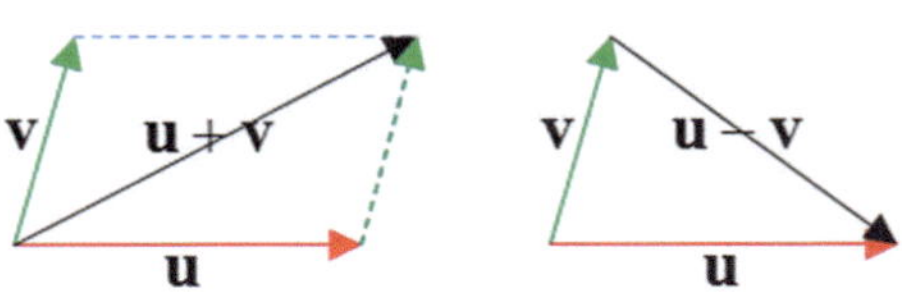

Fig. 7.1 Schematic of vector addition and subtraction

negative. Due to the reason,**u** and $-\mathbf{u}$ are vectors equal in the magnitude but opposite in the direction. Scalar multiplication obeys all of the three rules:

$$\alpha\mathbf{u} = \mathbf{u}\alpha \tag{7.9}$$

$$\alpha(\beta\mathbf{u}) = (\alpha\beta)\mathbf{u} \tag{7.10}$$

$$(\alpha + \beta)\mathbf{u} = \alpha\mathbf{u} + \beta\mathbf{u} \tag{7.11}$$

$$\alpha(\mathbf{u} + \mathbf{v}) = \alpha\mathbf{u} + \alpha\mathbf{v}. \tag{7.12}$$

As can be seen, these laws are identical to those in ordinary algebra.

Products of Vectors

The multiplication of vectors is different from that of scalars. Especially, two types of products are defined: dot product and cross product.

The dot or scalar product of two vectors **u** and **v**, written as $\mathbf{u} \cdot \mathbf{v}$, produces a real number $|\mathbf{u}| \cdot |\mathbf{v}| \cos(\mathbf{u}, \mathbf{v})$:

$$\mathbf{u} \cdot \mathbf{v} = |\mathbf{u}| \cdot |\mathbf{v}| \cos(\mathbf{u}, \mathbf{v}), \tag{7.13}$$

where $(\mathbf{u}, \mathbf{v})$ denotes the angle between the directions of **u** and **v**. As shown in Fig. 7.2, the dot product of two vectors thus can be understood as the elongated (according to the magnitude of the vector along the projection direction) projection of one vector on the direction of the other.

If **u** and **v** are at right angles, then $\mathbf{u} \cdot \mathbf{v} = 0$; and if neither **u** nor **v** is a zero vector, then a zero dot product indicates that the two vectors are perpendicular. If $\mathbf{u} = \mathbf{v}$ then $\cos(\mathbf{u} \cdot \mathbf{v}) = 1$, and $\mathbf{u} \cdot \mathbf{u} = |\mathbf{u}|^2$ gives the square of the length of **u**. It is also called scalar product in the context of tensor analysis.

The cross product can also be formulated using the full notation as follows:

$$\mathbf{u} \cdot \mathbf{v} = (u_i\mathbf{e}_i) \cdot (v_j\mathbf{e}_j) = (u_iv_j)\mathbf{e}_i \cdot \mathbf{e}_j. \tag{7.14}$$

In Cartesian coordinate systems, the axes are perpendicular to each other; therefore, the dot product of unit factors, $\mathbf{e}_i \cdot \mathbf{e}_j$, is 0 when they are different vectors and is 1 when they are identical. The Kronecker delta is defined as

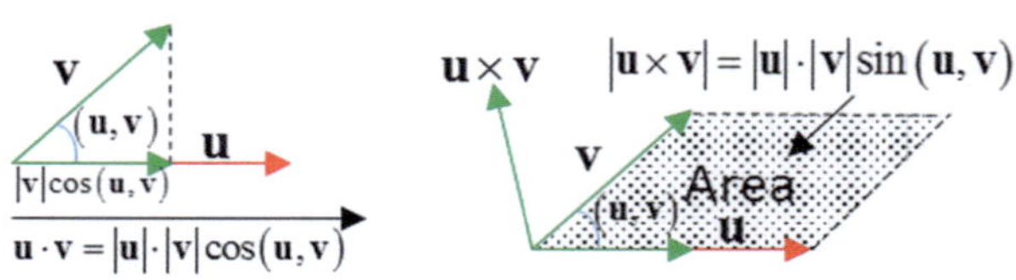

Fig. 7.2 Schematic of dot and cross products

$$\delta_{ij} = \mathbf{e}_i \cdot \mathbf{e}_j = \begin{cases} 0 & if \ \ i \neq j \\ 1 & \text{if} \ \ i = j \end{cases} \tag{7.15}$$

The vector form of the Kronecker delta is the identity matrix, which has ones on the main diagonal and zeros elsewhere. The symbols $\boldsymbol{\delta}$ and $\mathbf{I}$ are reserved for this matrix in this book.

Using the Kronecker delta, the dot product can be represented as

$$\mathbf{u} \cdot \mathbf{v} = (u_i v_j)\delta_{ij}. \tag{7.16}$$

In the above summation, any product with $i \neq j$ will be equal to 0; therefore, the above equation can be further simplified into

$$\mathbf{u} \cdot \mathbf{v} = u_i v_i. \tag{7.17}$$

This way of simplification to remove δ_{ij} is very useful and widely used in deductions in multiphysics. The above equation, in fact, gives out the definition of dot product using the array notation:

$$\mathbf{u} \cdot \mathbf{v} = [u_1 \ \ u_2 \ \ u_3] \begin{bmatrix} v_1 \\ v_2 \\ v_3 \end{bmatrix} = \sum u_i v_i. \tag{7.18}$$

It is identical to Eq. 7.17 considering Einstein's summation convention used with the full notation.

The commutative and distributive laws of elementary algebra are valid for the dot multiplication of vectors. The associative law does not apply here because the dot product of two vectors is a scalar, which is not able to form a dot product with another vector.

The cross or vector product of two vectors $\mathbf{u}$ and $\mathbf{v}$, written as $\mathbf{u} \times \mathbf{v}$, is defined as

$$\mathbf{u} \times \mathbf{v} = \mathbf{n}|\mathbf{u}| \cdot |\mathbf{v}| \sin(\mathbf{u}, \mathbf{v}), \tag{7.19}$$

where $\mathbf{n}$ is a vector of a unit length and a direction perpendicular to the plane of $\mathbf{u}$ and $\mathbf{v}$. Therefore, the cross product of two vectors generates another vector (Fig. 7.2). The direction of this vector can be obtained using the right-hand rule in which a right-handed screw that rotates from $\mathbf{u}$ toward $\mathbf{v}$ will advance in the direction of $\mathbf{n}$. If $\mathbf{u}$ and $\mathbf{v}$ are parallel, then $\mathbf{u} \times \mathbf{v} = 0$. The magnitude of $\mathbf{u} \times \mathbf{v}$ is identical to the area of the parallelogram which has $\mathbf{u}$ and $\mathbf{v}$ as two adjacent sides.

The cross product can be defined using the full notation as

$$\mathbf{u} \times \mathbf{v} = (u_i \mathbf{e}_i) \times (v_j \mathbf{e}_j) = (u_i v_j)\mathbf{e}_i \times \mathbf{e}_j. \tag{7.20}$$

In a Cartesian coordinate system, the special nature of $\mathbf{e}_i \times \mathbf{e}_j$ will lead to the definition of the Levi-Civita symbol, ε_{ijk}:

$$\varepsilon_{ijk} = \begin{cases} 0 & \text{if } i = j, \text{or } j = k, \text{or } k = i \\ +1 & \text{if}(i,j,k) \text{ is } (i,j,k), (i,j,k), \text{or } (i,j,k) \\ -1 & \text{if}(i,j,k) \text{ is } (i,j,k), (i,j,k), \text{or } (i,j,k) \end{cases} \tag{7.21}$$

Then the cross product can be simplified into

$$\mathbf{u} \times \mathbf{v} = u_i v_j \mathbf{e}_k \varepsilon_{ijk}. \tag{7.22}$$

Using the array notation, we will obtain the following definition of the cross product:

$$\mathbf{u} \times \mathbf{v} = \begin{vmatrix} \mathbf{e}_i & \mathbf{e}_j & \mathbf{e}_k \\ u_i & u_j & u_k \\ v_i & v_j & v_k \end{vmatrix}. \tag{7.23}$$

The associative law does not hold for the cross product operation. Instead, we can use Lagrange's formula and the Jacobi identity, which will be given out in the next section for the significant identities in tensor analysis. In fact, the cross product of three vectors is usually treated as a separate operation, which will be introduced later in this section. Although the associative law does not apply, the following equation is valid:

$$(\mathbf{A}\mathbf{u}) \times \mathbf{v} = \mathbf{A}(\mathbf{u} \times \mathbf{v}), \tag{7.24}$$

where $\mathbf{A}$ can be a tensor of any order: $\mathbf{A}\mathbf{a}$ or $\mathbf{A}(\mathbf{a} \times \mathbf{b})$ is a scalar product when $\mathbf{A}$ is a scalar and an outer product when $\mathbf{A}$ is a vector and higher-order tensors.

The distributive law is valid for cross products:

$$\mathbf{u} \times (\mathbf{v} + \mathbf{w}) = \mathbf{u} \times \mathbf{v} + \mathbf{u} \times \mathbf{w}. \tag{7.25}$$

However, the cross product is not commutative. This is because the rotation from $\mathbf{u}$ to $\mathbf{v}$ is opposite to that from $\mathbf{v}$ to $\mathbf{u}$:

$$\mathbf{u} \times \mathbf{v} = -\mathbf{v} \times \mathbf{u}. \tag{7.26}$$

Hence, a counter-commutative law is instead valid for the cross product.

Vector Calculus

Vector calculus is the mathematical study of the change of vectors. Similar to the calculus of scalars, vector calculus includes differential calculus and integral calculus. Changes of the vector, both differential and integral, are formulated through several vector operators. Vector operators are defined in terms of del. Del is a vector differential operator, usually represented by the nabla symbol ∇. In a three-dimensional Cartesian coordinate system with coordinates (x_1, x_2, x_3) and a standard basis $(\mathbf{e}_1, \mathbf{e}_2, \mathbf{e}_3)$, del is written as

$$\nabla = \left(\frac{\partial}{\partial x_1}, \frac{\partial}{\partial x_2}, \frac{\partial}{\partial x_3}\right) = \frac{\partial}{\partial x_1}\mathbf{e}_1 + \frac{\partial}{\partial x_2}\mathbf{e}_2 + \frac{\partial}{\partial x_3}\mathbf{e}_3 = \frac{\partial}{\partial x_i}\mathbf{e}_i, \tag{7.27}$$

where $\mathbf{e}_i$ is the unit vector along the direction of the coordinate axes i and i=1, 2, or 3.

Other common vector operators include gradient, divergence, and curl, which are defined using del in the following.

1. Gradient is a generalization of the concept of derivative of a function in one dimension into a function in several dimensions. The operator maps a scalar field to a vector field:

$$\begin{aligned} \text{grad} &= \nabla \\ \nabla u &= \frac{\partial u}{\partial x_1}\mathbf{e}_1 + \frac{\partial u}{\partial x_2}\mathbf{e}_2 + \frac{\partial u}{\partial x_3}\mathbf{e}_3 = \frac{\partial u}{\partial x_i}\mathbf{e}_i \\ \nabla \mathbf{u} &= \frac{\partial \mathbf{u}}{\partial x_1}\mathbf{e}_1 + \frac{\partial \mathbf{u}}{\partial x_2}\mathbf{e}_2 + \frac{\partial \mathbf{u}}{\partial x_3}\mathbf{e}_3 = \frac{\partial (u_j\mathbf{e}_j)}{\partial x_i}\mathbf{e}_i = \frac{\partial u_j}{\partial x_i}\mathbf{e}_i\mathbf{e}_j \end{aligned} \tag{7.28}$$

2. Divergence is a vector operator that measures the magnitude of the source or sink in a vector field at any given point, in terms of a signed scalar. The operator maps a vector field to a scalar field:

$$\begin{aligned} \text{div} &= \nabla\cdot \\ \nabla\cdot\mathbf{u} &= \frac{\partial (u_j\mathbf{e}_j)}{\partial x_i}\cdot\mathbf{e}_i = \frac{\partial u_j}{\partial x_i}\mathbf{e}_i\cdot\mathbf{e}_j = \frac{\partial u_j}{\partial x_i}\delta_{ij} = \frac{\partial u_i}{\partial x_i} \end{aligned} \tag{7.29}$$

3. Curl is a vector operator that describes the infinitesimal rotation of a three-dimensional vector field. The operator maps a vector field to another vector field:

$$\begin{aligned} \text{curl} &= \nabla\times \\ \nabla\times\mathbf{u} &= \begin{vmatrix} \mathbf{e}_i & \mathbf{e}_j & \mathbf{e}_k \\ \frac{\partial}{\partial x_i} & \frac{\partial}{\partial x_j} & \frac{\partial}{\partial x_k} \\ u_i & u_j & u_k \end{vmatrix} = \frac{\partial u_i}{\partial x_j}\mathbf{e}_k\varepsilon_{ijk} \end{aligned} \tag{7.30}$$

4. Laplacian measures the difference between the value of the scalar/vector field and its average on infinitesimal balls. The operator maps between scalar/vector fields:

$$\nabla^2 = \text{div grad} = \nabla\cdot\nabla = \Delta \tag{7.31}$$

$$\Delta u = \frac{\partial\left(\frac{\partial u}{\partial x_j}\mathbf{e}_j\right)}{\partial x_i}\cdot\mathbf{e}_i = \frac{\partial^2 u_i}{\partial x_i^2} \tag{7.32}$$

Therefore, these operators could change the order of the vectors that they act on. The order of tensors is a very useful tool to examine the consistency of governing equations and physical terms in multiphysics in a way similar to units.

Vector operators need to be placed right in front of a scalar or vector to take effect. Therefore ∇f represents the gradient of f while $f\nabla$ just produces another vector operator, which does not act on anything. Also, as just mentioned, a vector operator can operate on another vector operator to produce a new compound vector operator. Finally, the curl and divergence differ because the former uses a cross product while the latter uses a dot product.

Theorems

Three common theorems in calculus, i.e., Green's theorem, Stokes' theorem, and Gauss (divergence) theorem, are frequently encountered in the derivation of the equations in multiphysics. These theorems for the calculus of scalars can be generalized to higher-order tensors.

Green's theorem states that the integral of the scalar curl of a vector field over some regions in a plane equals the line integral of the vector field over the closed curve bounding the region oriented in the counterclockwise direction. This theorem can be formulated using vector operators as

$$\int_S \nabla \cdot \mathbf{F} dS = \oint_{\partial S} \mathbf{F} \cdot \mathbf{n} dL. \tag{7.33}$$

In Stokes' theorem, the integral of the curl of a vector field over a surface in R^3 equals the line integral of the vector field over the closed curve bounding the surface. The corresponding mathematical formulation is

$$\iint_{\sigma \subset R^3} \nabla \times \mathbf{F} \cdot d\boldsymbol{\sigma} = \oint_{\partial \sum} \mathbf{F} \cdot d\mathbf{r}. \tag{7.34}$$

The divergence theorem says that the integral of the divergence of a vector field over a volume equals the integral of the flux through the closed surface bounding the volume, which is formulated as

$$\iiint_{V \subset R^3} (\nabla \cdot \mathbf{F}) dV = \oiint_{\partial V} \mathbf{F} \cdot d\mathbf{S}. \tag{7.35}$$

It is worthwhile to point out that the above three theorems can be formulated using a general form:

$$\int_{\Omega}(\nabla * \mathbf{f})d\Omega = \int_{\partial\Omega}(\mathbf{n}^* \ \mathbf{f})d\Omega = \int_{\partial\Omega} d\boldsymbol{\Omega}^* \ \mathbf{f}, \tag{7.36}$$

where * is used to represent any tensor product, i.e., inner, outer, and cross products.

Basics of Tensor Analysis

Tensor analysis can be understood as the extension or generalization of the vector calculus to higher-order tensors. Therefore, it is concerned with relations or laws that remain valid regardless of the coordinate system for describing the quantities. In fact, tensor analysis includes the definition, formulation, operation, differentiation, and integration of tensors from scalars to higher-order tensors in various coordinate systems in a general sense. Multiphysics deals with physical fields, which involve the changes of tensors of different orders. Therefore, tensor analysis is an essential tool and necessary language in multiphysics. Tensor analysis is widely used in most science and engineering topics that require a continuum description. Due to time constraints, only the most relevant and critical content is included here. More details will be provided in later chapters whenever necessary.

Definition

Tensors are a generalization of scalars and vectors. Many different definitions from different viewpoints are available, such as multilinear algebra and differential geometry. But in a simple way, tensors can be just viewed as arrays of numbers or functions, which transform according to certain rules under a change of coordinates. Tensors can also be understood as geometric objects that describe linear relations between vectors, scalars, and other tensors. Examples of such relations include the dot product, the cross product, and the linear maps. In physics, tensors are used to characterize the properties or describe the state of a physical system.

For a simple and smooth introduction, let us first get back to the definition of first-order tensors, i.e., vectors. In a three-dimensional Euclidean space with an arbitrary basis ($\mathbf{e}_1$, $\mathbf{e}_2$, $\mathbf{e}_3$), a vector $\mathbf{u}$ can be expressed as

$$\mathbf{u} = u_1\mathbf{e}_1 + u_2\mathbf{e}_2 + u_3\mathbf{e}_3 = u_i\mathbf{e}_i. \tag{7.37}$$

Similarly, for a second-order tensor, two sets of bases will be needed to reflect the order of "two." If we use $\mathbf{m}_i$ and $\mathbf{n}_j$ for the two sets of basis, a general second-order tensor $\mathbf{A}$ can be written as

$$\mathbf{A} = a_{ij}\mathbf{m}_i\mathbf{n}_j. \tag{7.38}$$

Similar to vectors, we can also use two-dimensional arrays (matrices) to represent and visualize tensors. For example, a second-order tensor in this case can be visualized as a two-dimensional (or second rank) matrix:

$$\mathbf{A} = \begin{bmatrix} a_{11} & a_{12} & \dots & a_{1n} \\ a_{21} & a_{22} & \dots & a_{2n} \\ \dots & \dots & & \\ a_{m1} & a_{m2} & \dots & a_{mn} \end{bmatrix}. \tag{7.39}$$

The above notation only includes the components while omitting the basis. Therefore, the implication of the matrix notation is that each element will go with an invisible basis vector. An element is the projection of the tensor along the direction of its associated basis vector.

When the order of a tensor gets beyond 2, it is usually not easy to visualize tensors in terms of matrix anymore. But for scalars, vectors, and second-order tensors, which are extensively used in FDEs, matrices are still a good tool to help us understand the structures of tensors and the operations between them.

Tensor Product

The tensor product is a type of operation between tensors. This operation also applies to vectors, though we skipped it in the section for vectors. In linear algebra, the term outer product is typically used to refer to the tensor product of two vectors. In the dyadic context, dyadic product, outer product, and tensor product all share the same definition and meaning and thus are used synonymously. However, the tensor product is the most general and abstract term among them.

There are several equivalent terms and notations for this product:

1. The dyadic product of two vectors **u** and **v** is denoted by the juxtaposition of them.
2. The outer product of two column vectors **u** and **v** is denoted and defined as $\mathbf{u} \otimes \mathbf{v}$ or $\mathbf{u}\mathbf{v}^{\mathrm{T}}$, where T means transpose.
3. The tensor product of two vectors **u** and **v**is denoted by $\mathbf{u} \otimes \mathbf{v}$.

The above usages can be proven to be equivalent. Consider a three-dimensional Euclidean space, in which we have the two following vectors:

$$\mathbf{u} = u_1\mathbf{e}_1 + u_2\mathbf{e}_2 + u_3\mathbf{e}_3 \tag{7.40}$$

$$\mathbf{v} = v_1\mathbf{e}_1 + v_2\mathbf{e}_2 + v_3\mathbf{e}_3, \tag{7.41}$$

where $\mathbf{e}_1$, $\mathbf{e}_2$, and $\mathbf{e}_3$ are the standard basis vectors in this vector space. Then the dyadic product of **u** and **v** can be represented as a sum:

$$\begin{aligned}\mathbf{uv} = &u_1v_1\mathbf{e}_1\mathbf{e}_1 + u_1v_2\mathbf{e}_1\mathbf{e}_2 + u_1v_3\mathbf{e}_1\mathbf{e}_3 \\ &+ u_2v_1\mathbf{e}_2\mathbf{e}_1 + u_2v_2\mathbf{e}_2\mathbf{e}_2 + u_2v_3\mathbf{e}_2\mathbf{e}_3 \\ &+ u_3v_1\mathbf{e}_3\mathbf{e}_1 + u_3v_2\mathbf{e}_3\mathbf{e}_2 + u_3v_3\mathbf{e}_3\mathbf{e}_3.\end{aligned} \tag{7.42}$$

While using row and column vectors, the result of the outer product or tensor product of **u** and **v** is a 3 × 3 matrix:

$$\mathbf{uv} = \mathbf{u} \otimes \mathbf{v} = \mathbf{uv}^{\mathrm{T}} = \begin{pmatrix} u_1 \\ u_2 \\ u_3 \end{pmatrix} \begin{pmatrix} v_1 & v_2 & v_3 \end{pmatrix} = \begin{pmatrix} u_1v_1 & u_1v_2 & u_1v_3 \\ u_2v_1 & u_2v_2 & u_2v_3 \\ u_3v_1 & u_3v_2 & u_3v_3 \end{pmatrix}. \tag{7.43}$$

These two operations are thus essentially equivalent.

Unlike tensor products and outer products, which apply to the products of two tensors of any order, the dyad product is usually confined to the product of two vectors. In the dyad product, a dyad, e.g., u_1v_1, is a component of the dyadic **uv**. The dyad thus can be viewed as a monomial of the sum, equivalent entry of the matrix, or the juxtaposition of a pair of basis vectors scalar multiplied by a number.

Identities and Algebra

Tensor product is associative and distributive but not commutative.

As for the associative law, we have

$$(\mathbf{uv})\mathbf{w} = \mathbf{u}(\mathbf{vw}) = \mathbf{u}(\mathbf{vw}). \tag{7.44}$$

The associative low is compatible with scalar multiplication for any scalar α:

$$(\alpha\mathbf{u})\mathbf{v} == \alpha(\mathbf{uv}) = \mathbf{u}(\alpha\mathbf{v}). \tag{7.45}$$

The term dyadic algebra is introduced as an extension of vector algebra to include the dyadic product of vectors. In this case, dyadic algebra, dot product, and cross product may coexist. When working with dot and cross products, four operations can be defined on a vector and dyadic with the three forms of vector products. As shown in the following table, the associative law still applies for the coexistence of these products.

	Left	Right
Dot product	$\mathbf{c} \cdot \mathbf{ab} = (\mathbf{c} \cdot \mathbf{a})\mathbf{b}$	$\mathbf{ab} \cdot \mathbf{c} = \mathbf{a}(\mathbf{b} \cdot \mathbf{c})$
Cross product	$\mathbf{c} \times \mathbf{ab} = (\mathbf{c} \times \mathbf{a})\mathbf{b}$	$\mathbf{ab} \times \mathbf{c} = \mathbf{a}(\mathbf{b} \times \mathbf{c})$

When dealing with higher-order tensors, we can still take advantage of the above rules by treating a higher-order tensor as the juxtaposition of multiple vectors. A third-order tensor, $\mathbf{A} = A_{ijk}\mathbf{e}_i\mathbf{e}_j\mathbf{e}_k$, for example, can be treated as the juxtaposition, or tensor product, of three vectors:

$$\mathbf{A} = A_{ijk}\mathbf{e}_i\mathbf{e}_j\mathbf{e}_k = \mathbf{uvw} = (u_i\mathbf{e}_i)(v_j\mathbf{e}_j)(w_k\mathbf{e}_k), \tag{7.46}$$

in which $A_{ijk} = u_i v_j w_k$.

The tensor product operation obeys the distributive law:

$$\mathbf{u}(\mathbf{v} + \mathbf{w}) = \mathbf{uv} + \mathbf{uw}. \tag{7.47}$$

It is noted that the tensor product operation is not commutative, that is,

$$\mathbf{uv} \neq \mathbf{vu}. \tag{7.48}$$

Contraction and Inner Product

The inner product is another operator in need of further explanation for tensor algebra. The inner product is used interchangeably with the dot product in many occasions. Their difference lies in that the inner product generalizes the dot product to abstract vector spaces over a field of scalars, either real or complex numbers. The definition of inner product in this way is usually written as $\langle \mathbf{a}, \mathbf{b} \rangle$. In mathematics, the dot product (or called scalar product), or the more general inner product in the context of Euclidean space, is an algebraic operation that multiplies two equal-size arrays of numbers, e.g., coordinate vectors, and returns a single number. The dot product of two vectors

$$\mathbf{u} = (u_1, u_2, \ldots, u_n) \tag{7.49}$$

and

$$\mathbf{v} = (v_1, v_2, \ldots, v_n) \tag{7.50}$$

is defined as

$$\mathbf{u} \cdot \mathbf{v} = u_1 v_1 + u_2 v_2 + \ldots + u_n v_n = \sum_{i=1}^{n} u_i v_i. \tag{7.51}$$

It is seen from the symbols that the major difference between outer product and inner product is the symbol "." which is called contraction or tension contraction. The contract can be defined as follows:

$$\mathbf{e}_i \cdot \mathbf{e}_j = \delta_{ij}. \tag{7.52}$$

In a simple way, we can understand the contraction operation as relating two vectors on two sides of the operator using the same Cartesian coordinate system. Therefore, the two sets of bases associated with the two vectors degenerate into the Kronecker delta. So you can see that this operation reduces the order of a tensor by two. That is why the dot product makes the combination of two first-order tensors (vectors, 1 + 1) into a scalar (1 + 1–2 = 0). Besides dot product, two types of double-dot products are also useful in manipulating dyads: the vertical and horizontal double-dot products (Taber 2004):

$$\boldsymbol{\sigma}\mathbf{:}\boldsymbol{\tau} = \left(\sigma_{ij}\mathbf{e}_i\mathbf{e}_j\right)\mathbf{:}\left(\tau_{kl}\mathbf{e}_k\mathbf{e}_l\right) = \sigma_{ij}\tau_{kl}\left(\mathbf{e}_i \cdot \mathbf{e}_k\right)\left(\mathbf{e}_j \cdot \mathbf{e}_l\right) = \sigma_{ij}\tau_{kl}\delta_{ik}\delta_{jl} = \sigma_{il}\tau_{il} \tag{7.53}$$

$$\begin{aligned}\boldsymbol{\sigma} \cdot \cdot \boldsymbol{\tau} &= \left(\sigma_{ij}\mathbf{e}_i\mathbf{e}_j\right) \cdot \cdot \left(\tau_{kl}\mathbf{e}_k\mathbf{e}_l\right) \\ &= \sigma_{ij}\tau_{kl}\left(\mathbf{e}_i \cdot \mathbf{e}_l\right)\left(\mathbf{e}_j \cdot \mathbf{e}_k\right) = \sigma_{ij}\tau_{kl}\delta_{il}\delta_{jk} = \sigma_{ik}\tau_{ki}.\end{aligned} \tag{7.54}$$

These two operators are not commutative. The rule can be seen in the following operation:

$$\mathbf{abc} : \mathbf{def} = \mathbf{a}(\mathbf{bc} : \mathbf{de})\mathbf{f} = (\mathbf{b} \cdot \mathbf{d})(\mathbf{c} \cdot \mathbf{e})\mathbf{af}. \tag{7.55}$$

Important Identities in Vector Calculus

Operator and Order of Tensor

First, a few typical examples are given in the following to show the changes that different operators can cause to the order of the tensors that the operators act on.

$\mathrm{div}(\mathbf{u}) = \nabla \cdot \mathbf{u}$ is a scalar
$\mathrm{div}(\boldsymbol{\sigma}) = \nabla \cdot \boldsymbol{\sigma}$ is a vector
$\mathrm{curl}(\mathbf{u}) = \nabla \times \mathbf{u}$ is a vector
$\mathrm{grad}(\varphi) = \nabla\varphi$ is a vector
$\mathrm{grad}(\mathbf{u}) = \nabla\mathbf{u}$ is a second-order tensor

Special Combination of Multiple Operators

$$\nabla \times (\nabla\varphi) = 0 \tag{7.56}$$

$$\nabla \cdot (\nabla \times \mathbf{v}) = 0 \tag{7.57}$$

Distributivity over Vector Addition

$$\nabla \cdot (\mathbf{v}_1 + \mathbf{v}_2) = \nabla \cdot \mathbf{v}_1 + \nabla \cdot \mathbf{v}_2 \tag{7.58}$$

$$\nabla \times (\mathbf{v}_1 + \mathbf{v}_2) = \nabla \times \mathbf{v}_1 + \nabla \times \mathbf{v}_2 \tag{7.59}$$

Product of Scalar and Vector

$$\nabla(\psi\varphi) = \varphi\nabla\psi + \psi\nabla\varphi \tag{7.60}$$

$$\nabla \cdot (\varphi\mathbf{v}) = (\nabla\varphi) \cdot \mathbf{v} + \varphi(\nabla \cdot \mathbf{v}) \tag{7.61}$$

$$\nabla \times (\varphi\mathbf{v}) = (\nabla\varphi) \times \mathbf{v} + \varphi\nabla \times \mathbf{v} \tag{7.62}$$

Scalar Triple Product

The scalar triple product (also called the mixed or box product) is defined as the dot product of one vector and the cross product of the other two:

$$\mathbf{u} \cdot (\mathbf{v} \times \mathbf{w}).$$

As for the scalar triple product, there exists an important relationship:

$$\mathbf{u} \cdot (\mathbf{v} \times \mathbf{w}) = \mathbf{v} \cdot (\mathbf{w} \times \mathbf{u}) = \mathbf{w} \cdot (\mathbf{u} \times \mathbf{v}). \tag{7.63}$$

The above relationship leads to another important deduction:

$$\mathbf{u} \cdot (\mathbf{u} \times \mathbf{v}) = \mathbf{u} \cdot (\mathbf{v} \times \mathbf{u}) = \mathbf{u} \cdot (\mathbf{v} \times \mathbf{v}) = \mathbf{u} \cdot (\mathbf{u} \times \mathbf{u}) = 0. \tag{7.64}$$

The gradient operator, ∇, can be treated as a normal vector in the scalar triple product; thus

$$\nabla \cdot (\mathbf{u} \times \mathbf{v}) = \mathbf{u} \cdot (\mathbf{v} \times \nabla) = \mathbf{v} \cdot (\nabla \times \mathbf{u}). \tag{7.65}$$

Vector Triple Product

While the scalar triple product produces a scalar, the vector triple product will lead to a vector. The vector triple product is defined as the cross product of one vector and the cross product of the other two. The following relationship holds:

$$\mathbf{u} \times (\mathbf{v} \times \mathbf{w}) = \mathbf{v}(\mathbf{u} \cdot \mathbf{w}) - \mathbf{w}(\mathbf{u} \cdot \mathbf{v}). \tag{7.66}$$

This is known as the triple product expansion, or Lagrange's formula, although the latter name is also used for several other formulae. Its right-hand side can be remembered by using the mnemonic "BAC-CAB," provided one keeps in mind which vectors are dotted together.

Similar to the scalar triple product, the gradient vector can be treated as a normal vector in the vector triple product. A vector triple product with two gradient vectors in it leads to

$$\nabla \times (\nabla \times \mathbf{u}) = \nabla(\nabla \cdot \mathbf{u}) - \nabla^2 \mathbf{u}. \tag{7.67}$$

If we replace only one of the vectors with the gradient, then we can obtain

$$\nabla \times (\mathbf{u} \times \mathbf{v}) = \mathbf{u}(\nabla \cdot \mathbf{v}) - \mathbf{v}(\nabla \cdot \mathbf{u}) \tag{7.68}$$

$$\nabla(\mathbf{u} \cdot \mathbf{v}) = \mathbf{u} \times (\nabla \times \mathbf{v}) + \mathbf{v}(\nabla \cdot \mathbf{u}). \tag{7.69}$$

In addition, Lagrange's formula can lead to another important relationship, the Jacobi identity:

$$\mathbf{u} \times (\mathbf{v} \times \mathbf{w}) + \mathbf{v} \times (\mathbf{w} \times \mathbf{u}) + \mathbf{w} \times (\mathbf{u} \times \mathbf{v}) = 0. \tag{7.70}$$

Chapter 8
Index Notation and Tensor Notation

Introduction

The full notation and array notation are very helpful when introducing the operations and rules in tensor analysis. However, tensor notation and index notation are more commonly used in the context of partial differential equations and tensor analysis. The tensor notation just requires the utilization of different symbols for tensors of different orders and the use of appropriate symbols as operators connecting these tensors. The tensor notation thus enables us to write PDEs in a concise way, which is also independent of the adopted coordinate system. But in many cases, the index notation is preferred as it is proven to be much more powerful for occasions such as derivations. In this chapter, we will start from the basic rules of the index notation, then move to the use of the index notation for tensor algebra, and finally reach the calculus in terms of the index notation. At the end of the chapter, two examples will be given to show the algebraic manipulations, i.e., derivations, using the index notation.

Rules of Index Notation

In the index notation, indices are categorized into two groups: free indices and dummy indices. A free index means an "independent dimension" or an order of the tensor whereas a dummy index means summation. The following three basic rules must be met for the index notation:

1. The same index (subscript) may not appear more than twice in a product of two (or more) vectors or tensors. Thus

Z. Liu, *Multiphysics in Porous Materials*,
https://doi.org/10.1007/978-3-319-93028-2_8

$$A_{ik}u_k, A_{ik}B_{kj}, A_{ij}B_{jk}C_{nk}$$

are valid, but

$$A_{kk}u_k, A_{ik}B_{kk}, A_{ij}B_{ik}C_{ik}$$

are meaningless.

2. Free indices on each term of an equation must agree. Thus

$$u_i = v_i + w_i \tag{8.1}$$

$$u_i = A_{ik}B_{kj}v_j + C_{ik}w_k \tag{8.2}$$

are valid, but

$$u_i = A_{ij} \tag{8.3}$$

$$u_j = A_{ik}u_k \tag{8.4}$$

$$u_i = A_{ik}v_k + w_j \tag{8.5}$$

are meaningless.

3. Free and dummy indices may be changed without altering the meaning of an expression under the condition that Rules 1 and 2 are not violated. Thus

$$u_j = A_{jk}v_k \Leftrightarrow u_i = A_{ik}v_k \Leftrightarrow u_j = A_{ji}v_i.$$

Tensor Algebra in Index Notation

First, according to the meaning of free dimension, a vector **u** and a second-order tensor **A** should be written as u_i and A_{ij}, respectively. The free indices can be changed to other symbols. The basic operations in tensor algebra can be expressed using the index notation as follows:

Addition	$u_i = v_i + w_i \Leftrightarrow \mathbf{u} = \mathbf{v} + \mathbf{w}$	
Dot product	$\lambda = \mathbf{u} \cdot \mathbf{v} \Leftrightarrow \lambda = u_i v_i$	
Vector product	$\mathbf{u} = \mathbf{v} \times \mathbf{w} \Leftrightarrow u_i = \varepsilon_{ijk} v_j w_k$	
Dyadic product	$\mathbf{A} = \mathbf{u}\mathbf{v} \Leftrightarrow A_{ij} = u_i v_j$	
Addition	$\mathbf{C} = \mathbf{A} + \mathbf{B} \Leftrightarrow C_{ij} = A_{ij} + B_{ij}$	
Transpose	$\mathbf{A} = \mathbf{B}^T \Leftrightarrow A_{ij} = B_{ji}$	
Scalar products	$\lambda = \mathbf{A} : \mathbf{B} \Leftrightarrow \lambda = A_{ij}B_{ij}$	$\lambda = \mathbf{A} \cdot \cdot \mathbf{B} \Leftrightarrow \lambda = A_{ij}B_{ji}$

Inner product of a tensor and a vector	$\mathbf{u} = \mathbf{A}\mathbf{v} \Leftrightarrow u_i = A_{ij}v_j$
Inner product of two tensors	$\mathbf{C} = \mathbf{A} \cdot \mathbf{B} \Leftrightarrow C_{ij} = A_{ik}B_{kj}$
Determinant	$\lambda = \det\mathbf{A} \Leftrightarrow \lambda = \frac{1}{6}\varepsilon_{ijk}\varepsilon_{lmn}A_{li}A_{mj}A_{nk} = \varepsilon_{ijk}A_{l1}A_{j2}A_{k3} \Leftrightarrow \varepsilon_{lmn}$ $\lambda = \varepsilon_{ijk}A_{li}A_{mj}A_{nk} = \varepsilon_{ijk}A_{il}A_{jm}A_{kn}$
Inverse	$\mathbf{A}_{ji}^{-1} = \frac{1}{2\det(\mathbf{A})}\varepsilon_{ipq}\varepsilon_{jkl}A_{pk}A_{ql}$

Calculus Using Index Notation

The derivative with respect to one coordinate in the index notation can be represented by putting the corresponding symbols for the coordinate after a comma in the indices. The gradient of a scalar u thus can be written as

$$\nabla u = u_{,i} \tag{8.6}$$

where i is the free index. Therefore, $u_{,i}$ is still a vector since it has one free index like u_i. Accordingly, the gradient of a vector $\mathbf{u}$ is written as

$$\nabla \mathbf{u} = u_{i,j}, \tag{8.7}$$

$u_{i,\,j}$ has two free indices; hence, it is a second-order tensor.

The divergence of a vector, which can be viewed as the combination of the gradient vector and contraction, requires the contraction of two indices into one via Kronecker delta. Therefore, we will have this operation in the index notation as

$$\nabla \cdot \mathbf{u} = u_{i,i}. \tag{8.8}$$

The divergence of a second-order tensor $\mathbf{A}$ is

$$\nabla \cdot \mathbf{A} = A_{ij,j}. \tag{8.9}$$

Likewise, the curl of a vector $\mathbf{u}$ is

$$\nabla \times \mathbf{u} = \varepsilon_{ijk}u_{j,k}. \tag{8.10}$$

The Laplacian of a quantity involves both the gradient and the divergence. For a scalar u, we apply the gradient and divergence sequentially, accordingly,

$$\Delta u = u_{,ii}. \tag{8.11}$$

For a vector $\mathbf{u}$, we will have

$$\Delta \mathbf{u} = u_{i,jj}. \tag{8.12}$$

The derivative $\partial x_i/\partial x_j$ can be deduced by noting that $\partial x_i/\partial x_j = 1$when $i = j$ and $\partial x_i/\partial x_j = 0$ when $i \neq j$. Therefore $\frac{\partial x_i}{\partial x_j} = \delta_{ij}$.

The same argument can be used for higher-order tensors

$$\frac{\partial A_{ij}}{\partial A_{kl}} = \delta_{ik}\delta_{jl}. \tag{8.13}$$

Examples of Algebraic Manipulations Using Index Notation

1. Let **a**, **b**, **c**,**d** be vectors. Prove that

$$(\mathbf{a} \times \mathbf{b}) \cdot (\mathbf{c} \times \mathbf{d}) = (\mathbf{a} \cdot \mathbf{c})(\mathbf{b} \cdot \mathbf{d}) - (\mathbf{b} \cdot \mathbf{c})(\mathbf{a} \cdot \mathbf{d}). \tag{8.14}$$

Solution

Let us first express the left-hand side of the equation using index notation. Please check the rules for cross products and dot products of vectors to see how this is done.

$$(\mathbf{a} \times \mathbf{b}) \cdot (\mathbf{c} \times \mathbf{d}) = \varepsilon_{ijk} a_j b_k \varepsilon_{imn} c_m d_n \tag{8.15}$$

Recall the identity

$$\varepsilon_{ijk}\varepsilon_{imn} = \delta_{jm}\delta_{kn} - \delta_{jn}\delta_{mk}. \tag{8.16}$$

Then we will have

$$\varepsilon_{ijk} a_j b_k \varepsilon_{imn} c_m d_n = \left(\delta_{jm}\delta_{kn} - \delta_{jn}\delta_{mk}\right) a_j b_k c_m d_n. \tag{8.17}$$

Multiply out, and note that

$$\delta_{jm} a_j = a_m$$

and

$$\delta_{kn} a_k = a_n.$$

That is, multiplying a Kronecker delta has the effect of switching indices; hence

$$\left(\delta_{jm}\delta_{kn} - \delta_{jn}\delta_{mk}\right) a_j b_k c_m d_n = a_m b_n c_m d_n - a_n b_m c_m d_n \tag{8.18}$$

Finally, note that $a_m c_m = \mathbf{a} \cdot \mathbf{c}$ and similarly for other products with the same index so that

$$\begin{aligned} a_m b_n c_m d_n - a_n b_m c_m d_n &= a_m c_m b_n d_n - b_m c_m a_n d_n \\ &= (\mathbf{a} \cdot \mathbf{c})(\mathbf{b} \cdot \mathbf{d}) - (\mathbf{b} \cdot \mathbf{c})(\mathbf{a} \cdot \mathbf{d}) \end{aligned} \tag{8.19}$$

2. The stress-strain relation for linear elasticity may be expressed as

$$\sigma_{ij} = \frac{E}{1+v}\left(\varepsilon_{ij} + \frac{v}{1-2v}\varepsilon_{kk}\delta_{ij}\right) \tag{8.20}$$

where σ_{ij} and ε_{ij} are the components of the stress and strain tensor and E and v are Young's modulus and Poisson's ratio, respectively. Find an expression for strain in terms of stress.

Solution
Set $i = j$ to see that

$$\sigma_{ii} = \frac{E}{1+v}\left(\varepsilon_{ii} + \frac{v}{1-2v}\varepsilon_{kk}\delta_{ii}\right). \tag{8.21}$$

Recall that $\sigma_{ii} = 3$, and notice that we can replace the remaining ii by kk, so that

$$\sigma_{kk} = \frac{E}{1+v}\left(\varepsilon_{kk} + \frac{v}{1-2v}3\varepsilon_{kk}\right) = \frac{v}{1-2v}\varepsilon_{kk} \tag{8.22}$$

and

$$\varepsilon_{kk} = \frac{1-2v}{E}\sigma_{kk}. \tag{8.23}$$

Now, substitute for ε_{kk} in the given stress-strain relation and we obtain

$$\sigma_{ij} = \frac{E}{1+v}\left(\varepsilon_{ij} + \frac{v}{E}\sigma_{kk}\delta_{ij}\right) \tag{8.24}$$

and

$$\varepsilon_{ij} = \frac{1+v}{E}\left(\sigma_{ij} - \frac{v}{1+v}\sigma_{kk}\delta_{ij}\right). \tag{8.25}$$

Chapter 9
Partial Differential Equations

Introduction

Partial differential equations (PDEs) are another mathematical language required for expressing multiphysics in addition to tensors. The chapter starts the introduction to PDE with the role of PDE in the numerical simulation. Basic definitions including dependent variables, coordinates, domain, boundary, and numerical methods will be briefly introduced to lay down the background for the use of PDE in multiphysics. The common classification of PDEs will be discussed next. Later, the PDEs that we would possibly encounter in science and engineering applications, including linear, nonlinear, and PDE systems, will be presented. Finally, boundary conditions, which are needed for the solution of PDEs, will be introduced. Effort will be made to detail the common types of boundary conditions, confusion about boundary conditions, boundary conditions in 3D problems and computer treatments, and initial conditions.

Numerical Simulation and PDEs

A computer simulation is a numerical experiment to reproduce the behavior of a system. Computer simulations can be categorized into stochastic and deterministic simulations. Stochastic simulations adopt a collection of random variables to represent the evolution of some systems of random values over time, which will not be discussed in this book. By contrast, a deterministic algorithm is an algorithm which, given a particular input, will always produce the same output, which is multiphysics simulation in porous materials subordinate to. A deterministic algorithm aims at obtaining a mathematical function,

$$u = u(t, x_1, x_2, \ldots, x_N), \tag{9.1}$$

Z. Liu, *Multiphysics in Porous Materials*,
https://doi.org/10.1007/978-3-319-93028-2_9

which has a unique value for any input in its domain. The algorithm gives out a way that produces this particular value as output (or solution). u is the unknown(s) and it is the function of both temporal coordinate t and spatial coordinate x_i. N is the number of dimensions (spatial) associated with the problem of interest. Because an analytical solution, which is a continuous function, is usually difficult to obtain or even unavailable, numerical methods are usually adopted to seek a discrete form of the solution. In other words, we can get the values of u at some specific positions and times, which depend on pre-defined grid/mesh sizes and time steps. For the points between the grid points or time steps, interpolations are usually employed to estimate the function values.

In order to obtain a discrete solution (function), a collection of equations needs to be solved numerically. There are a variety of numerical methods, within which the finite difference method (FDM), finite volume method (FVM), and finite element method (FEM) are the most popular ones and thus will be introduced in this course.

As introduced in the Introduction part, the equations to be solved can be categorized based on their role in the description of the problem as governing equations, auxiliary equations, and boundary/initial conditions. A typical example is the Cauchy problems for the Laplace equation,

$$\frac{\partial^2 u}{\partial x^2} + \frac{\partial^2 u}{\partial y^2} = A, \tag{9.2}$$

with boundary conditions,

$$u(x, 0) = 0, \tag{9.3}$$

$$\frac{\partial u}{\partial x}(x, 0) = \frac{\sin(nx)}{n}, \tag{9.4}$$

and an auxiliary equation,

$$A = x^2 + y^2, \tag{9.5}$$

where n is a real number defining the Cauchy problem.

This example is not time-dependent as time, t, is not involved, so we often call it a boundary value problem (BVP) in mathematics. But in engineering applications, when the equations are endowed with physical meanings, we usually call it a steady-state problem. When time is involved, we call it an Initial-Boundary Value Problem (IBVP) and a time-dependent problem in mathematics and engineering applications, respectively. Initial value problems, in which no spatial coordinates are involved, are not quite common as what we target at in engineering is usually a domain, 1D, 2D, or 3D, rather than a point.

The governing equations are usually partial differential equations, which describe the major governing mechanisms underlying a physical process. When there is a single type of phenomenon for a single type of material/object, only one governing equation is necessary. Otherwise, multiple equations will be introduced. Initial conditions provide some information for the function at the initial time point, while boundary conditions lay down constraints on the boundaries. Auxiliary equations could be the relationships between any two or more of the materials properties, dependent variables, and independent variables. In this book, we will primarily focus on governing equations, especially partial differential equations. Initial and boundary conditions are relatively easy to deal with. Auxiliary relationships are more closely related to the research in individual disciplines. In any of the aforementioned numerical methods, the whole equation system will be discretized over a computational domain, Ω, leading to an algebraic equation system,

$$\mathbf{K} \cdot \mathbf{u} = \mathbf{b}, \tag{9.6}$$

where **K** is the stiffness matrix, **u** is the matrix of unknowns, and **b** is the matrix of forces.

In PDEs, it is also common to denote partial derivatives using subscripts, that is,

$$u_x = \frac{\partial u}{\partial x} \tag{9.7}$$

and

$$u_{xy} = \frac{\partial^2 u}{\partial x \partial y} = \frac{\partial u}{\partial y}\left(\frac{\partial u}{\partial x}\right). \tag{9.8}$$

Especially in physics, del (∇) is often used for spatial derivatives and $\dot{u}$ and $\ddot{u}$ the first and second temporal derivatives, respectively.

Classification of Partial Differential Equations

... partial differential equations are the basis of all physical theorems.
------Bernhard Riemann (1826–1866)

A partial differential equation (PDE) is an equation consisting of a multivariable function(s) and partial derivatives of the function(s). Similar to ordinary differential equations (ODE), PDEs are also a sub-category of differential equations (DE). The difference between ODEs and PDEs lies in the fact that ODEs contain a single variable and its derivatives while PDEs deal with multiple dependent variables and their partial derivatives. A general form of a second-order PDE for the function $u(x_1, x_2, \cdots, x_n)$ is

$$F\left(\frac{\partial^2 u}{\partial x_1 \partial x_1}, \ldots, \frac{\partial^2 u}{\partial x_1 \partial x_n}, \ldots, \frac{\partial^2 u}{\partial x_n \partial x_n}, \frac{\partial u}{\partial x_1}, \ldots, \frac{\partial u}{\partial x_n}, x_1, \ldots x_n\right) = 0, \tag{9.9}$$

where x_i's are general coordinates including both spatial and temporal coordinates here. It is noted that x_i's are general coordinates in the context of mathematics but are spatial coordinates in the context of physics.

If F is a linear function of u and its derivatives, then the PDE is called linear. Otherwise, it is nonlinear. In mathematics, nonlinear PDEs can be further categorized according to the relationships between u and its derivatives, especially the highest order of the derivatives. However, this detailed categorization will not be discussed here due to its less significant role in numerical simulations. The highest order of derivatives determines the order of the PDE. In numerical simulations, the second-order PDEs are of the major interest.

Assuming $\frac{\partial u^2}{\partial x \partial y} = \frac{\partial u^2}{\partial y \partial x}$, the general form of a second-order PDE with two independent variables are as follows:

$$Au_{xx} + 2Bu_{xy} + Cu_{yy} + \ldots (\text{lower order terms}) = 0, \tag{9.10}$$

where the coefficients A, B, and C may depend upon x and y. If $A^2 + B^2 + C^2 > 0$ over a region of the xy plane, the PDE is second-order in that region. This form is analogous to the equation for a conic section by replacing ∂x with x and likewise for other variables:

$$Ax^2 + 2Bxy + Cy^2 + \ldots = 0. \tag{9.11}$$

Similar to the way of classifying conic sections and quadratic forms into parabolic, hyperbolic, and elliptic functions based on the discriminant $B^2 - AC$, we can categorize a second-order PDE into the three types using the discriminant as well:

1. Elliptic PDEs: $B^2 - AC < 0$:

$$\text{e.g., } \frac{\partial^2 u}{\partial x^2} + \frac{\partial^2 u}{\partial y^2} = 0, \tag{9.12}$$

in which $A = 1$, $B = 0$, $C = 1$, and $B^2 - AC = -1$

2. Parabolic PDEs: $B^2 - AC = 0$

$$\text{e.g., } \frac{\partial u}{\partial t} - \frac{\partial^2 u}{\partial x^2} = 0, \tag{9.13}$$

in which, $A = 1$, $B = 0$, $C = 0$, and $B^2 - AC = 0$

3. Hyperbolic PDEs: $B^2 - AC > 0$

$$\text{e.g.,}\ \frac{\partial^2 u}{\partial t^2} - \frac{\partial^2 u}{\partial x^2} = 0, \tag{9.14}$$

in which $A = 1$, $B = 0$, $C = -1$, and $B^2 - AC = 1$

The above classification method may not be directly used when there are more than two independent variables. In this case, we first need to reduce the quadratic form of the second-order PDE, i.e.,

$$\sum a_{ij} \frac{\partial^2 u}{\partial x_i \partial x_j} + \text{lower order terms} = 0 \tag{9.15}$$

into the canonical form

$$\sum c_i \frac{\partial^2 u}{\partial x_i^2} + \text{lower order terms} = 0 \tag{9.16}$$

by means of an appropriate linear nondegenerate transformation (Polyanin 1997). The following criterion can be used:

1. The PDE is elliptic if all the c_i's are nonzero and have the same sign.
2. The PDE is parabolic if at least one of the c_i's is zero.
3. The PDE is hyperbolic if all the c_i's are nonzero and have different signs.

Common PDEs in Engineering Applications

Linear Equations

1. **Laplace's equations:** $\Delta u = 0$. This equation is widely used to describe the electric, gravitational, and fluid potentials in the equilibrium state.
2. Helmholtz's equation (involves eigenvalues):$-\Delta u = \lambda u$. This equation a time-independent form of the wave equation, resulting from applying the technique of separation of variables to reduce the complexity of the analysis. It is widely used for applications such as electromagnetic radiation, seismology, and acoustics.
3. **First-order linear transport equation:** $u_t + c\nabla u = 0$. This is the general time-dependent transport equation, e.g., for mass, momentum, and energy, with a diffusive term.
4. **Heat or diffusion equation:** $u_t - \Delta u = 0$. This is a simplified version of the above linear transport equation.
5. Schrodinger's equation: $iu_t + \Delta u = 0$. This is the time-dependent version of the Schrodinger's equation for quantum mechanical systems.
6. **Wave equation:** $u_{tt} - c^2\Delta u = 0$. The wave equation can be used to describe both mechanical waves and electromagnetic waves.

7. Telegraph equation: $u_{tt} + du_t - u_{xx} = 0$. The telegraph equation is proposed to formulate changes of the voltage and current on an electrical transmission line with distance and time.

Nonlinear Equations

1. Eikonal equation: $|\nabla u| \cdot f(x) = 1$. This equation is proposed to solve problems of wave propagation when the wave equation is approximated using the WKB (Wentzel-Kramers-Brillouin) theory. Derivable from Maxwell's equations, the equation provides a link between physical (wave) optics and geometric (ray) optics.
2. **Nonlinear Poisson equation:** $-\nabla u = f(u)$. This equation is similar to the Laplace equation but with a source term that relies on the dependent variable u. It turns into the linear Poisson equation if the function f is not dependent on u.
3. Burgers' equation:$u_t + uu_x = 0$. This equation provides a mathematical description for some very special problems in fluid dynamics.
4. Minimal surface equation: $\nabla \cdot \left[\frac{\nabla u}{\left(1+|\nabla u|^2\right)^{1/2}}\right] = 0$. This equation is the key equation in the minimal surface theory. The theory is used to solve the variational problem of finding the surface of least area stretched across a given closed contour.
5. Monge-Ampère equation: $\det(\nabla \nabla u) = f$. The Monge-Ampère equation arises in several problems in Riemannian geometry and conformal geometry. One of the simplest of applications is the prescribed Gauss curvature.
6. Korteweg-de Vries equation (KdV):$u_t + f(u)u_x + u_{xxx} = 0$. This equation presents a mathematical model of waves on shallow water surfaces.
7. Reaction-diffusion equation: $u_t - \Delta u = f(u)$. This is a transport equation with a diffusive term and nonlinear source term.

System of Partial Differential Equations

1. Evolution equation of linear elasticity.
2. $\mathbf{u}_{tt} - \mu\Delta\mathbf{u} - (\lambda + \mu)\nabla(\nabla \cdot \mathbf{u}) = 0$. This is the governing equation of the linear stress-strain problems.
3. System of conservation laws: $\mathbf{u}_t + \nabla \cdot F(\mathbf{u}) = 0$. This is the general form of the conservation equation with multiple scalar quantities of the dependent variables.
4. Maxwell's equations in vacuum:

$$\nabla \times \mathbf{E} = -\mathbf{B}_t$$

$$\nabla \times \mathbf{B} = \mu_0 \varepsilon_0 \mathbf{E}_t$$

$$\nabla \cdot \mathbf{B} = 0$$
$$\nabla \cdot \mathbf{E} = 0$$

Maxwell's equation for describing electromagnetic fields in vacuum.

5. Reaction-diffusion system:

$$v_t - \Delta v = f(v)$$
$$\mathbf{u}_t + \mathbf{u} \cdot \nabla \mathbf{u} = -\nabla p$$
$$\nabla \cdot \mathbf{u} = 0$$

The equation systems consider both the fluid movement and the solute balance involving both transient transport and chemical reactions in the fluid.

6. Euler's equations for incompressible and inviscid fluid:

$$\mathbf{u}_t + \mathbf{u} \cdot \nabla \mathbf{u} = -\nabla p$$
$$\nabla \cdot \mathbf{u} = 0$$

A simplified version of the Navier-Stokes equations for impressible and inviscid fluids.

7. Navier-Stokes equations for incompressible viscous fluid:

$$\mathbf{u}_t + \mathbf{u} \cdot \nabla \mathbf{u} = -\nabla p + \Delta \mathbf{u}$$
$$\nabla \cdot \mathbf{u} = 0$$

The Navier-Stokes equations for incompressible viscous fluids.

The equations in bold are those we encounter more often in engineering applications. It is interesting to mention that the above equations can be summarized into a generic form. For example, the generic PDE used by COMSOL is a good generalization of common PDEs:

$$e_\alpha \frac{\partial^2 u}{\partial t^2} + d_a \frac{\partial u}{\partial t} - \frac{\partial}{\partial x_i}\left(c\frac{\partial u}{\partial x_i} + \alpha u - \gamma\right) + \beta \frac{\partial u}{\partial x_i} + au = 0. \tag{9.17}$$

Boundary Conditions and Initial Conditions

Common Types of Boundary Conditions

Boundary conditions, which exist in the form of mathematical equations, exert a set of additional constraints to the problem on specified boundaries. The concept of boundary conditions applies to both ordinary and partial differential equations. There are five types of boundary conditions: Dirichlet, Neumann, Robin, Mixed, and Cauchy, within which Dirichlet and Neumann are predominant. To understand the difference, let us take a look at an ordinary differential equation:

$$\frac{du}{dx} + u = 0 \text{ in the domain } [\mathrm{a}, \mathrm{b}]$$

We have the Dirichlet boundary condition when the boundary prescribes a value to the dependent variable(s). A Dirichlet boundary condition for the above ODE looks like

$$y(\mathrm{a}) = A \tag{9.18}$$

$$y(\mathrm{b}) = B. \tag{9.19}$$

For example, in a 1D heat transfer problem, when both ends of a wire are maintained in a water bath with constant temperatures, the above boundary condition will be appropriate.

In the Neumann boundary condition, the derivative of the dependent variable is known in all parts of the boundary:

$$y'(\mathrm{a}) = \alpha \tag{9.20}$$

and

$$y'(\mathrm{b}) = \beta. \tag{9.21}$$

In the above heat transfer example, if heaters exist at both ends of the wire, via which energy would be added at a constant rate, the Neumann boundary condition will apply.

The Robin boundary condition is a weighted combination of the Dirichlet boundary and the Neumann boundary condition in all the parts of the boundary:

$$\chi_1 \cdot y(\mathrm{a}) + \chi_2 \cdot y'(\mathrm{a}) = A_\alpha \tag{9.22}$$

$$\chi_1 \cdot y(\mathrm{b}) + \chi_2 \cdot y'(\mathrm{b}) = B_\beta \tag{9.23}$$

where χ_i's are constants representing the weights. The Dirichlet, Neumann, and Robin are also called the first-type, second-type, and third-type boundary condition, respectively.

The mixed boundary condition refers to the cases in which Dirichlet boundary conditions are prescribed in some parts of the boundary while Neumann boundary conditions exist in the others. Accordingly, for the above ODE, the following is a typical mixed boundary condition:

$$y(\mathrm{a}) = A \tag{9.24}$$

$$y'(\mathrm{b}) = \beta. \tag{9.25}$$

In the above 1D heat transfer problem, this corresponds to the condition that one end of the wire is placed in a water bath while the other end is connected to a heater with constant heat transfer rate.

While being less common, Cauchy boundary conditions are also used in second-order differential equations, in which one may specify the value of the function y and the value of the derivative y' at a given point:

$$u(\mathrm{a}) = A \tag{9.26}$$

$$u'(\mathrm{a}) = \alpha. \tag{9.27}$$

Therefore, the Cauchy boundary conditions correspond to imposing Dirichlet and Neumann boundary conditions simultaneously.

Confusion About Boundary Conditions

In the above introduction to boundary conditions, we determine the boundary type by assessing the constraints applied on the whole boundary of a computational domain, which may consist of multiple segments. A rectangular (2D) domain, for example, has four boundary segments. In many cases, we do not call the top side of the rectangular area "top boundary segment," but instead, we call it "top boundary" directly. This may cause confusion. In many software packages, we need to assign a boundary condition to this "top boundary," e.g., Dirichlet and Neumann boundary conditions. In this case, the Dirichlet and Neumann boundary conditions just correspond to the assignment of value to the dependent variable and its derivative, respectively, on the "top boundary." As a result, we may have different boundary conditions on different boundary segments. In this use, we mostly talk about Dirichlet, Neumann, and Robin or other physical boundary conditions rather than mixed boundary conditions, which need to be defined with respect to the whole boundary. The use of boundary conditions in this way is somewhat in conflict with the boundary condition categorization in the previous section. However, we possibly need to understand and keep both of them in mind due to their wide use.

Another common confusion is associated with the level of boundary conditions. The definition and categorization of boundary conditions introduced in the previous section were laid down on the mathematical level. That is, we define and classify boundary conditions by evaluating whether the dependent variable or its derivative is equal to a prescribed value. However, boundary conditions can also be defined and categorized on other levels, such as geometric and physical levels. Common examples of boundary condition defined on the geometric level are symmetric boundary conditions and periodic boundary conditions. Taking monolithic transport in the 2D Cartesian coordinate system, for example, a symmetric boundary implies that the derivative along the direction normal to the boundary is zero, which can be explained using a Neumann boundary condition defined on the mathematical level. However, symmetric boundary conditions for more complicated physical fields may imply more complicated mathematical boundary conditions. In the FVM platform OpenFOAM, the boundary conditions defined on the geometric level is called "primitive" boundary conditions. In addition, boundary conditions can also be defined on the physical level. That is, they are defined from the perspective of physical meanings. Typical examples are those used in computational fluid dynamics packages: slip, nonslip, outflow, and so on. These boundary conditions are defined based on the description of the physical processes on the boundaries, and they can be formulated using the more fundamental mathematical level boundary conditions. The definitions and categorization of physical boundary conditions depend on the targeted physical fields. In addition, different software packages may provide different physical boundary conditions, though some physical boundary conditions have been accepted across packages. These physical boundary conditions serve as preset combinations of mathematical boundary conditions for boundary segments to save users' efforts at repeatedly implementing these combinations themselves. Due to the above reason, this book will mostly focus on mathematical boundary conditions. The physical boundary conditions will be introduced for specific physical fields if necessary.

Boundary Conditions in 3D Domains

When moving from 1D to 2D or 3D problems, the Dirichlet boundary condition requires no extra treatment as the dependent variable or its components can be assigned a value directly. Some quantities such as velocity that appear as a scalar in 1D application will retrieve its vector nature in 2D or 3D. The treatment of a Dirichlet boundary condition of this type of quantity just changes from assigning one component in 1D to two and three components in 2D and 3D, respectively.

However, the Neumann boundary condition will call for additional treatment. This is because Neumann boundary conditions deal with a derivative. In 2D and 3D, there is more than one direction, and the derivative is generalized into the gradient as introduced in previous sections. Therefore, theoretically, we now have the freedom to assign a value to each component of the gradient. This also happens in real simulations. However, a more common scenario in multiphysics simulation is that we introduce the directional derivative, which defines the projection of the gradient in a prescribed direction:

$$\frac{\partial u}{\partial \mathbf{n}} = (\nabla u) \cdot \mathbf{n}, \tag{9.28}$$

where u is the dependent variable which can be a tensor of any order and $\mathbf{n}$ is the prescribed direction. In general, this boundary condition is utilized to represent the flux across the boundary. The flux can usually be correlated to the gradient of the dependent variable. Therefore, $\mathbf{n}$ is usually the outward normal direction of the boundary. It is noticed that $\mathbf{n}$ is not necessarily unchanged from point to point. Detailed examples will be given in the chapters for monolithic physics.

Boundary Condition Treatments in Computer Programs

The treatment of the Neumann boundary condition in FEM software packages desires special attention, such as MATLAB PDE Toolbox and FlexPDE used in this book. The FEM usually starts the consideration of the Neumann boundary condition in the derivation of the weak form of the governing equation. In this process, terms with second or higher derivatives of the dependent variable(s) will be integrated by parts so that the order of the derivatives can be lowered. Taking the diffusion term, $\nabla \cdot (D\nabla u)$, for example, the integration by parts produces $(-D\nabla u) \cdot \mathbf{n}$. As $\mathbf{J} = -D\nabla u$ is the flux, $(-D\nabla u) \cdot \mathbf{n}$ defines the normal outflux on the boundary. Accordingly, when we define the Neumann boundary condition by specifying a value β, we, in fact, assign this value to the normal outflux

$$\mathbf{n} \cdot (-D\nabla u) = \beta. \tag{9.29}$$

The software package will substitute the value into the stiffness matrix for calculation. Likewise, if the second or higher derivative exists in the curl operator as $\nabla \times (D\nabla u)$, the software package will carry out the following replacement

$$\mathbf{n} \times (-D\nabla u) = \beta. \tag{9.30}$$

There is a significant ramification of this treatment. That is, other parameters in these terms containing second or higher derivatives will also be included in the definition. In FlexPDE, any parameters, regardless of its position relative to the differential operator, will be included. That is, if we multiply the two sides of the PDE with a constant, such as 2, then the Neumann boundary condition, taking diffusion for example, will change accordingly,

$$\mathbf{n} \cdot (-2D\nabla u) = \beta. \tag{9.31}$$

This means, if we still assign the same β value to the boundary, then the outflux will be one half of the original case without multiplying the PDE by 2 as

$$\mathbf{n} \cdot \mathbf{J} = \frac{\beta}{2}. \tag{9.32}$$

Initial Condition

The treatment of the initial condition is usually straightforward. If the initial condition for a field is a constant, then the dependent variable will be equal to this constant at all the points of discretization in the domain at the beginning of the transient process to be simulated. Otherwise, a function may be given for the dependent variable to define its initial distribution. In that case, we just need to discretize the function to find out the values of the dependent variables at all the points of discretization. All the other parameters containing the dependent variable will be calculated with the dependent variable values calculated above.

Chapter 10
Numerical Solution of PDEs

Introduction

As stated before, multiphysics can be viewed as a fusion of mathematics, physics, applications, and numerical analysis. For a normal multiphysics end user, researcher, or developer, it is suggested to cover necessary mathematics, understand the physics on an application basis, and touch upon numerical analysis to an appropriate depth, just like the arrangement of this book. The content of numerical analysis, especially the discretization methods and computing techniques, could be overwhelming. However, what many multiphysics users and researchers would be interested in is not very likely how to develop a new numerical method or develop an iterative solver for a problem but instead to mathematically describe the problem using multiphysics and then to solve it using existing numerical methods and numerical solvers.

Following the above viewpoint, a handy tool would be a computer program containing preprocessing for geometry development, numerical solver, and postprocessing as well as a necessary interface to assist the setup of the model. If a multiphysics description of a problem has already been well established, then the remaining task is to create and solve the problem using the way prescribed in the problem. Currently, there are many commercial and open-source programs that can satisfy this need, such as COMSOL, Elmer, MOOSE, FiPy, OpenFOAM, FEniCS, and FEATool. This book adopts the MATLAB PDE Toolbox and FlexPDE which are very friendly to starters and helpful in exhibiting and implementing the content of this book. The MATLAB PDE Toolbox will be introduced and used in the first few chapters considering that it has a convenient graphical user interface to create models. FlexPDE will be used as the major tool considering its simplicity, close relationship to the essence of multiphysics, capacity, and complimentary licenses for education and evaluation purposes.

In the following sections, we will focus on two tools that will be used in the later implementations of monolithic physics and multiphysics problems in this book:

Z. Liu, *Multiphysics in Porous Materials*,
https://doi.org/10.1007/978-3-319-93028-2_10

MATLAB PDE Toolbox and FlexPDE. Step-by-step directions will be given for the use of these two tools to solve a simple PDE with typical boundary conditions. It is expected that readers will get a real feeling of numerical implementations with these two tools with distinctly different styles: one with a graphical user interface for pre- and postprocessing and the other with script files for model setup. Advanced numerical implementations of multiphysics with self-developed code using the finite difference method, finite volume method, and finite element method will be introduced in the last three chapters of this book.

Solve PDEs with MATLAB PDE Toolbox

The MATLAB PDE Toolbox is a tool for solving 2D partial differential equations with the finite element method. An example is provided in this section to guide you through the use of the graphical user interface for interactive use. Here we use MATLAB R2014b to demonstrate the use of the PDE Toolbox.

At the MATLAB prompt, type “pdetool,” and you should see a window like the one below popping out (Fig. 10.1).

We will show the process of solving Laplace’s equation with Dirichlet boundary conditions using the PDE Toolbox. The solutions to Laplace’s equation are harmonic functions, which describe situations in which forces are balanced, for example, in electrostatics or thermal equilibrium. This is an example which might be first

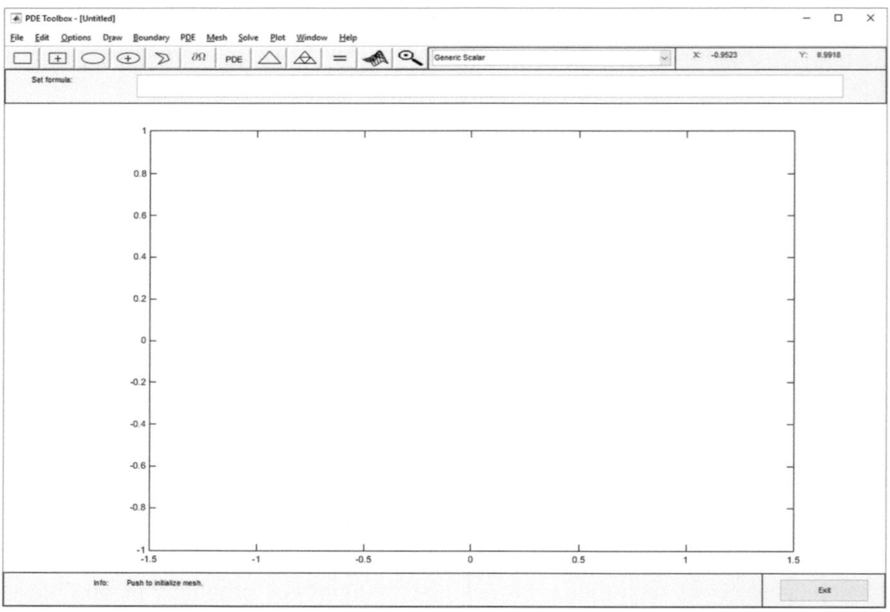

Fig. 10.1 Main window of PDE Toolbox

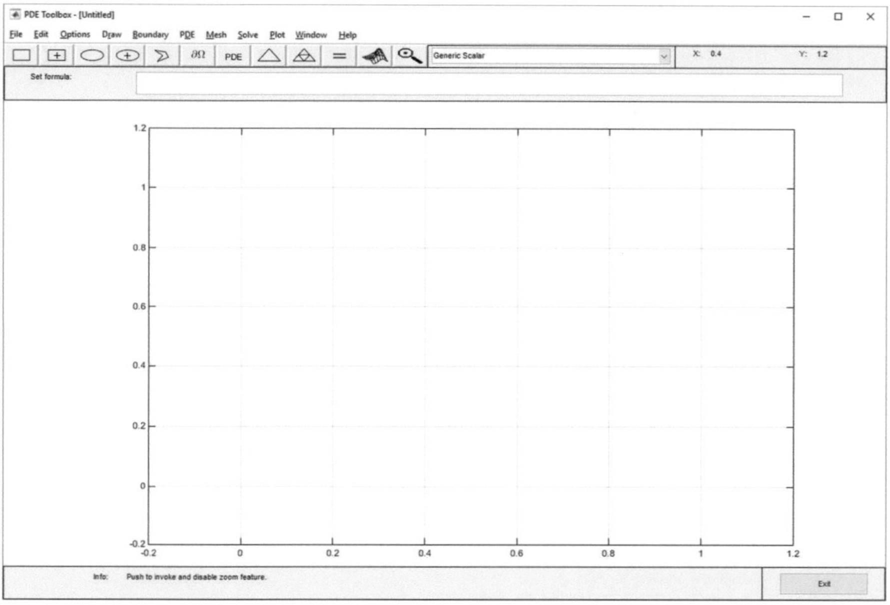

Fig. 10.2 Setup of canvas

encountered and be solved using Fourier series methods by hand in an elementary PDE course. You will find that you obtain the solution faster with the PDE Toolbox than the Fourier series methods. Moreover, the solution can be visually represented for a better understanding of the problem and the solution. We here show an example which might not be easily solved by hand.

1. Draw the region of interest: suppose we want the rectangle $(x, y) = [0, 1] \times [0, 1]$.
   - In the Options menu, select Grid and Snap; then set the Axes' Limits to [−0.2 1.2] and [0.2 1.2]. When finished, click Apply and Close the Axes' Limits window (Fig. 10.2).
   - In the Draw menu, select Rectangle/Square (not centered).
   - In the drawing region, click and hold with the cursor at the grid point (0, 0), and drag it to the point with coordinates (1, 1). The region will be shaded and marked SQ1 (Fig. 10.3).
2. Set up the boundary conditions: the dependent variable u is -1 on the top and bottom and + 1 on the sides.
   - In the Boundary menu, select Boundary Mode. The window changes to this (Fig. 10.4):
   - In the drawing region, double-click on the left edge of the computational domain (SQ1). The boundary condition window pops up. The default Dirichlet condition, $h*u = r$, is OK; the default $h = 1$ is OK; change the value of r to +1 (Fig. 10.5).

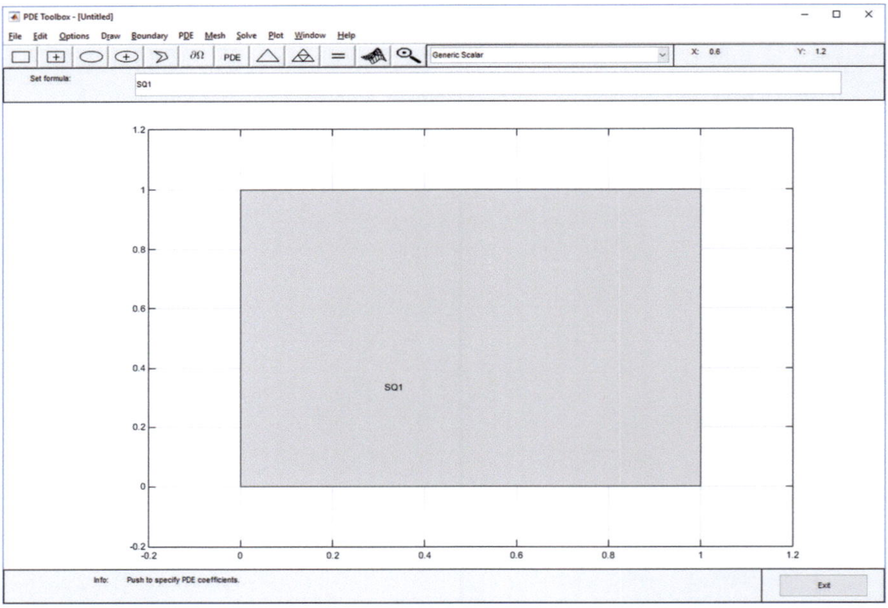

Fig. 10.3 Development of geometry (computational domain)

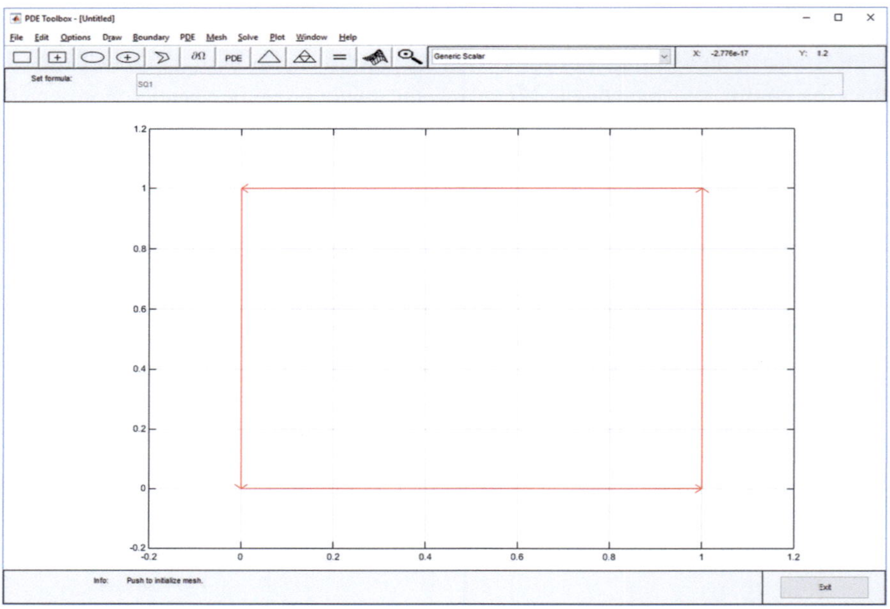

Fig. 10.4 Boundary mode

- Do the same thing with the right edge: double-click and set the value. On the top and bottom edges, set r to -1. Be patient and careful. When the boundary conditions are correctly set, you should be able to click on any edge, and see the correct value for *r*.

3. Specify the particular PDE of interest: $\Delta u = 0$.

   - In the PDE menu, select PDE specification. The PDE specification window pops up; we are doing the default type of problem, elliptic, of the form –div (c*grad(u)) + a*u = f, where $a = 0, f = 0$, and $c = 1$. Change the values as needed, and click on OK (Fig. 10.6).

4. Generate the mesh with the automatic mesh generator:

   - In the Mesh menu, select Initialize Mesh. This time we will use the default mesh (Fig. 10.7).

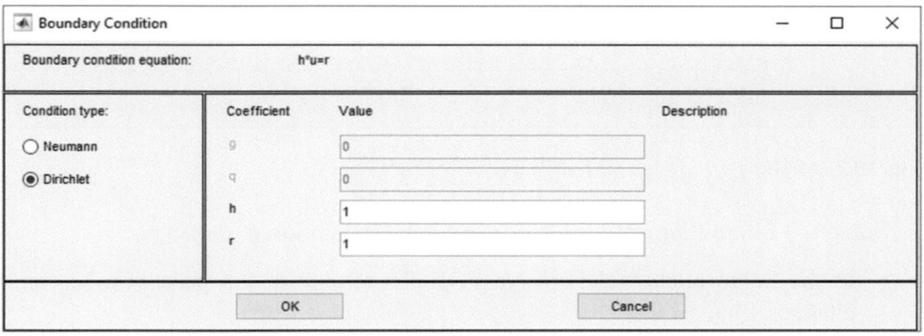

Fig. 10.5 Setup of boundary conditions

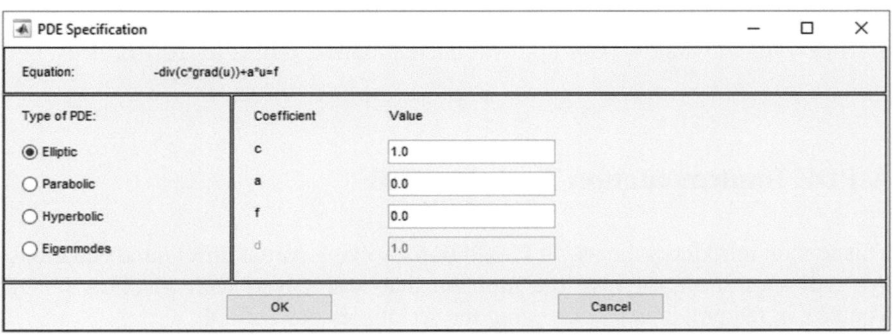

Fig. 10.6 Setup of governing equation

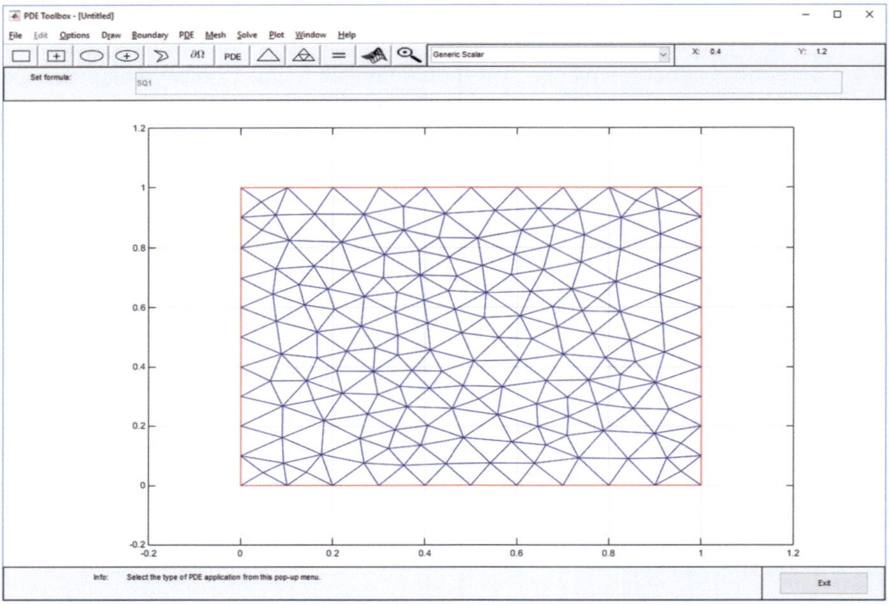

Fig. 10.7 Meshing

5. Solve the Finite Element problem using default solution parameters.
   - In the Solve menu, select Solve PDE. The figure in the main drawing area changes (Fig. 10.8).
6. Generate other plots:
   - In the Plot menu, select Parameters. The plot parameter window pops up; click on the checkboxes for color, contour, and height and then on the Plot button (Fig. 10.9).

   A new figure window pops up, with the appropriate plot (Fig. 10.10).

A PDE Implementation Tool: FlexPDE

This section introduces the use of FlexPDE for solving partial differential equations. We will show how to solve the problem that was solved with MATLAB PDE Toolbox in the previous section using the script files of FlexPDE.

FlexPDE can solve mathematical systems consisting of first- or second-order partial differential equations in 1D, 2D, or 3D Cartesian geometry, in 1D spherical or cylindrical geometry, or in axisymmetric 2D geometry. Other geometries can be supported by including the proper terms in the PDE. The system may be steady-state

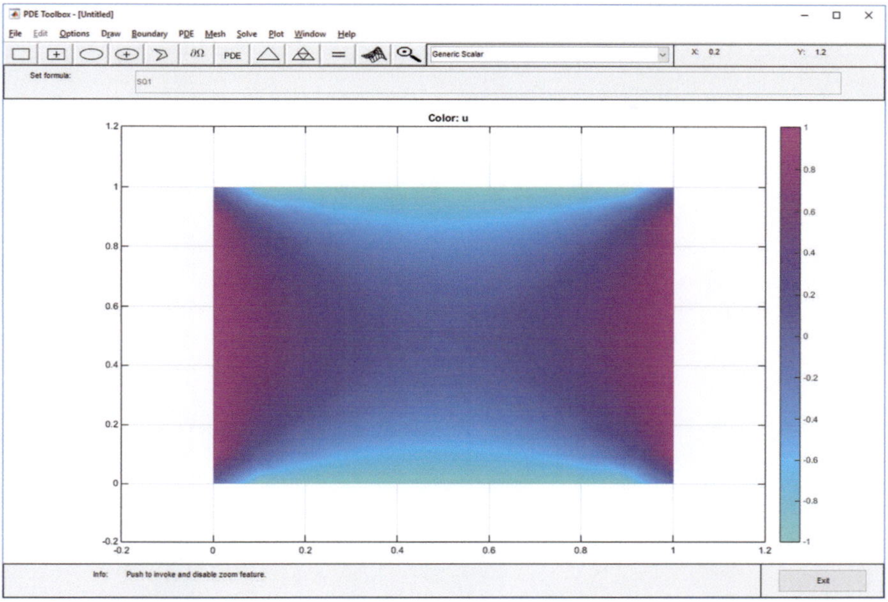

Fig. 10.8 Solution and results

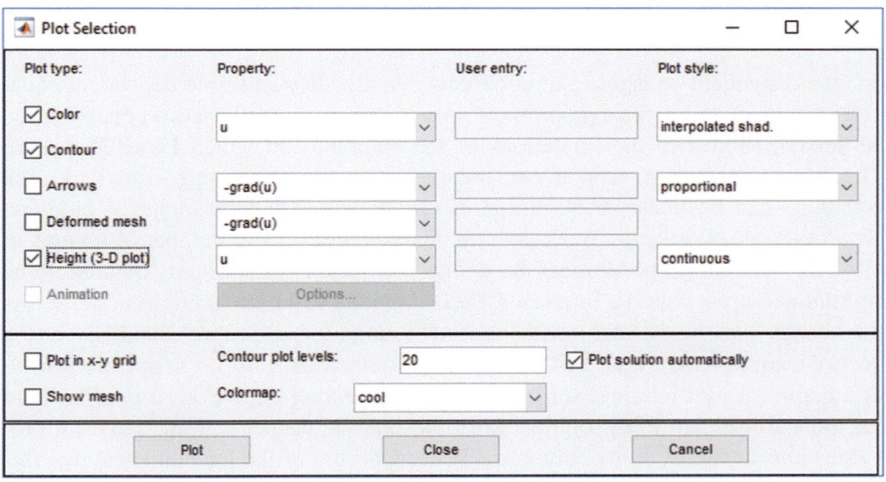

Fig. 10.9 Setup for postprocessing

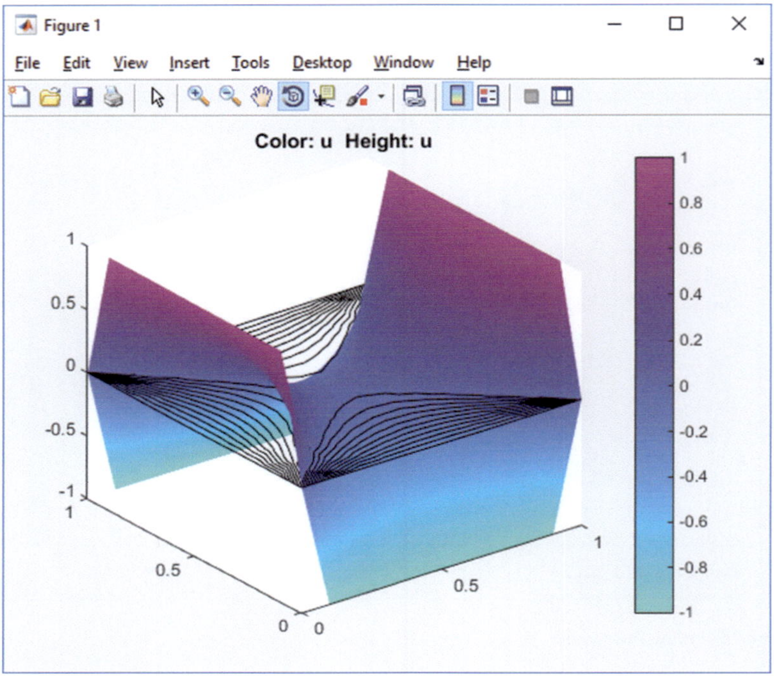

Fig. 10.10 Visualization of results

or time-dependent or eigenvalue problems. Steady-state and time-dependent equations can be mixed in a single problem. Any number of simultaneous equations can be solved, subject to the limitations of the computer on which FlexPDE is run. The education/student version can handle up to five governing equations. The equations can be linear or nonlinear. FlexPDE automatically applies a modified Newton-Raphson iteration process in nonlinear systems. Any number of regions of different material properties may be defined. Modeled variables are assumed to be continuous across material interfaces. Discontinuous variables or jumps in the across boundaries can be handled with special boundary conditions. FlexPDE has a deeper learning curve than MATLAB FDE Toolbox as it uses a script file instead of a graphical user interface for pre- and postprocessing. However, it is still one of the most convenient PDE solution tools. Despite the deeper learning curve, it provides more flexibility in modeling and better tailoring of the mathematical description. By full use of its power, FlexPDE can be used for very complex numerical analysis.

A problem description script is a readable text file. The contents of the file consist of a number of sections, each identified by a header. The fundamental sections are listed in the following.

TITLE	A descriptive label for the output
SELECT	User controls that override the default behavior of FlexPDE
VARIABLES	Here the dependent variables are named
DEFINITIONS	Useful parameters, relationships, or functions are defined
EQUATIONS	Each variable is associated with a partial differential equation
BOUNDARIES	The geometry is described by walking the perimeter of the domain, stringing together line or arc segments to bound the figure
MONITORS and PLTS	Desired graphical output is listed, including any combination of CONTOUR, SURFACE, ELEVATION, or VECTOR plots
END	Complete the script

There are several other optional sections for describing special aspects of the problem. Some of these will be introduced later in this document. Detailed rules for all sections are presented in the FlexPDE Problem Descriptor Reference chapter "The Sections of a Descriptor." Comments can be placed anywhere in a script to describe or clarify the work. Two forms of comment are available:

- {Anything inside curly brackets is a comment.}
- ! from an exclamation to the end of the line is a comment

The script file is not case sensitive. But some people prefer to use capital letters for the keywords reserved by FlexPDE.

To demonstrate the use of FlexPDE, here we will solve the same problem that we just solved using PDE Toolbox with FlexPDE 6.32s (a free student version). The complete script file is listed in the following.

Example
A simple diffusion equation on a square might look like this:

```
TITLE '2D Laplace Demo'    { the problem identification }
! COORDINATES cartesian2 { coordinate system, 1D,2D,3D, etc; the
default is cartesian2: 2D Cartesian}
VARIABLES       { system variables }
   u        { choose your own names }
! SELECT       { method controls }
! DEFINITIONS   { parameter definitions }
! INITIAL VALUES
EQUATIONS       { PDE's, one for each variable }
   DIV(GRAD(u))=0 { one possibility }
! CONSTRAINTS   { Integral constraints }
BOUNDARIES      { The domain definition }
   REGION 1     { For each material region }
     START(0,0)  { Walk the domain boundary }
     VALUE(u)=-1
     LINE TO (1,0)
     VALUE(u)=+1
     LINE TO (1,1)
     VALUE(u)=-1
     LINE TO (0,1)
     VALUE(u)=+1
     LINE TO CLOSE
```

```
! TIME 0 TO 1   { if time-dependent }
MONITORS       { show progress }
PLOTS        { save result displays }
  CONTOUR(u)
  SURFACE(u)
END
```

In the script file, VARIABLEs and EQUATIONS define the PDEs and BOUNDARIES give out the computational domain and boundary conditions. In the simplest scenario, as shown above, a domain is plotted by assigning the start and end points of a line and then connecting these lines to form a closed 2D domain. Boundary conditions need to be specified for each boundary segment, with VALUE reserved for Dirichlet boundary conditions and NATURAL reserved for Neumann boundary conditions. Auxiliary relationships can be given in the DEFINITIONS section. MONITORS and PLOTS are designed for showing the process and final results, respectively. Common plot types include CONTOUR (2D contour), SURFACE (3D contour), ELEVATION (curves for spatial distribution), and HISTORY (temporal variation). The keyword END is needed to close a script file. Many sections have default settings, for example, a Neumann boundary condition will be used if no boundary conditions are assigned to a boundary segment.

To solve the above Laplace's equation example, we just need to click the "New Script" tab in the Menu drop-down list. Then we paste the script file into the editable area. After saving the file before the first time we run it, we can obtain the following results by clicking the "run" button in the quick access tool area in the editor mode. After clicking the "run" button, the program will automatically enter the "Plot" mode to show the solution process and results (Figs. 10.11 and 10.12).

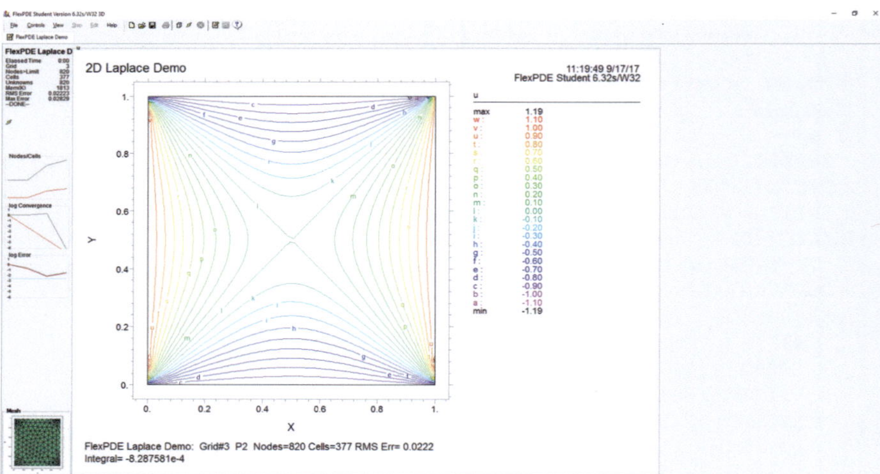

Fig. 10.11 Simulation result 1: contour

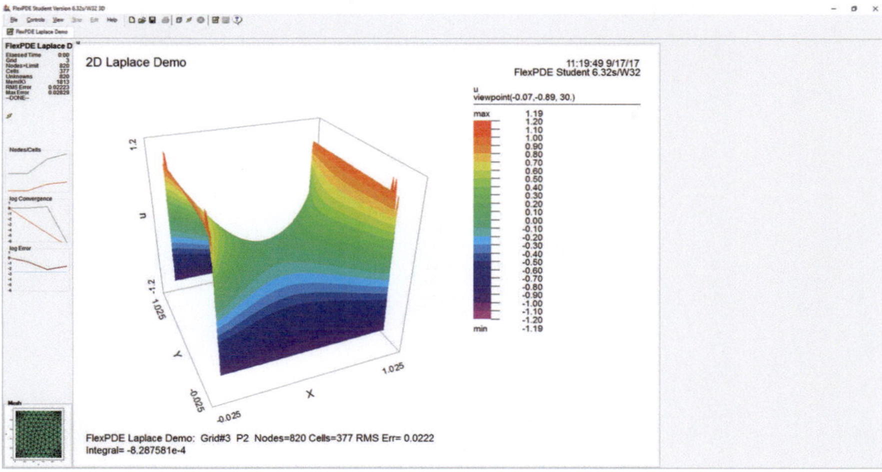

Fig. 10.12 Simulation result 2: 3D contour

Part III
Monolithic Physics

Chapter 11
Thermo: Heat Transfer

Introduction

Heat transfer is a great example to show how a fundamental monolithic physical phenomenon can be mathematically formulated. In this chapter, the fundamentals of heat transfer for continua and mixtures including porous materials will be described consecutively. We will start with the concept of the representative elementary volume (REV) and conservation laws. Based on that, the general governing equation of conservation problems will be derived. Following that, we will derive the governing equation of the heat transfer in continuous materials. The validity of the REV assumption and conservation laws will help us extend the basics of continuum mechanics to the theories in porous materials. Emphasis will be placed on the heat transfer in porous materials which represents a much more complicated and general framework for heat transfer. Boundary conditions for heat transfer will be discussed from both a mathematical and a physical perspective. Finally, a practice problem will be solved using the MATLAB PDE Toolbox and compared against the analytical solution.

General Background for Continuum Mechanics

There are two types of methods to implement multiphysics: continuous methods and particle methods. Continuous methods directly deal with the PDEs which describe the physical processes in the continuous materials. There are rare examples of perfect continuous materials as imperfections widely exist in natural materials. However, if we assume that the scale of these imperfections, such as vacancies, dislocations, cracks, surfaces, and pores, is much smaller than that of the domain of interest, in which the processes occur, then we can still use the knowledge based on the continuum mechanics. The assumption of REV will be introduced in the following

Z. Liu, *Multiphysics in Porous Materials*,
https://doi.org/10.1007/978-3-319-93028-2_11

paragraphs. Porous materials are an extreme example of imperfect materials, in which imperfections are more of normality than an exception. But the knowledge of continuum mechanics still applies as long as the assumption of REV can be ensured. This type of method includes most of the numerical simulation methods that we are familiar with such as FEM, FVM, FDM, element free Galerkin, boundary element method, discontinuous Galerkin method, and collocation method, which is thus the interest of this book. The second type of method deals with multiphysics from a discontinuous perspective, in which the material is usually assumed to be an ensemble of particles. Typical examples include the discrete element method, smoothed particle hydrodynamics, lattice Boltzmann method, and molecular dynamics, which will not be discussed in this book. It is noted that methods such as smoothed particle hydrodynamics are more of a hybrid of the above two groups than any of them because of the employment of both PDEs from continuum mechanics and particles in the method.

Representative Elementary Volume

The representative elementary volume, which can be abbreviated as REM, is also called representative volume element (RVE). In continuum mechanics for a heterogeneous material, as shown in Fig. 11.1, an RVE can be considered as a volume Ω that represents a composite statistically. The volume should effectively include a sampling of all microstructural heterogeneities occurring in the composite, such as grains, inclusions, voids, and fibers (Kanit et al. 2003). This volume, in the meanwhile, must stay small enough to be considered as a volume element of the bulk continuum. Several types of boundary conditions can be prescribed on the boundary of the volume, $\partial\Omega$, such as a given mean strain or mean stress in the mechanical field. The word "volume" and the symbol Ω here are general terms which apply to 1D, 2D, and 3D problems. The REV has a volume of V in 3D, a surface area of S in 2D, and a length of L in 1D.

In the theory of composite materials, the REV is believed to be the smallest volume over which a value representative of the whole can be obtained. Porous materials can be treated as a special and very complicated case of composite materials. Let us revisit the sampling of materials to understand the REV in porous materials. We usually collect and measure samples of a porous medium to acquire

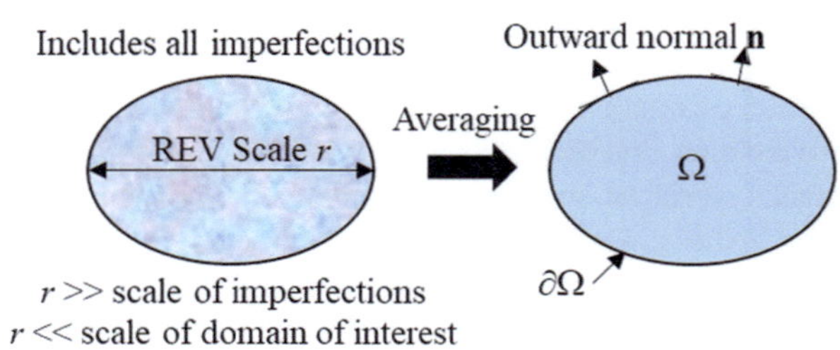

Fig. 11.1 Representative elementary volume

the properties of the material. If the sample is too small, the readings tend to oscillate. As we increase the sample size, the oscillations begin to dampen out. Eventually, the sample size will become large enough that we begin to get consistent readings. This sample size is referred to as the representative elementary volume. If we continue to increase the sample size, the measurement will remain stable until it becomes comparable to the size of the domain of interest. However, the sample needs to be much smaller than the size of the domain (whole sample); otherwise, sampling will be meaningless. The validity of the REV assumption is thus the foundation based upon which you can apply the PDEs developed for continuous materials to "imperfect" materials such as porous materials.

Conservation Law

The conservation laws, i.e., mass, momentum, or energy, define the way that how a physical process of a conservation nature is formulated with a PDE(s). Conservation, in fact, accounts for a major part of physical processes that we encounter in multiphysics, including heat transfer. The derivation of the PDE for conservation problems starts from the global or integral balance laws as the following equation:

$$\frac{d}{dt}\int_{\Omega} u d\mathbf{x} + \oint_{\partial\Omega} \mathbf{J}\cdot\mathbf{n} d\mathbf{x} = \int_{\Omega} q d\mathbf{x}, \tag{11.1}$$

where Ω corresponds to the REV in continuum mechanics, u is the generic conserved quantity defined by unit mass, $\mathbf{x}$ is the coordinate vector, $\mathbf{J}$ is the flux of this conserved quantity, and q is the source term. Ω is a more general representation of the computational domain. Material points occupying $\Omega(0)$ may move under the influence of inertia and applied loads, so that at any time, t, they may occupy a volume $\Omega(t)$ which is different from $\Omega(0)$. The function $\Omega(t)$ is a material volume which is defined by a set of material points rather than by any explicit spatial coordinates.

After applying the Gauss theorem, Reynolds transport theorem, and DuBois-Reymond lemma theorem (Allen 1988), we can obtain the local balance equation:

$$\frac{\partial u}{\partial t} + \nabla\cdot(u\mathbf{v}) + \nabla\cdot\mathbf{J} = q, \tag{11.2}$$

where the flux, $\mathbf{J}$, can be induced by different mechanisms such as diffusion, dispersion, electromagnetic field, and so on. This flux term can be formulated as Eq. 11.3:

$$\mathbf{J} = f(t, \mathbf{x}, u, \nabla u) = -K\nabla u - D\nabla u + \ldots \tag{11.3}$$

By substituting this flux term into the local balance equation, we then obtain the general conservation equation for transport phenomena (Eq. 11.4). More mechanisms responsible for flux across the boundary can be added by extending the flux term:

$$\underbrace{\frac{\partial u}{\partial t}}_{\text{Accumulation}} \underbrace{+\nabla \cdot (u\mathbf{v})}_{\text{Advection}} \underbrace{-\nabla \cdot (K\nabla u)}_{\text{Diffusion (Conduction)}} \underbrace{-\nabla \cdot (D\nabla u)}_{\text{Dispersion}} = \underbrace{q}_{\text{Source}}. \quad (11.4)$$

The conservation equation states that the accumulation of conserved quantity is balanced with the conserved quantity advected into the control volume via mobile phases and diffused directly as diffusion and indirectly as dispersion (Deen 1998) into the control volume via the aqueous and gas phases, plus the mass generation associated with sources.

Heat Transfer in Continua

The governing equation for heat transfer can be easily obtained based on the conservation equation introduced below:

$$\frac{\partial e}{\partial t} + \nabla \cdot (e\mathbf{v}) + \nabla \cdot \mathbf{J}_{\mathrm{T}} = q_{\mathrm{T}}, \quad (11.5)$$

where the subscript, T, denotes the conservation of energy, which is the extended definition of thermal fields. The conservation of quantities and the corresponding physical fields are used interchangeably throughout this book.

The above governing equation can be further transformed to obtain the solution. Generally, terms such as energy density, heat flux, and source can be formulated as functions of the dependent variables and material properties. In such a way, the equation only contains the dependent variables, materials properties depending on the dependent variables, and constants. One governing equation may contain multiple dependent variables, but the total numbers of dependent variables and that of the governing equations should be equal to guarantee a unique solution to the equation system. Auxiliary relationships, which are necessary for the internal energy and heat flux, mainly refer to the heat capacity and thermal conductivity and will be discussed in next section. The auxiliary relationships for the source term will be considered in the chapter for the chemical field (Chap. 14). In order to get a heat transfer equation that we are more familiar with, let us start from the formation of the energy density term as Eq. 11.6:

$$e = \underbrace{\left(\int_{T_0}^{T} \rho C_v T + U_0 \right)}_{\text{Internal Energy}} + \underbrace{\left(\frac{1}{2} \rho \mathbf{v} \cdot \mathbf{v} \right)}_{\text{Kinetic Engergy}} + \underbrace{(\rho g z)}_{\text{Potential Energy}}, \qquad (11.6)$$

where C_v is the gravimetric heat capacity at a constant volume and U_0 is the internal energy per unit volume at the reference temperature T_0.

The internal energy is part of the accumulation term in addition to the kinetic energy and potential energy. But in general, the internal energy is dominant over the other energy types. As a result, the energy density is frequently represented by only the internal energy density as in Eq. 11.7, in which the pressure term vanishes for solids:

$$e = \int_{T_0}^{T} \rho C_v T + U_0 = \int_{T_0}^{T} \rho C_p T - \int p dV + U_0, \qquad (11.7)$$

where C_p is the heat capacity at a constant pressure.

Next, the flux term is linked to the temperature by Fourier's law of heat conduction as

$$\mathbf{J}_T = -\lambda \nabla T, \qquad (11.8)$$

where λ is the thermal conductivity. Substituting the above equations for energy density and heat flux into the energy balance equation, we can obtain the heat equation including advection and source terms as:

$$\frac{\partial (\rho C_v T)}{\partial t} + \nabla \cdot (\rho C_v T \mathbf{v}) - \nabla \cdot (\lambda \nabla T) = q_T. \qquad (11.9)$$

As can be seen, this equation is a comprehensive version of the Fourier's equation for heat transfer. The above deduction shows how to obtain the equation from the basic heat conservation law.

Heat Transfer in Porous Materials

Energy balance is mostly considered for the whole system instead of individual phases. This treatment is based on the assumption that thermodynamic equilibrium is maintained between different phases. Therefore, most existing models for porous materials employed one governing equation for heat transfer for the whole system. Governing equations are established for individual phases/components only in a few cases. The general form of the governing equation for heat transfer of the whole system is as follows:

$$\frac{\partial(\sum \rho_i \theta_i C_{\mathrm{v},i} T)}{\partial t} + \sum \nabla \cdot (\rho_i \theta_i C_{\mathrm{v},i} T \mathbf{v}_i) - \nabla \cdot (\lambda \nabla T) = q_{\mathrm{T}}. \quad (11.10)$$

But in most cases, researchers prefer simple treatment in which phases and components are not differentiated. In such a case, the material properties are the overall properties of the porous materials instead of that of any individual phase. In even more simplified treatments, these material properties are moved out of the operators, leading to Eq. 11.11:

$$\rho C_{\mathrm{v}} \frac{\partial T}{\partial t} + \rho C_{\mathrm{v}} \nabla \cdot (T \mathbf{v}) - \lambda \nabla \cdot (\nabla T) = q_{\mathrm{T}}. \quad (11.11)$$

However, a considerable error may be involved when these materials' properties have spatial or temporal variabilities as they are functions of dependent/independent variables or other materials properties.

Heat Capacity

Heat capacities of individual components are mostly used to directly construct the accumulation term in Eq. 11.11 based on the formulation of the internal energy (Eq. 11.7). This formulation, which is the weighted average of the heat capacities of the components with respect to mass, has long been used for soils (de Vries 1963; Campbell 1985). This mass-weighted average method was also used in many other mixtures as

$$C_{\mathrm{v}} = \sum (\rho_i \theta_i C_{\mathrm{v},i}) / \sum (\rho_i \theta_i). \quad (11.12)$$

The above formulation is based on the definition of the thermal energy and can be derived based on thermodynamics.

A constant value is also frequently used for the average heat capacity of porous materials for simplicity. There are also more complex relationships that allow more effects on the heat capacity to be considered. For example, the capacity of gas can be described as a function of temperature:

$$C_{\mathrm{g}} = C_0 + C_1 \cdot T + C_2 \cdot T^2 + C_3 \cdot T^3, \quad (11.13)$$

where C_0 is 1238.79, C_1 is 3.1303, C_2 is 7.905×10^{-4}, C_3 is -6.858×10^{-7}, and C_i is the fitting constants throughout this section. When the phase changes of water occur, especially for freezing and thawing, the latent heat is typically considered with the concept of apparent heat capacity. This term helps ensure the computational stability (Anderson et al. 1973). In this treatment, the released or absorbed energy is incorporated into the heat capacity term using Eq. 11.14:

$$C_{\mathrm{a}} = C + L_{\mathrm{f}} \frac{d\theta_i}{dT}, \tag{11.14}$$

where C_{a} is the apparent gravimetric heat capacity and L_{f} is the latent heat of phase transition.

Thermal Conductivity

The formulation of the thermal conductivity is not as straightforward as that of the heat capacity. The average thermal conductivity of a composite is usually used. In many simulation models, this average thermal conductivity was assumed to be a constant. This treatment was adopted to reduce the nonlinearities of the models or to enable analytical solutions. However, a volume-weighted value of thermal conductivities was more frequently employed in the existing numerical models. The mathematical formulation of this volume-weighted average value is as follows:

$$\lambda = \sum \theta_i \lambda_i / \sum \theta_i. \tag{11.15}$$

The formulation in Eq. 11.15, which has been extensively used, is essentially an empirical equation. There are still other ways for the calculation of the average thermal conductivity, such as a geometric average as the following:

$$\lambda = \left(\prod \lambda_i^{\theta_i} \right)^{1/\sum \theta_i}. \tag{11.16}$$

Russell's equation (Perry and Chilton 1973) has also been applied to the formulation of the thermal conductivity by Scott et al. (2006):

$$\lambda = \lambda_{\mathrm{m}} \sum \frac{[\theta_i/(1-\theta_{\mathrm{m}})]\left[(1-\theta_{\mathrm{m}})^{2/3}(\lambda_i/\lambda_{\mathrm{m}}-1)+1\right]}{\left[(1-\theta_{\mathrm{m}})^{2/3}-(1-\theta_{\mathrm{m}})\right](\lambda_i/\lambda_{\mathrm{m}}-1)+1}, \tag{11.17}$$

where m indicates solid matrix and *i* denotes the phases other than the solid matrix.

Thermal Diffusivity

In many documents about PDE, especially those used in mathematics, the thermal diffusivity is frequently used. The adoption of the thermal diffusivity was also common in early simulation models. As thermal diffusivity is the combination of the heat capacity and thermal conductivity, the purpose for using this term is to simplify the governing equation and the solution process. So in these methods, a constant value of the thermal diffusivity is usually employed. More recent research such as Rempel and Buffett (1997) claimed that the effective thermal diffusivity should include dispersive effects.

Boundary Conditions

The boundary conditions for heat transfer is just materialization of the aforementioned basic types of boundary conditions for general PDEs. But as we are moving into real physical problems, the basic boundary condition types will be detailed and extended to allow for physical mechanisms. The following boundary conditions can be specified on outward and inner boundaries of a region.

The Dirichlet boundary condition is used to handle the condition when the temperature on a boundary is known. For the purpose, a known value of temperature T_{env} is assigned to the vertex or the edge of the computational domain. This is common in chemical reactions when we have a water bath with a constant temperature in contact with the boundary. The T_{env} value at the edge can be specified as a linear function of coordinates. The function parameters can vary from one edge to another but have to be adjusted to avoid discontinuities at edges' junction points. This boundary condition sometimes is called the boundary condition of the first kind.

The Neumann boundary condition is used to consider the condition when the heat flux across a boundary is prescribed. This type of boundary condition can be applied to external boundaries or internal boundaries between regions (Fig. 11.2):

$$
\begin{aligned}
-\mathbf{n}\cdot\mathbf{J} &= -F_n && \text{at external boundaries,}\\
-\mathbf{n}_+\cdot\mathbf{J}_+ - \mathbf{n}_-\cdot\mathbf{J}_- &= F_n && \text{at internal boundaries.}
\end{aligned}
$$

For external boundaries, F_n is the outward normal component of heat flux. For the internal boundary, "+" and "–" superscripts denote quantities to the left and to the right side of the boundary by assuming the axis is pointing to the right. The left region is "+" because it has a boundary outward normal vector pointing in the positive direction. On the internal boundary, F_n is the generated power per unit area or equivalently the surface source.

If F_n on the boundary is zero, the boundary condition is called homogeneous. The homogeneous condition at the outward boundary indicates vanishing of the heat flux across the surface. This kind of boundary condition is used when the boundary is thermally insulated or on a symmetry plane when heat transfer across the plane is obviously absent. In a few cases, a heat source in terms of heat flux can exist on the

Fig. 11.2 External and internal boundaries

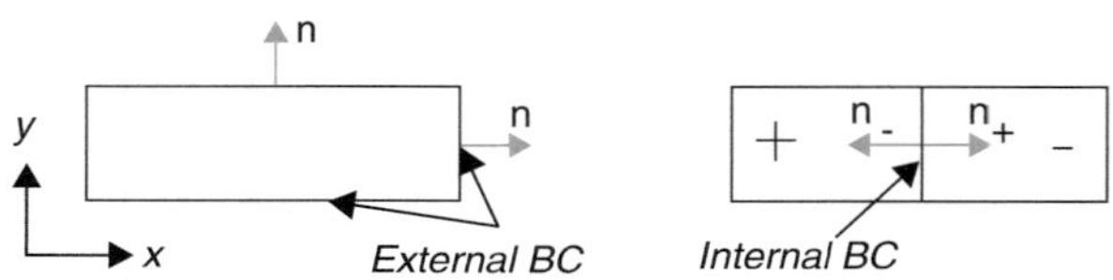

symmetry plane; then this boundary can be simulated as an external boundary, but the surface power has to be divided based on the symmetry. A homogeneous condition for an internal boundary indicates that there is no source on the boundary, and the flux across the internal boundary is continuous at least along the normal direction of the boundary.

Two other physical boundary conditions can be derived from the classic Neumann boundary condition by incorporating the real heat transfer mechanisms across boundaries: convection boundary condition and radiation boundary condition. Technically, they are still the Neumann boundary condition, but the heat flux density is related to temperature and material properties by considering the physical mechanisms.

The convection boundary condition describes convective heat transfer. The convective heat transfer can be described using Newton's law of cooling, in which the heat flux is proportional to the difference between the temperature of the material adjacent to the boundary and the ambient temperature. This type of boundary condition including Newton's law of cooling is defined as

$$F_n = h(T - T_{\text{env}}), \tag{11.18}$$

where h is the coefficient of convective heat transfer and T_{env} is the temperature of contacting fluid medium. Parameters h and T_{env} may differ from part to part of the boundary.

The Neumann boundary condition can also be extended to consider radiative heat transfer across boundaries:

$$F_n = \beta k_{\text{SB}}\left(T^4 - T_{\text{env}}^4\right), \tag{11.19}$$

where k_{SB} is a Stephan-Boltzmann constant ($5.67032 \cdot 10^{-8}$ W/m^2/K^4) and β is an emissivity coefficient. Parameters β and T_0 may differ from part to part of the boundary.

Practice Problem

Problem Description

In rectangular Cartesian coordinates (Fig. 11.3), the 2D sourceless heat equation can be simplified into the following form when the density, heat capacity, and thermal conductivity are constants:

$$\frac{\partial w}{\partial t} = a\left(\frac{\partial^2 w}{\partial x^2} + \frac{\partial^2 w}{\partial y^2}\right), \tag{11.20}$$

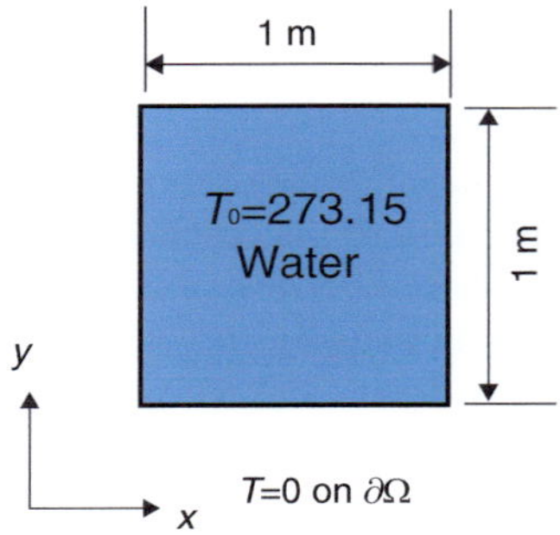

Fig. 11.3 Computational domain

where w is the dependent variable, which is the temperature in this case; and a is the thermal diffusivity. The material filling the domain is water.

1. The problem to be solved corresponds to a 2D square region (1 m by 1 m). The initial temperature is 273.15 K throughout the interior of the region. The temperature on all the boundaries is fixed as 0 K. The question is how the temperature in the domain changes in the first hour. Please describe the problem using mathematical terms we learned in the past chapters.
2. Please conduct simulations using the PDE toolbox in MATLAB. Explain the process step by step (print your screen and add some text to explain what you do).
3. Compare the numerical results with the analytical solution as follows:

$$w = \frac{16w_0}{\pi^2}\left\{\sum_{n=0}^{\infty}\frac{1}{2n+1}\sin\left[\frac{(2n+1)\pi x}{l_1}\right]\exp\left[-\frac{\pi^2(2n+1)^2 at}{{l_1}^2}\right]\right\} \times \left\{\sum_{n=0}^{\infty}\frac{1}{2m+1}\sin\left[\frac{(2m+1)\pi y}{l_2}\right]\exp\left[-\frac{\pi^2(2m+1)^2 at}{{l_2}^2}\right]\right\}. \tag{11.21}$$

To save your time, a MATLAB program has been developed for implementing the analytical solution. But you still need to extract values from both numerical and analytical solutions so that you can do a quantitative evaluation. For example, you can try to compare the temperature distributions along $y = 0$ at $t = 3600$. For the numerical solution, you will need to export mesh and solution from the toolbox GUI first. Then can you use a MATLAB command tri2grid to get the solution on a straight line, i.e., tri2grid(p,e,u(:,3600),-0.5:0.01:0.5,0). As for the distributions in the analytical solution, you can extract the values using a couple of commands as long as you know the structure of the solution (spend a few minutes to understand the code in the appendix).

4. Refine the mesh in the PDE toolbox and run the simulation again. Please show whether the accuracy of the solution is improved.

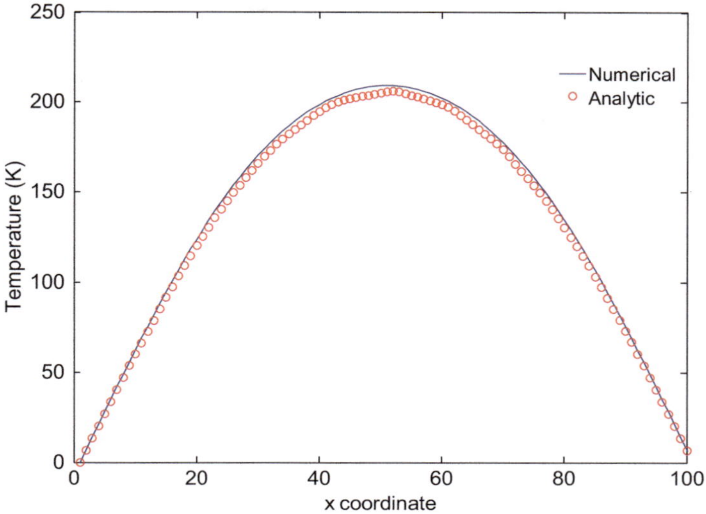

Fig. 11.4 Comparison between numerical and analytical results

Solution

Typical results are shown in Fig. 11.4.

Appendix: MATLAB Code

```
L1=1;
L2=1;
t=0:1:3600;
a=1e-5;
u0=273.15;
x=0:0.01:1;
y=0:0.01:1;
[xx,yy]=meshgrid(x,y);
solution=zeros(length(x),length(y),length(t));
A_temp=zeros(length(x),length(y));
A=zeros(length(x),length(y));
B_temp=zeros(length(x),length(y));
B=zeros(length(x),length(y));
time_step=0;
for temp=t
  time_step=time_step+1
  A=zeros(length(x),length(y));
  B=zeros(length(x),length(y));
  for n=0:100
```

```
    A_temp=1/(2*n+1)*sin((2*n+1)*pi*xx/L1).*exp(-pi^2*a*temp*
(2*n+1).^2/L1^2);
    A=A+A_temp;
    B_temp=1/(2*n+1)*sin((2*n+1)*pi*yy/L2).*exp(-pi^2*a*temp*
(2*n+1).^2/L2^2);
    B=B+B_temp;
  end
  solution(:,:,time_step)=16*u0/pi^2*A.*B;
  figure(1)
  surf(xx,yy,solution(:,:,time_step))
end
%surf(xx,yy,solution(:,:,time_step))
```

Chapter 12
Hydro: Pore Water Movement

Introduction

Groundwater is the water located beneath the Earth's surface in soil pore spaces and in the fractures of rock formations. Usually, the topsoil or rock layer is unsaturated unless it is right close or beneath the surface water. However, a groundwater table may be encountered if we go deeper. The boundary between the unsaturated layer and the completely saturated layer is called the water table. A rock body or an unconsolidated deposit is called an aquifer when it can yield a usable quantity of water. Groundwater is closely related to surface water which it is recharged from and eventually flows to. Natural discharge often occurs at springs and seeps, possibly forming oases or wetlands. Groundwater is also often withdrawn for agricultural, municipal, and industrial use via extraction wells. The study of the distribution and movement of groundwater is hydrogeology, also called groundwater hydrology. Besides serving as an essential resource, groundwater can also be a harmful factor in various natural hazards such as landslides and earthwork construction. Therefore, groundwater movement is also of great interest in water resources and geotechnical engineering. Hydro is used to represent water in the general context and water movement in the context of multiphysics.

Water moves in porous materials in terms of saturated flow or unsaturated flow. In the hydrologic cycle, as illustrated in Fig. 12.1, water flows both in saturated and unsaturated porous media (soils, rock formations). The vadose zone, also termed unsaturated zone or zone of aeration elsewhere, is the Earth portion connecting the atmosphere and the phreatic zone where the groundwater presents. The vadose zone contains, at least sometimes, air and water in the pores, leading to an unsaturated condition. The thickness of vadose zone ranges from zero, as when a lake or marsh is at the surface, to hundreds of meters, as is common in arid regions. The upper part of the vadose zone commonly includes the plant root zone. The phreatic zone, also called saturated zone, is the part below the groundwater table. This part is where we can locate aquifers. It is worthwhile to mention the water flows in the unsaturated

Z. Liu, *Multiphysics in Porous Materials*,
https://doi.org/10.1007/978-3-319-93028-2_12

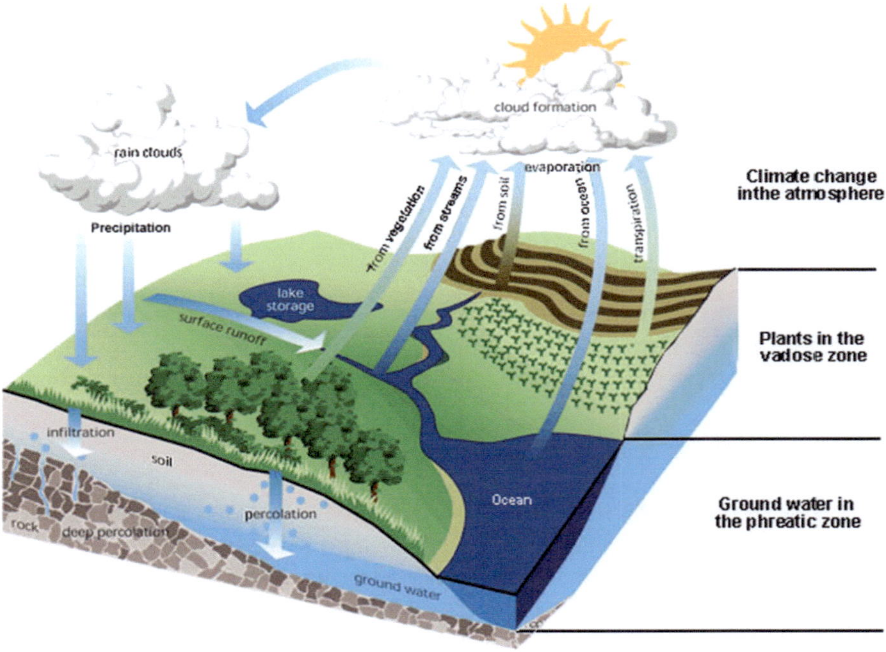

Fig. 12.1 Schematic of vadose zone and the hydrologic cycle

and saturated zones occur according to significantly different mechanisms and thus are described in distinct ways. The water in the unsaturated zone is mostly called water flows in the vadose zone or just unsaturated flow and is formulated using the Richards equation. The water in the saturated zone is usually called groundwater flow and is formulated using the groundwater flow equation.

In the following sections, the groundwater flow will be introduced first, including mass conservation and Darcy's law. The introduction to these laws and their corresponding mathematical formulations will be conducted for both the saturated flow and unsaturated flow. Next, brief information regarding surface water will also be provided to lay down a complete framework for the water movement related to porous materials. Finally, a practice problem for coupled saturated-unsaturated flow will be presented, which can be solved with FlexPDE and validated against documented testing results.

Groundwater Movement

The groundwater can be described using the groundwater flow equation. This equation is extensively used in hydrogeology as the mathematical formulation for the flow of groundwater through an aquifer. The transient flow corresponds to a flow field changing with time. That is, the water head/pressure at any point could change with time. If dye is placed in the porous media, then the flow net generated by the migratory dye with water movement will also change with time. A parabolic PDE will be used for this transient field. This PDE is also used for many other types of transient conservation problems, such as the heat conduction and the particle diffusion. For an equilibrium (or steady-state) problem, the time-dependent term or accumulation term drops out, and the governing equation will reduce into an elliptic PDE, i.e., Laplace's equation.

The groundwater flow equation can be derived using the general REV concept. The assumption of REV still needs to be ensured in which the properties of the medium are assumed to be constants. In this chapter, we will introduce a mass balance between the water flowing in and out of a small control volume. The flux terms in the balance are usually expressed using head via a constitutive equation called Darcy's law, whose validity requires that the flow is slow.

Mass Balance

The scientific fact underlying the groundwater equation is the mass balance. This balance is analogous to the energy balance used in heat transfer, which leads to the heat equation. The mass balance in a porous material states that, for a given control volume, aside from sources or sinks, mass cannot be created or destroyed. Alternatively, the conservation of mass defines that, for a given increment of time Δt, the difference between the mass flowing in across the boundaries, the mass flowing out across the boundaries, and the sources within the volume is the change in storage. Mathematically, we would have

$$\frac{\Delta M_{\text{stor}}}{\Delta t} = \frac{M_{\text{in}}}{\Delta t} - \frac{M_{\text{out}}}{\Delta t} - \frac{M_{\text{gen}}}{\Delta t}. \tag{12.1}$$

Applying the above equation to a REV, we obtain

$$\frac{1}{V}\frac{d}{dt}\int_{\Omega} m dV = -\oint_{\partial\Omega} \rho \mathbf{J}_{\text{H}} \cdot \mathbf{n} dS + \int_{\Omega} \rho q_{\text{H}} dV. \tag{12.2}$$

Applying the divergence theorem and turning the integral form into the differential form, we will obtain

$$\frac{1}{V}\frac{\partial V}{\partial t} = -\nabla \cdot \mathbf{J}_{\text{H}} + q_{\text{H}}. \tag{12.3}$$

Darcy's Law

Narrow Definition for Special Case

Darcy's law at a constant elevation is a simple proportional relationship between the instantaneous discharge rate through a porous medium, the viscosity of the fluid, and the pressure drop over a given distance:

$$Q = \frac{-kA}{\mu}\frac{(p_b - p_a)}{L}. \tag{12.4}$$

The total discharge, Q (units of volume per time, e.g., m^3/s), is equal to the product of the intrinsic permeability of the medium, k (m^2); the cross-sectional area to flow, A (units of area, e.g., m^2); and the total pressure drop ($p_b - p_a$), (Pa), all divided by the viscosity, μ (Pa·s), and the length over which the pressure drop is taking place (m). The negative sign is needed because fluid flows from a location with a high pressure to that with a low pressure. It is noted that the elevation head must be taken into account if the inlet and outlet are at different elevations. If the change in pressure is negative (where $p_a > p_b$), then the flow will be in the positive direction. Dividing both sides of the equation by the area and using more general notation leads to

$$\mathbf{J}_H = -\frac{k}{\mu}\nabla p, \tag{12.5}$$

where $\mathbf{J}_H$ is the flux (discharge per unit area per unit time, m/s) and ∇p is the pressure-gradient vector (Pa/m). It is also very common to formulate Darcy's law using hydraulic conductivity and water head:

$$\mathbf{J}_H = -K\nabla h \tag{12.6}$$

where

$$K = \frac{\rho g k}{\mu}, \tag{12.7}$$

$$p = \rho g h. \tag{12.8}$$

Equation 12.7 implies the intrinsic permeability k is a property of the solid matrix while hydraulic conductivity K is a property of the whole porous material. The misuse of these two terms is a common mistake in technical literature for the pore water movement.

General Definition for Porous Materials

A more general Darcy's law requires a better understanding of the energy state of pore water. The water head h is a measure of the energy state of pore water in terms of the height of an equivalent water column. In porous materials, the total energy of the pore water has three components: pressure, elevation, and velocity. Accordingly, the water head also has these three components. In normal conditions, the water velocity in porous materials is low due to the small pore size. Therefore, the total head usually contains pressure head and elevation head:

$$h \approx h_p + h_z, \tag{12.9}$$

where h_p and h_z are pressure and elevation head, respectively. However, the following symbols will be used to render a consistent symbol system for the flow in both saturated and unsaturated porous materials:

$$h = \psi + z, \tag{12.10}$$

where ψ is the pressure head, and z is the elevation head as the coordinate z is usually reserved for the gravitational direction.

The flux $\mathbf{J}_\mathrm{H}$, which is often referred to as the Darcy flux or velocity, is not the velocity at which the fluid travels through the pores $\mathbf{v}$. However, this true fluid velocity $\mathbf{v}$ is related to the Darcy flux $\mathbf{J}_\mathrm{H}$ by the porosity ϕ. The flux is divided by porosity to account for the fact that only a fraction of the total formation volume is available for the fluid flow. The true fluid velocity would be the velocity that a conservative tracer would experience when it is carried by the moving fluid through the porous material, that is,

$$\mathbf{v} = \frac{\mathbf{J}_\mathrm{H}}{\phi_\mathrm{eff}}, \tag{12.11}$$

where ϕ_eff is the effective porosity, which is defined as the ratio between the flow area and the total sectional area available to the flow. It is noted that the porosity of the porous material is defined as the ratio between the pore volume and the total volume. Therefore, the effective porosity and porosity are defined using area and volume, respectively. Their values may be close despite their different definitions.

The Darcy's law holds when the velocity of pore water is slow, which is the case for most cases in porous materials. The Darcy's law can be derived from the Navier-Stokes equations under the condition of very slow liquid flow velocity, i.e., creep flow.

Derivation of Governing Equation

Mass is the product of the density and volume. Under creep flow conditions in porous materials, water can be considered incompressible: the density does not depend on the pressure. The mass fluxes across the boundaries then become volume fluxes as are used in Darcy's law. Using Taylor series to represent the in- and outflux terms across the boundaries of the control volume, and using the divergence theorem to turn the flux across the boundary into a flux over the entire volume, the final form of the groundwater flow equation (in differential form) is

$$\frac{1}{V}\frac{\partial V}{dt} = -\nabla \cdot \mathbf{J}_{\mathrm{H}} + q_{\mathrm{H}}. \tag{12.12}$$

In reality, we define a material property as a specific storage, which is the volume of water released from storage per unit decline in the hydraulic head:

$$S_{\mathrm{s}} = \frac{\partial h}{\partial V/V}. \tag{12.13}$$

Then we will have

$$S_{\mathrm{s}}\frac{\partial h}{\partial t} = -\nabla \cdot \mathbf{J}_{\mathrm{H}} + q_{\mathrm{H}}. \tag{12.14}$$

This parabolic PDE states that the change in the hydraulic head with time (left-hand side) equals the negative divergence of the flux $\mathbf{J}_{\mathrm{H}}$ and the source term q_{H}. This equation contains two unknowns, the head h and flux $\mathbf{J}_{\mathrm{H}}$, while q_{H} is usually known for a given problem. As Darcy's law relates the flux to the head, we can substitute it into the above equation to eliminate the flux $\mathbf{J}_{\mathrm{H}}$, leading to

$$S_{\mathrm{s}}\frac{\partial h}{\partial t} = -\nabla \cdot (-K\nabla h) - q_{\mathrm{H}}. \tag{12.15}$$

Now if the hydraulic conductivity K is spatially uniform and isotropic can be formulated using a scalar rather than a tensor or a function, then this term can be taken out of the spatial derivative, further simplifying the above equation into

$$S_{\mathrm{s}}\frac{\partial h}{\partial t} = K\nabla^2 h - q_{\mathrm{H}}. \tag{12.16}$$

In some cases, the above equation is further simplified by dividing both sides of the equation by the specific storage S_{s}, leading to a new term, hydraulic diffusivity $\alpha = K/S_{\mathrm{s}}$ on the right-hand side. The hydraulic diffusivity is proportional to the speed at which a finite pressure pulse will propagate through the system. The groundwater flow equation for transient flow then becomes

$$\frac{\partial h}{\partial t} = \alpha\nabla^2 h - \frac{q_{\mathrm{H}}}{S_{\mathrm{s}}}, \tag{12.17}$$

where the sink/source term, q_{H}, is divided by the appropriate storage term.

The steady-state equation for groundwater movement is obtained by eliminating the transient term in the above equation, which has the form of Poisson's equation:

$$\nabla^2 h = q_{\mathrm{H}}/S_{\mathrm{s}}. \tag{12.18}$$

As the source terms usually do not exist, the above equation is reduced to Laplace's equation, which serves as the theoretical basis of equilibrium groundwater problems and engineering methods for solving such problems such as the flow net method:

$$\nabla^2 h = 0. \tag{12.19}$$

The boundary conditions of either the transient or steady-state governing equations can be easily dealt with. For transient flow, either Dirichlet or Neumann boundary conditions can be applied. For the steady-state, it seems that Neumann boundary condition may not be included because an equilibrium state may be impossible unless with exactly equal influx and outflux. The Dirichlet boundary condition assigns the value of the head on the boundary, while the Neumann boundary condition assigns the derivative of the head to define the flux through the boundary via Darcy's law:

$$(-K\nabla h)\cdot(-\mathbf{n}) = \text{influx}, \tag{12.20}$$

where **n** is the unit outward normal vector of the boundary.

Water Flow in Unsaturated Zones

The Richards equation represents the movement of water in unsaturated soils and was formulated by Lorenzo A. Richards in 1931. It is a nonlinear PDE, which is often difficult to approximate since it does not have a closed-form analytical solution.

Derivation

We can derivate the governing equation for water flow in unsaturated zones in a way similar to that is used for saturated flow. The major difference lies in both the water accumulation and Darcy's law. For the water accumulation, the difference is that the water volume is no more majorly attributed to the volume expansion, $\delta V/V$, caused by the change in volume pressure. This is because, as the porous material is unsaturated, the change in the water content $\delta\theta$ can now be well accommodated by the partially filled pore space. Therefore, the governing equation can be written as

$$\frac{\partial \theta}{\partial t} = -\nabla \cdot \mathbf{J}_{\mathrm{H}} + q_{\mathrm{H}}. \tag{12.21}$$

Darcy's law still holds in unsaturated flow as the velocity of water movement is still proportional to the energy gradient:

$$\mathbf{J}_{\mathrm{H}} = -K\Delta h. \tag{12.22}$$

However, in unsaturated porous materials, the water pressure head now turns into a negative value, which indicates that its energy is lower than the reference energy, i.e., the pressure head of bulk water under standard conditions (1 atmosphere).

Substituting $\mathbf{J}_{\mathrm{H}}$ into the equation above, we get

$$\frac{\partial \theta}{\partial t} = \nabla \cdot (K\nabla h) + q_{\mathrm{H}}. \tag{12.23}$$

Substituting $h = \psi + z$ into the above equation, then we have

$$\frac{\partial \theta}{\partial t} = \nabla \cdot [K(\nabla \psi + \nabla z)] = \nabla \cdot (K\nabla \psi) + \nabla \cdot (K\mathbf{e}_k), \tag{12.24}$$

where $\mathbf{e}_k$ is the unit vector along the gravitational (z) direction. The above equation contains two unknowns: water content and pressure head. Due to the reason, it is called the mixed form of the Richards equation.

Formulations

The Richards equation is used extensively in hydrology, hydrological engineering, and geotechnical engineering. It also appears in pure mathematical journals due to the fact that it has nontrivial solutions. Three forms of the equation can be found in the literature. The mixed form contains the pressure and the degree of saturation as introduced above. It can also appear in two other forms: the head-based formulation and saturation-based formulation.

Head Based

$$C(h)\frac{\partial \psi}{\partial t} = \nabla \cdot [K(h)\nabla h], \tag{12.25}$$

where $C(h)$ [1/L] is a function describing the rate of the change of saturation with the change of the pressure (suction) head:

$$C(h) = \frac{\partial \theta}{\partial \psi}. \tag{12.26}$$

This function is called "specific moisture capacity" in the literature and could be determined for different soil types using curve fitting and laboratory experiments measuring the rate of infiltration of water into a soil column as described in van Genuchten (1980).

Saturation Based

$$\frac{\partial \theta}{\partial t} = \nabla \cdot [D(\theta) \nabla \theta], \tag{12.27}$$

where $D(\theta)$ [L^2/T] is "the soil-water diffusivity":

$$D(\theta) = \frac{K(\theta)}{C(\theta)} = K(\theta) \frac{\partial h}{\partial \theta}. \tag{12.28}$$

A concept called the soil-water characteristic curve needs to be introduced before we further investigate C and K. This SWCC (mostly used in geotechnical engineering), which is also called water retention curve (in soil science) and saturation-capillary pressure curve (in petroleum engineering), is a fundamental constitutive relationship in porous materials. The SWCC defines a unique relationship between the water content, or degree of saturation, and the suction (pressure head or potential) if hysteresis is neglected. Both C and K are functions of the SWCC in unsaturated porous materials. The hydraulic conductivity can be estimated by integration of the SWCC (Mualem 1976; Fredlund and Xing 1994). Both van Genuchten's equation (van Genuchten 1980) and Fredlund and Xing's equation (1994) are common for formulating the SWCC. Different approaches were then adopted to predict the hydraulic conductivity based on SWCC.

$$S_e = \frac{\theta - \theta_r}{\theta_s - \theta_r} = (1 + |\alpha\psi|^n)^{-m}, \tag{12.29}$$

where S_e is the effective saturation, θ_s and θ_r are the saturated and residual water content, respectively; and α, m, and n are empirical parameters.

The hydraulic conductivity in unsaturated porous materials can be viewed as the product of the saturated hydraulic conductivity and the relative hydraulic conductivity. The relative conductivity K_r quantifies how the hydraulic conductivity in unsaturated porous materials changes with the water content or the degree of saturation. One widely adopted relationship is proposed by van Genuchten (1980):

$$K(S_e) = K_s K_r = K_s S_e \left[1 - \left(1 - S_e^{\frac{1}{m}}\right)^m\right]^2. \tag{12.30}$$

The treatment of the boundary condition for the Richards equation is also straightforward, similar to those in the groundwater movement equation.

Mixed Saturated-Unsaturated Flow

In the real world, water moves in both saturated and unsaturated zones which may coexist as contagious saturated and unsaturated zones, whose boundary may change over time. So the numerical simulation of such a case requires the simultaneous implementations of the groundwater flow equation and the Richards equation. A standard formulation of the Richards equation that also includes groundwater (saturated zone) flow (Richards 1931) is

$$\frac{S_s \theta(h)}{\phi} \frac{\partial h}{\partial t} + \frac{\partial \theta}{\partial t} = \nabla \cdot [K_s K_r \nabla (h + z)] + q_H, \tag{12.31}$$

where h is the pressure head, z is the vertical coordinate, K_s is the saturated hydraulic conductivity tensor, K_r is the relative hydraulic conductivity, S_s is the specific storage coefficient, and ϕ is the porosity. The detailed implementation of the equation will be practiced in a later practice problem.

Surface Flow

Surface flow is much different from the flow in porous materials; the latter can be termed subsurface flow in hydrogeology. However, it is very common that these two types of flow interfere with each other. In the water cycle, when precipitation falls on the earth's land, some of the water flows on the surface forming streams and rivers. The remaining water, through infiltration, penetrates the soil traveling underground, hydrating the vadose zone soil, recharging aquifers, with the excess flowing in the subsurface runoff. Water flows from areas where the water table is higher to areas where it is lower. This flow can be either surface runoff in rivers or streams or subsurface runoff infiltrating rocks and soils. The amount of runoff reaching surface and groundwater can vary significantly, depending on rainfall, soil moisture, permeability, groundwater storage, evaporation, upstream use, and frost depth.

Subsurface water may return to the surface as groundwater flow, such as from a spring, seep, or a water well, rivers, and oceans. Water returns to the land surface at a lower elevation than the place where infiltration occurs, under the force of gravity or gravity-induced pressures. Groundwater tends to move slowly and be replenished slowly, so it can remain in aquifers for thousands of years. Mainly, water flows through the ground and eventually to the ocean where the cycle begins again.

Despite the difference, the mathematical description of the surface flow resembles that of the flow in porous materials. Mass conservation for surface flow is

$$\frac{\partial h}{\partial t} = \nabla \cdot (\mathbf{v}h) + q_e + q_r, \tag{12.32}$$

where q_e is the exchange flux between surface and subsurface domains, and q_r is a general source/sink term. Manning's equation is often used to establish a flow depth-discharge relationship, where the velocity vector $\mathbf{v}$ in may be written as follows:

$$\mathbf{v} = \frac{\sqrt{\mathbf{S}_f}}{n} h^{2/3}, \tag{12.33}$$

where $\mathbf{S}_f$ [L] is the friction slope and n [TL$^{-1/3}$] is Manning's coefficient. Other approaches are available to take the place of Manning's equation (Nepf 1999).

Practice Problem

Experiment Description

A lab test for mixed saturated-unsaturated flow was conducted and documented by Luthin et al. (1975). Figure 12.2 shows the schematic of a tank used in their test. Initially, the sector tank was filled with the sand. A gravel-filled reservoir was placed at the wide end of the tank to simulate a constant horizontal replenishment. Please simulate the experiment and validate your simulation against the lab test.

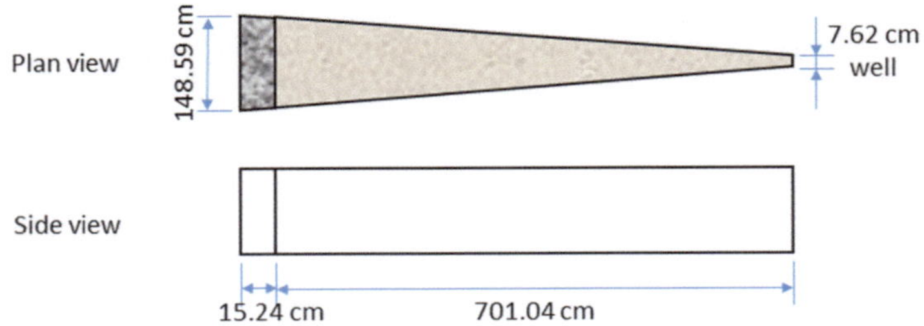

Fig. 12.2 Schematic of the sector tank used and the locations of pressure measuring points

Model Development and Implementation

This is a typical mixed saturated-unsaturated flow problem, in which the boundary between the two regions may change with time. To describe the flow, we need to implement both the groundwater flow equation and the Richards equation, which can be formulated with either the total head or the pressure head:

Total head:

$$\frac{S_s\theta(h)}{\phi}\frac{\partial h}{\partial t}+C(h)\frac{\partial(h-z)}{\partial t}=\nabla\cdot[K_sK_r\nabla h] \tag{12.34}$$

Pressure head:

$$\frac{S_s\theta(\psi+z)}{\phi}\frac{\partial(\psi+z)}{\partial t}+C(\psi)\frac{\partial\psi}{\partial t}=\nabla\cdot[K_sK_r\nabla(\psi+z)] \tag{12.35}$$

where S_s is the specific storage, θ is the water content, h is the total water head, K_s is the saturated hydraulic conductivity, ϕ is the porosity, and K_r is the relative hydraulic conductivity. It is worthwhile to mention that the boundary conditions and auxiliary relationships such as SWCC and unsaturated (relative) hydraulic conductivity will take different forms depending on the choice of the dependent variable.

The specific storage is assumed to be 10^{-5} as a dense sand was used in the test. *C*(*h*) is the derivative of the function of water content with respect to the water head. The saturated hydraulic diffusivity is 0.946 cm/min. The relative hydraulic conductivity and SWCC were obtained via curve fitting using the experimental data in Fig. 12.3. The porosity is 0.6 and the specific gravity is 2.7.

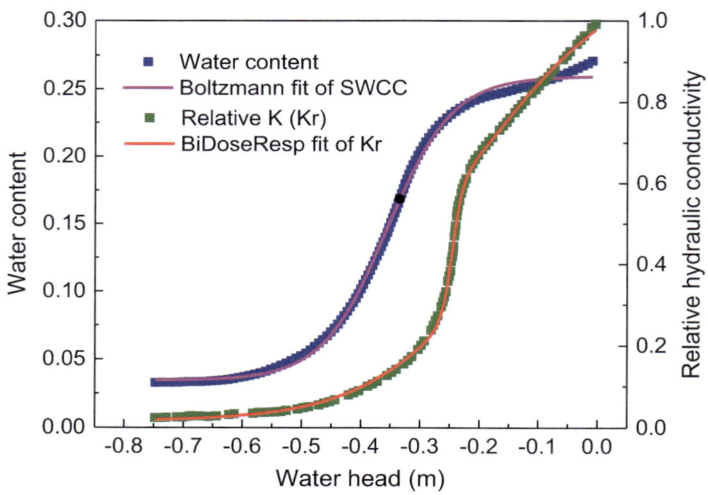

Fig. 12.3 Moisture content and hydraulic conductivity vs soil moisture tension

The computational domain takes a cylindrical coordinate system (r and z). The governing equation needs to be formatted accordingly if we do not use the nabla symbol.

$$S/\text{phi}^{*}dt(H) = K/r^{*}dr(H+z) + dr(K^{*}dr(H+z)) + dz(K^{*}dz(H+z))$$

Alternatively, we can use the nabla symbol

$$S/\text{phi}^{*}dt(H) = K^{*}\text{div}(\text{grad}(H+z))$$

where S is the water content, H is the pressure head, and K is the general hydraulic conductivity for both saturated and unsaturated soils. The "IF…THEN…" command needs to be used to define K, H, and the coexistence of the two accumulation terms on the left-hand side of the governing equation.

Boundary conditions are key factors influencing the behavior of simulations. In this case, the boundary conditions were established to approximate the experimental conditions:

1. Bottom of the tank: no flux
2. Left (reservoir): constant total head
3. Top of the tank: no flux
4. Right (well): constant total head from the bottom to the free water surface; no flux in the rest of the boundary

Here is the mesh of the computational domain (Fig. 12.4).

After the model is established, cases with different heights of water in the well are simulated with different time steps. The definition of each parameter is illustrated in the codes in the Appendix.

Typical Numerical Results

The major purpose of this practice problem is to guide the readers through the development of numerical models using FlexPDE and the validation of the model against experimental results. In the previous chapter, we used one practice problem to show how to create a model using MATLAB PDE Toolbox and validate it against analytical solutions. Figure 12.5 plots the water table (boundary between saturated and unsaturated zones) and pressure head distribution at the end of 5 min in the test, which are comparable to the experimental results in Fig. 12.6.

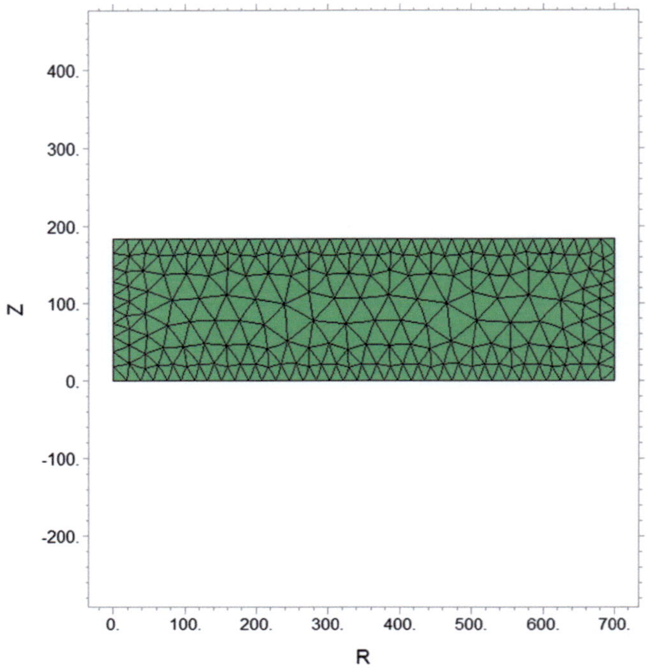

Fig. 12.4 Meshing with default settings

Appendix: FlexPDE Script File

```
    { POROUS.PDE
    This problem describes the flow through an a soil tank, Luthin
and Orhun, Water Resrouces Research, Vol. 11, No. 6, 1975
    }
TITLE 'Coupled saturated and unsaturated flow'
SELECT
  CHANGELIM=200
  fixdt = OFF
  ngrid=10
  regrid=ON
  order=3
COORDINATES YCYLINDER
VARIABLES
  H !Pressure Head
DEFINITIONS
  K0=0.946/60/100
  Ss=0.22473/0.0547*(1+exp((H+0.35549)/0.0547))^(-2)*exp((H
+0.35549)/0.0547)
```

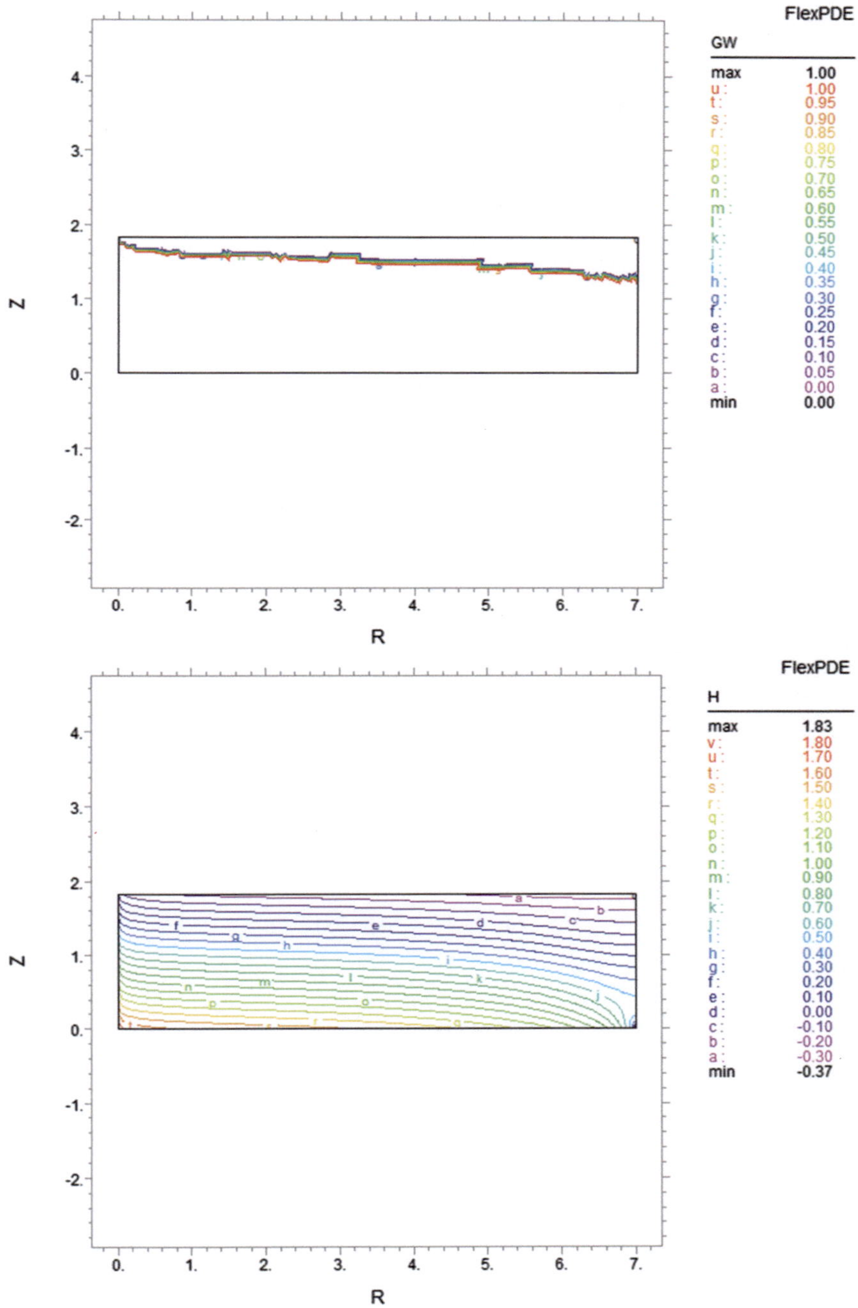

Fig. 12.5 Numerical solutions at 5 min for water table (upper) and pressure head (lower)

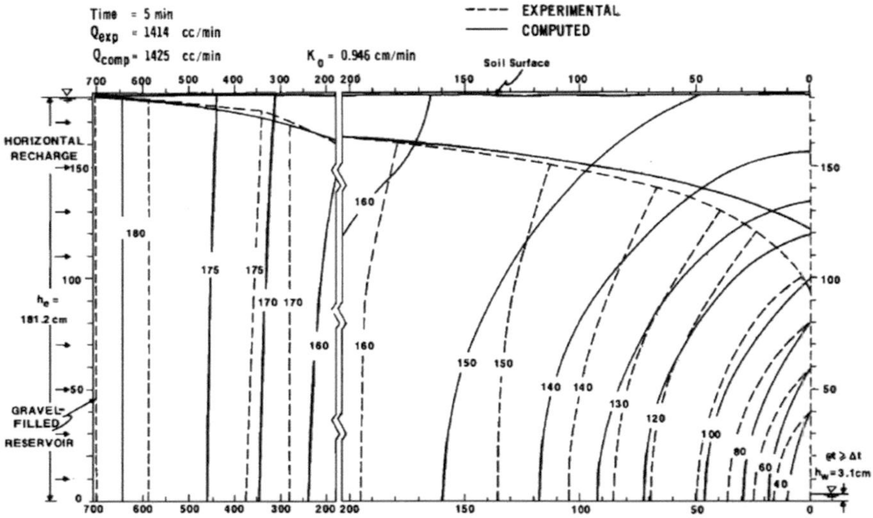

Fig. 4. Computed and experimental equipotentials, free water surface, and well discharge for t = 5 min for the case of a water level of 3.1 cm in the well.

Fig. 12.6 Experimental results at 5 min (Luthin et al. 1975)

```
  Kr=0.01628+1.09848*0.70914/(1+10^((-0.17064-H)/4.03296))
+1.09848*(1-0.70914)/(1+10^((-0.24585-H)/50.58359))
  K= if H>=0 then K0 else K0*Kr
  S= if H>=0 then 1e-8 else Ss
  GW=if H>=0 then 1 else 0
  Discharge=LINE_INTEGRAL(NORMAL(-k*grad(H)),'outlet')
  phi=0.6
EQUATIONS
  H : S/phi*dt(H)=K/r*dr(H+z)+dr(K*dr(H+z))+dz(K*dz(H+z))
INITIAL VALUES
  H = 1.8288-z
BOUNDARIES
  REGION 1
    START 'outlet' (0,0) NATURAL(H)=0
    LINE TO  (7.0104,0)   VALUE(H)=0.031-z
    LINE TO  (7.0104,0.031) NATURAL(H)=0
    LINE TO (7.0104,1.8288) NATURAL(H)=0
    LINE TO (0,1.8288)   VALUE(H)=1.8288-z
    LINE TO CLOSE
TIME 0 TO 300 BY 0.001
MONITORS
  FOR cycle=10
    CONTOUR(GW)
    CONTOUR(H)
PLOTS
  FOR TIME=10 BY 10 TO endtime
    CONTOUR(GW)
    CONTOUR(H)
```

```
    !surface(H)
    ELEVATION (H) FROM (0,0) TO (7.0104,0)
    ELEVATION (H) FROM (0,1.8288) TO (7.0104,1.8288)
    HISTORY (H) AT (7,1.8)
    HISTORY (Discharge)
END
```

Chapter 13
Concentrato: Transport of Dispersed Mass

Introduction

This chapter first clarifies the different definitions of concentrations. Following the clarification on the basic terms, the formulation of the general governing equation of the concentration field will be given out. Additional effort will be made to introduce the three major dispersed mass transport mechanisms in detail, i.e., advection, diffusion, and dispersion. The introduction will cover the definitions, origins in physics, mathematical formulations, material properties, and applications in porous materials. A practice problem developed from a documented test will be presented at the end to exemplify the implementation of the concentration field.

Definitions and Mechanisms

The root "concentrato" is newly coined in this book to represent any process involving the evolution or/and equilibrium of the concentration field. In other words, it is used as a generic term for various mechanisms underlying mass transport of dispersed phases in the context of multiphysics. Though the mass transport of dispersed phases in a mixture (Fig. 13.1), i.e., solution, colloid, and suspension, is a very basic monolithic physical process, a representative root such as concentrato has seldom been used. This fact is possibly due to a few inconspicuous reasons related to the nature and historical development of this topic.

First, mass transport includes three mechanisms: diffusion, dispersion, and advection. One unique feature of concentrato is that these three mechanisms are mutually independent but not exclusive of each other. This feature enables any of them to constitute a governing equation for the mass transport in a system in which the corresponding mechanism is the only (considered) mechanism or the dominant one therein, such as diffusion:

Z. Liu, *Multiphysics in Porous Materials*,
https://doi.org/10.1007/978-3-319-93028-2_13

Fig. 13.1 Dispersed phases in different types of mixtures

$$\frac{\partial c}{\partial t} = \nabla \cdot (D\nabla c), \tag{13.1}$$

where c is the concentration and D is the diffusion coefficient. Also, as they are not exclusive, they can work together to consider any combined mass transport mechanisms, such as combined diffusion and advection:

$$\frac{\partial c}{\partial t} = \nabla \cdot (D\nabla c) - \mathbf{v}\nabla c, \tag{13.2}$$

where $\mathbf{v}$ is the velocity, which is associated with the dispersed phase or the bulk phase (or dispersion phase) that the dispersed phase is distributed (dispersed) in. It seems that, in applications, these three individual mechanisms have been more of interest than the ensemble of the three. Accordingly, roots such as "diffuso" and "advecto" are still present in the literature though not common, while "concentrato" is rarely seen.

Second, mass transport is a topic in need of further integration. Mass transport could refer to two apparently different types of processes. One is the mass transport of a continuum, which includes an accumulation term and an advection term. The equation describes that the mass accumulation of a material in a REV, usually a fluid, is caused by the mass advected by the flux of the fluid into the REV. This mass transport of the fluid occurs due to its own inertial, viscous, gravitational, and other types of forces. In the meanwhile, we are also familiar with the second type of mass transport such as the contaminant transport. The distinction between these two makes mass transport a loose and confusing term. The solution is to treat the aforementioned single-phase case as a special case of the mixture and resort to the definition of concentration.

Concentration can be described in four different ways. The mass concentration ρ_i is defined as the mass of a constituent m_i divided by the volume of the mixture V:

$$\rho_i = \frac{m_i}{V}. \tag{13.3}$$

The SI unit is kg/m^3 (equal to g/L). This use is common in environmental engineering such as the study of sulfur dioxide in the air and that of dissolved oxygen in the water.

The molar concentration c_i is defined as the amount of a constituent n_i (in mole) divided by the volume of the mixture, V:

$$c_i = \frac{n_i}{V}. \tag{13.4}$$

The SI unit is mol/m^3, while the use of mol/L (mol/dm^3) is also common. This expression is convenient when water is the dispersion fluid and chemical reactions occur in the fluid.

The number concentration C_i is defined as the number of entities of a constituent N_i in a mixture divided by the volume of the mixture V:

$$C_i = \frac{N_i}{V}. \tag{13.5}$$

The SI unit is 1/m^3. This expression is helpful when dealing with large objects contained in a bulk phase such as suspensions.

The volume concentration θ_i is defined as the volume of a constituent V_i divided by the volume of the mixture V:

$$\theta_i = \frac{V_i}{V}. \tag{13.6}$$

Being dimensionless, the volume concentration is expressed as a number, e.g., 0.18 or 18%. This expression is frequently adopted in multiphase and multicomponent problems, such as the water content concept in soil science. The mass transport of a single continuum can be described using the first definition of concentration.

Third, the terms of transport, e.g., diffusion and dispersion, have different meanings in different disciplines; and even within multiphysics, they could mean different things. In multiphysics, transport and diffusion could mean processes involving mass, energy, or momentum. Therefore, in a broad sense, diffusion includes both heat diffusion, mass diffusion, and momentum diffusion. Besides, terms such as dispersion are widely used in chemistry, geology, materials science, and physics for totally different things. This fact impedes the wide adoption of these terms in multiphysics and prevents the use of these terms as a purely mass transport concept. It is proposed in this book that these terms are reserved for mass transport only. Accordingly, heat diffusion will be replaced with conduction, and momentum diffusion mainly is referred to as momentum transfer via the Reynolds stress.

Finally, the physical nature of the three mechanisms blurs the distinctions between them. Transport mechanisms need to be discussed with respect to the mixture type. Mixtures can be roughly divided into suspension, colloid, and solution, which correspond to suspended particles (>1 μm), colloids (between 1 nm and 1 μm), and solutes (dissolved molecule, atoms, or ions). The scale boundaries are in fact not as clear as noted. We know that diffusion includes both molecular diffusion and turbulent diffusion which occur due to different reasons and at different scales. Usually, molecular diffusion occurs in solutions and colloids; however, turbulent diffusion can be associated with all three types of mixtures. In general, molecular diffusion is far better known than turbulent diffusion. This explains why in many places diffusion is introduced as a microscopic phenomenon, which possibly means molecular diffusion only. Another example is dispersion, which has a very similar mathematical formation to diffusion. However, dispersion is a macroscopic phenomenon caused by changes in velocities, which is totally different from the causes and mechanisms of diffusion.

A general equation for the dispersed mass transport or concentration field, including diffusion, dispersion, and advection, is formulated as Eq. 13.7. A combination of any two of them, such as diffusion and advection, is named with these two mechanisms, that is, diffusion-advection equation:

$$\frac{\partial c}{\partial t} = \underbrace{\nabla \cdot (D_{\mathrm{diff}} \nabla c)}_{\text{diffusion}} + \underbrace{\nabla \cdot \left(D_{\mathrm{disp}} \nabla c\right)}_{\text{dispersion}} - \underbrace{\mathbf{v} \cdot \nabla c}_{\text{advection}}, \tag{13.7}$$

where c can be replaced by the other definitions of the concentration. But such a replacement requires to change the coefficients for diffusion D_{diff} and dispersion D_{disp} accordingly.

In porous materials, the fact that only part of the volume is occupied by fluids needs to be considered. Thus porosity and tortuosity need to be incorporated into the formulation of the material properties or even the definition of the concentration. Detailed mathematical formations for advection, diffusion, and advection will be discussed in the following sections.

Advection

Definition

In a broader sense, advection is a transport mechanism of a substance or a conserved property by a fluid due to the bulk motion of the fluid. An example of advection is the transport of pollutants and sediments by the bulk water flow. Here the fluid may be any material that contains thermal energy, such as water or air. In general, any substance or conserved quantity can be advected by a fluid that can carry the substance or quantity.

In advection, a fluid transports some conserved quantities or dispersed materials in it via its bulk motion. The motion of the fluid is described mathematically as a vector field, whereas the transported materials are described by scalar fields showing the distribution of the materials over space, i.e., concentration fields. Therefore, advection requires the presence of currents; and as a result, it cannot happen in rigid solids. The term advection sometimes serves as a synonym for convection. However, technically, convection covers the transport attributed to both diffusion and advection.

Mathematical Description

Advection is the transport of mass due to the bulk fluid motion. A closer experimental investigation can easily observe that the convective flux vector $\mathbf{J}_{\mathrm{adv}}$ is proportional to the fluid velocity $\mathbf{v}$, so it has the same direction as the velocity. The motion of a bulk fluid, therefore, contributes to the flux of its dispersed phases, in addition to the flux due to diffusion. This flux can be formulated as follows:

$$\mathbf{J}_{\mathrm{adv}} = \mathbf{v} \cdot \nabla c, \tag{13.8}$$

where $\mathbf{J}_{\mathrm{adv}}$ is the advective flux, $\mathbf{v}$ is the velocity field, and c is the concentration of the dispersed material. One assumption underlying this equation is that the movement of the dispersed phase is identical to that of the bulk phase carrying it. The advection equation can be obtained with this flux, which is a special case of the comprehensive mass transport equation:

$$\frac{\partial c}{\partial t} = -\mathbf{v} \cdot \nabla c, \tag{13.9}$$

where $\mathbf{v} \cdot \nabla$ is the advection operator, and in Cartesian coordinates the advection operator is

$$\mathbf{v} \cdot \nabla = v_x \frac{\partial}{\partial x} + v_y \frac{\partial}{\partial y} + v_z \frac{\partial}{\partial z}, \tag{13.10}$$

where $\mathbf{v} = (v_x, v_y, v_z)$ is the velocity field in Cartesian coordinates. Both the advection operator and the above advection equation have a much more general sense than mass transport. The advection equation for a conserved quantity described by a scalar field ψ can be expressed mathematically using the following general continuity equation:

$$\frac{\partial \psi}{\partial t} + \nabla \cdot (\psi \mathbf{v}) = 0 \tag{13.11}$$

Frequently, it is assumed that the flow is incompressible; as a result, the velocity field satisfies the following equation to ensure continuity (mass balance):

$$\nabla \cdot \mathbf{v} = 0, \tag{13.12}$$

and $\mathbf{v}$ is said to be solenoidal. If this is the fact, the above equation can be rewritten as

$$\frac{\partial \psi}{\partial t} + \mathbf{v} \nabla \cdot \psi = 0. \tag{13.13}$$

The above is the general convection equation, a time-dependent, first-order partial differential equation. The concept of the solenoidal field will be encountered in the future chapters.

In particular, if the flow is steady, then

$$\mathbf{v}\nabla \cdot \psi = 0, \tag{13.14}$$

which shows that ψ is a constant along a streamline. Hence, $\partial\psi/\partial t = 0$, so ψ does not vary in time.

If a vector quantity $\mathbf{u}$, such as a magnetic field, is advected by a solenoidal velocity field $\mathbf{v}$, the above advection equation becomes

$$\frac{\partial \mathbf{u}}{\partial t} + \mathbf{v}\nabla \cdot \mathbf{u} = 0. \tag{13.15}$$

Advection in Porous Materials

The advection in porous materials is much different from that in bulk fluids as only part of the volume is occupied by the pore fluids. Taking this fact into consideration, the advection equation can be modified into

$$\phi \frac{\partial c}{\partial t} = -\mathbf{v}_{\mathrm{Darcy}} \cdot \nabla c, \tag{13.16}$$

where ϕ is the porosity and $\mathbf{v}_{\mathrm{Darcy}}$ is the Darcy velocity in Darcy's law. $\mathbf{v}_{\mathrm{Darcy}}$ is related to $\mathbf{v}$ via the effective porosity ϕ_{eff} as $\mathbf{v}_{\mathrm{Darcy}} = \mathbf{v}\phi_{\mathrm{eff}}$. ϕ_{eff} is defined using areas instead of volume as used for ϕ. ϕ_{eff} is close to ϕ in many cases; therefore, the above equation for porous materials is also close to the equation for bulk fluids.

Physical Implications of the Mathematical Description

The velocity in the above equation appears straightforward but could be hard to deal with in real applications. The velocity of a molecule undergoing mass transfer incorporates both an advective and a diffusive component. For a dilute species (qualitative definition, species with low concentrations), the understanding is quite clear as the dilute species is the diffusive component and the solvent is the advective component. Correspondingly, the velocity is the velocity of the solvent or carrier material. However, there are a few special cases in which the identification of the velocity is not that easy. First, if a flow involves multiple phases, which occupy comparable fractions, and all are discontinuous to some extent, then it may be hard to identify an advective phase and its velocity. Second, for the transport of concentrated species (qualitative definition, species with high concentrations), a more careful definition of the bulk fluid velocity is required. For example, if the dispersed phase is large and heavy conglomerates, then its velocity may not be identical to that of the advective phase.

The boundary and relation between advection and diffusion also deserve consideration. We understand advection as the mass transport due to the average velocity of all molecules and diffusion as mass transport due to the instantaneously varying and randomized velocity of individual molecules with reference to the average velocity of the whole fluid. Hence, the choice of the reference velocity will affect the definition of both the diffusive flux and the convective flux. For the transport of dilute species, where one component dominates the momentum of the system, such as a solvent or carrier gas, it is normal to take the velocity of the dominant component as the reference velocity.

Natural convection, which is also called free convection or simply buoyant convection, also deserves attention. It happens in the case where density variations are caused by temperature and salinity differences. Without a forced flow, a buoyancy-driven flow may still arise in the presence of temperature gradients because of the density differences. For fluid parcels with dimensions larger than a few milliliters in normal ambient conditions, convection currents usually cannot achieve a steady state. This implies that this convection will contribute to mass transport even when no forced convection is applied. A typical example is the convection caused by gradients in salinity in oceans, which is usually referred to as gravitational convection.

Diffusion

Diffusion is the process by which a substance is moved from one place to another under the action of random fluctuations. On the molecular level, the cause is the perpetual agitation of molecules; while on a higher level, its occurrence is attributed to the turbulent eddies of the carrying fluid. Both types of diffusion are determined by concentrations, leading to a proportionality between the mass flux and the concentration gradient. This proportionality is characterized as a material property and a key quantity in diffusion problems. Both types of diffusion can be formulated using Fick's laws. In the following paragraphs, the general formulation of Fick's laws will be introduced first. Then special cases including the dependence of the diffusion coefficient on various factors, diffusion in porous materials, multicomponent diffusion, and turbulent diffusion will be introduced consecutively.

The driving force for molecular diffusion is the dynamic oscillation of molecules, which also contributes to the macroscopic thermal state. As we know, molecules stay in motion at temperatures above absolute zero. This motion leads to their kinetic energy. The motion and collision of the particles are random at the microscopic scale, however, follow thermodynamic rules at the macroscopic scale. These collisions are common even in the air at an atmospheric pressure which can be hardly viewed as a dense fluid. If the concentration of a species is initially not uniform, e.g., the concentration is greater in one region of a fluid than another, then, diffusion will enable mass transfer in the direction toward a more uniform concentration in the fluid.

Turbulent diffusion is a special category of diffusion pertaining to turbulent fluid systems. In turbulent diffusion, the variation of the concentration occurs when the turbulent fluid system reaches critical conditions in response to shear flow. Such a condition is usually caused by high velocities while it can also result from steep concentration gradients, steep density gradients, or their combinations. The rate of turbulent diffusion is usually much higher than molecular diffusion. Therefore, it is necessary to take turbulent diffusion into consideration for diffusion in fluids where turbulent flow patterns are dominant.

Mathematical Description: Fick's Laws

Fick's laws provide a simple and universal mathematical description of diffusion. These physical laws were reported by Adolf Fick in the nineteenth century. Two different laws exist:

1. The first Fick's law states that the molar flux due to diffusion is proportional to the concentration gradient.
2. The second law states that the rate of the change in the concentration at any point is proportional to the second spatial derivative of the concentration at that point.

Fick's First Law of Diffusion

Fick's first law of diffusion can be written in a modern mathematical form as

$$\mathbf{J}_{\mathrm{diff}} = -\mathbf{D}_{\mathrm{diff}} \nabla c, \tag{13.17}$$

where $\mathbf{J}_{\mathrm{diff}}$ is the mass flux, $\mathbf{D}_{\mathrm{diff}}$ is the diffusion coefficient, and c is the concentration. $\mathbf{D}_{\mathrm{diff}}$ is a vector here to allow for anisotropy. Other definitions of the concentration can also be used, but consistent units need to be used for the diffusion and flux.

Fick's Second Law of Diffusion

From the continuity equation for mass balance, we have

$$\frac{\partial c}{\partial t} + \nabla \cdot \mathbf{J}_{\mathrm{diff}} = 0. \tag{13.18}$$

Then we can derive Fick's second law directly:

$$\frac{\partial c}{\partial t} = \nabla \cdot (\mathbf{D}_{\mathrm{diff}} \nabla c) = D_{\mathrm{diff}} \nabla^2 c, \tag{13.19}$$

where D_{diff} is moved out of the Laplacian operator if $\mathbf{D}_{\text{diff}}$ is a constant along different directions. Usually, the assumption is also a good approximation for the diffusion in solids, diffusion of chemicals in a liquid solvent, and diffusion of dilute species in the gas phase, such as carbon dioxide in the air. However, this may not be the case in porous materials, where anisotropy is common.

Understanding of Diffusion Coefficient

In dilute species transport, the flux due to diffusion, as formulated by Fick's first law, only depends on the diffusion coefficient, which quantifies the interaction between the solute and the solvent. The diffusion coefficient is the magnitude of the molar flux through a unit surface area under a unit concentration gradient. Therefore, it is analogous to the thermal diffusivity in heat transfer, hydraulic diffusivity in groundwater flow, and electric conductivity in electric currents in conductors. When moving to the mixtures with multiple dispersed phases/species, the above Fick's law will also be valid:

$$\mathbf{J}_{\text{diff},i} = -\mathbf{D}_{\text{diff},i} \nabla c_i. \tag{13.20}$$

Or in 1D,

$$J_{\text{diff},i} = -D_{\text{diff},i} \nabla c_i, \tag{13.21}$$

so

$$D_{\text{diff},i} = \frac{|J_{\text{diff},i}|}{|\nabla c_i|}. \tag{13.22}$$

However, in multicomponent diffusion, the diffusion coefficient is possibly not a constant or composition-independent anymore. Especially, the intermolecular dependencies cannot be overlooked as the interactivity of the molecules of different species becomes prevalent. To allow for the interactivity, it is more accurate to formulate the diffusion coefficient using a tensor. Also, the equation for diffusion needs to be altered to relate the mass flux of one chemical species to the concentration gradients of all chemical species. Such a mechanism can be formulated using the Maxwell-Stefan theory of diffusion (Krishna and Wesselingh 1997). The Maxwell-Stefan theory is a more comprehensive version of the "classical" Fick's diffusion theory.

Fick's laws also apply to turbulent diffusion. The diffusion coefficient associated with turbulent diffusion is called the eddy diffusion coefficient or eddy diffusivity.

The magnitude of the diffusion coefficient in molecular diffusion varies significantly from mixture to mixture. In gaseous solutions, a typical diffusion coefficient for a molecule lies in the range of 10^{-6} to 10^{-5} m^2/s. By contrast, the diffusion coefficient for molecules dissolved in liquids is much slower. In aqueous solution such as water, typical diffusion coefficients range from 10^{-10} to 10^{-9} m^2/s. For example, the diffusion coefficient for salt (NaCl) diffusing into water is about 1.5×10^{-9} m^2/s. Accordingly, it could take more than 1 month for sugar in a cup with a height of 5 cm to become uniformly distributed. By contrast, the turbulent diffusion coefficient in typical rivers could be 0.01 m^2/s along the horizontal direction and 0.1 m^2/s along the vertical direction (Roberts and Webster 2002)

The molecular diffusion coefficient can be predicted in some simple cases. In an ideal gas, the following relationship can be derived from the Maxwell-Boltzmann distribution by considering the mean free path and average velocity of the molecule:

$$\mathbf{D}_{\mathrm{diff}} \propto \frac{T^{\frac{3}{2}}}{p}, \tag{13.23}$$

where T and p are the temperature and pressure, respectively. According to this equation, diffusion would be faster in hotter and more rarefied gasses. In a viscous fluid, the diffusion coefficient of particles or large molecules obeys the Stokes-Einstein equation (Logan 2012):

$$\mathbf{D}_{\mathrm{diff}} = \frac{kT}{6\pi\mu r}, \tag{13.24}$$

where k is the Boltzmann constant, μ is the solvent viscosity, and r is the radius of the diffusing particle. The equation is obtained by calculating the drag force from Stokes' law.

Diffusion Coefficient in Porous Materials

In a porous material, the fluid phase must travel a tortuous path that winds through the pores perforating the solid skeleton. Therefore, the effective diffusion coefficient alters from the theoretical diffusion coefficient for a bulk solution due to various effects (Fig. 13.2). For example, the available cross section for diffusion in the porous material is smaller than for the bulk solution. This effect can be considered using the effective porosity defined in terms of areas, which can be approximated using the porosity of the porous material defined in terms of volumes. Additionally, the distance that a molecule must travel from one point to another in the porous material is greater than the distance between these points because the molecule must navigate between the solid skeleton. As a result, the real concentration gradient is less than the apparent concentration gradient.

This effect is considered by multiplying the diffusion coefficient by the tortuosity ($\tau > 1$). With the above considerations, Fick's first law can be modified into

$$\mathbf{J}_{\mathrm{diff}} = -\mathbf{D}_{\mathrm{diff}}^{\mathrm{eff}} \nabla c, \tag{13.25}$$

where $\mathbf{D}_{\mathrm{diff}}^{\mathrm{eff}}$ is the effective diffusion coefficient. This effective diffusion coefficient is smaller than that in a continuous liquid phase due to consideration of the porosity ϕ and tortuosity τ

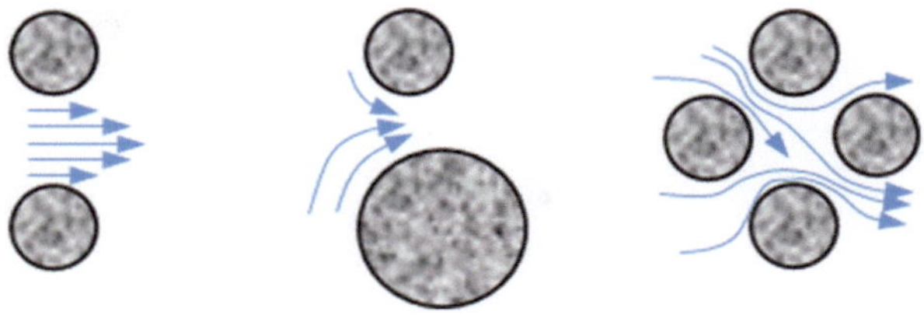

Fig. 13.2 Flow and particle transport in porous materials affected by (**a**) velocity variation, (**b**) particle size distribution, and (**c**) tortuosity

$$\mathbf{D}_{\text{diff}}^{\text{eff}} = \frac{\phi}{\tau}\mathbf{D}_{\text{diff}}, \tag{13.26}$$

where tortuosity is defined as the ratio of the length of the real traveling path to the distance between the two ends of the path.

Here the flux $\mathbf{J}_{\text{diff}}$ and concentration gradient ∇c are still defined with respect to the solution where diffusion actually occurs instead of the whole porous material. It is worthwhile to mention that the concentration and flux defined with respect to the whole porous material are also used for porous materials.

The tortuosity is easy to obtain. Some correlations relating the tortuosity to the porosity are available for rough calculations. Typical correlations are the Millington-Quirk equation (Campbell 1974)

$$\tau = \phi^{-\frac{1}{3}}, \tag{13.27}$$

and the Bruggeman equation (Brakel and Heertjes 1974)

$$\tau = \phi^{-\frac{1}{2}}. \tag{13.28}$$

It is also very common that the effective diffusion coefficient is anisotropic in the porous material. In this case, the rate of diffusion depends on the direction of the concentration gradient. The corresponding mathematical treatment is to formulate the diffusivity using a vector.

Dispersion

Diffusion can be significant or even dominant in the vapor transport in unsaturated zones of porous materials. However, diffusion is usually not a significant mass transport mechanism in saturated pore liquid flow where dispersion could be the dominant one. Dispersion, or dispersive mass transfer, is also the spreading of dispersed mass from highly concentrated areas to less concentrated areas. However, this dispersed mass transport mechanism is caused by non-ideal flow patterns like turbulent diffusion. The difference is that the turbulent diffusion is still one type of random mixing which is related to the turbulence (perturbation) in the dispersion liquid while dispersion relies heavily on the average velocity. Another major

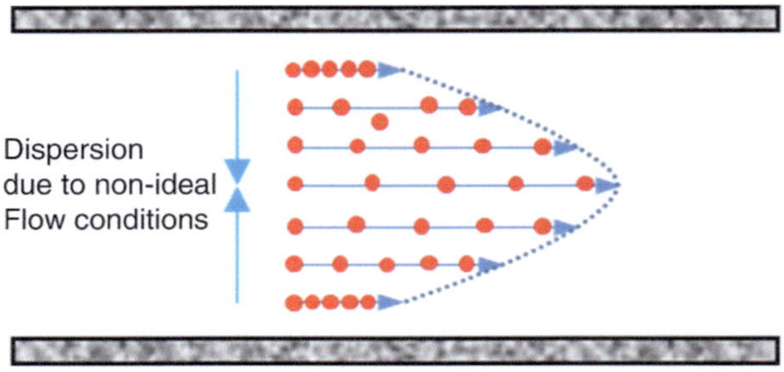

Fig. 13.3 Dispersion and non-ideal flow conditions

difference is that dispersion is a macroscopic phenomenon while diffusion occurs at much smaller scales because diffusion is related to the vibration of molecules, i.e., Brownian motion.

The non-ideal flow patterns refer to the deviations from a “plug flow” in a conceptual pipe housing the flow, in which the velocity of the fluid is assumed to be a constant across any cross section perpendicular to the axis of the pipe. The following example can be used to understand the deviation from such ideal flow patterns (Deen 1998). If the flow velocity is uniform everywhere, i.e., plug flow (Fig. 13.3), then molecules at different places in the flow will move with the same convective velocity. In that case, only the diffusion rates differentiate between them. However, in a non-ideal flow, velocity gradients exist along the radial direction of the pipe – the fluid at the center of the pipe moves faster than that adjacent to the walls. As a result, different streamlines have different concentrations. Molecules then tend to go across streamlines, leading to diffusion in the radial directions. This special transport mechanism, which results from this smearing out or enhanced mass transfer due to the combined effect of radial diffusion and velocity differences, is interpreted as dispersion. Surprisingly, dispersion can decrease with an increase in diffusion coefficients. This is because the molecules will not get enough time to be transported far along each streamline when the diffusion is too strong along the radial direction.

Based on the above description, dispersive mass flux is analogous to diffusion, and it can also be described using Fick’s first law:

$$\mathbf{J}_{\text{disp}} = -\mathbf{D}_{\text{disp}} \nabla c, \tag{13.29}$$

where c is the mass concentration of the dispersed species and $\mathbf{D}_{\text{disp}}$ is the dispersion coefficient. Dispersion is usually more significant than diffusion in typical convection-diffusion problems.

The dispersion coefficient is frequently modeled as the product of a characteristic length scale, α, and the fluid velocity, $\mathbf{v}$:

$$\mathbf{D}_{\text{disp}} = \alpha \mathbf{v}. \tag{13.30}$$

Dispersion in porous materials is usually appreciable because of the complicated structure of the solid skeleton which can significantly deviate the flow patterns from the ideal flow situation and thus enhance dispersion therein significantly. The dispersion coefficient or the characteristic length scale can be obtained directly via experiments (Fischer 1966; Irwin et al. 1996).

Practice Problem

Problem Description

An experiment and its analytical solution for typical concentration fields were reported by Massabo et al. (2011). As shown in Fig. 13.4, the experiment was conducted in a Perspex box, which was 0.28 m long, 0.2 m wide, and 0.01 m thick. The box was filled with transparent glass beads to simulate a porous material. There is a constant flow from the left to the right with a velocity of 2.9×10^{-4} m/s along the length direction (horizontal). The combined diffusion and dispersion coefficients considering the porosity and tortuosity of the porous material are 3.322×10^{-7} m^2/s and 2.672×10^{-7} m^2/s along the horizontal direction and vertical direction, respectively. Dye was injected into the box at the location that is 0.05 m from the left side and 0.1 m from the top. Unfortunately, the accurate formulation of the pulse source is hard to determine based on the description in the paper. Here we can assume it is a square pulse with a magnitude of 0.1 between $t = 0$ s and $t = 1$ s in a circular area with a radius of 0.001 m. The initial concentration is zero across the domain. All the boundaries have no mass influx or outflux. Please simulate this experiment, and compare the simulation result against the analytical solution at $t \cdot v/L = 1/50$ (18.96 s) and $t \cdot v/L = 1/10$ ($t = 94.8$ s).

For simplicity, the effective porosity is assumed to be the same as the porosity.

Results

Direct comparisons between the simulation results and the analytical (Fig. 13.5) and experimental results from Massabo et al. (2011) are difficult due to the unknown source function. In Fig. 13.6, we normalized the concentration for a quantitative comparison for the concentration distributions along x symmetry axis at two different times (Figs. 13.5 and 13.6).

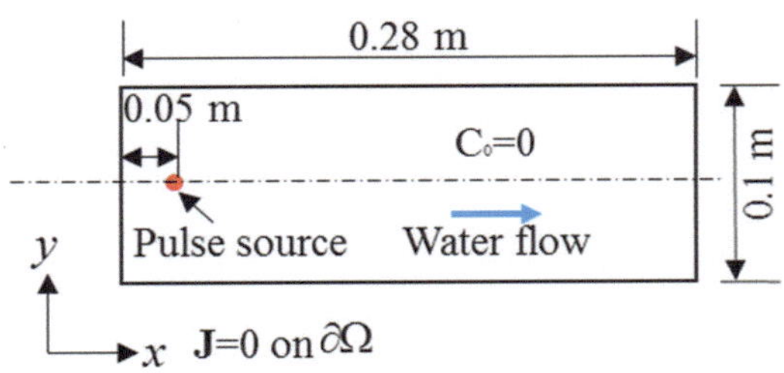

Fig. 13.4 Schematic of the experiment

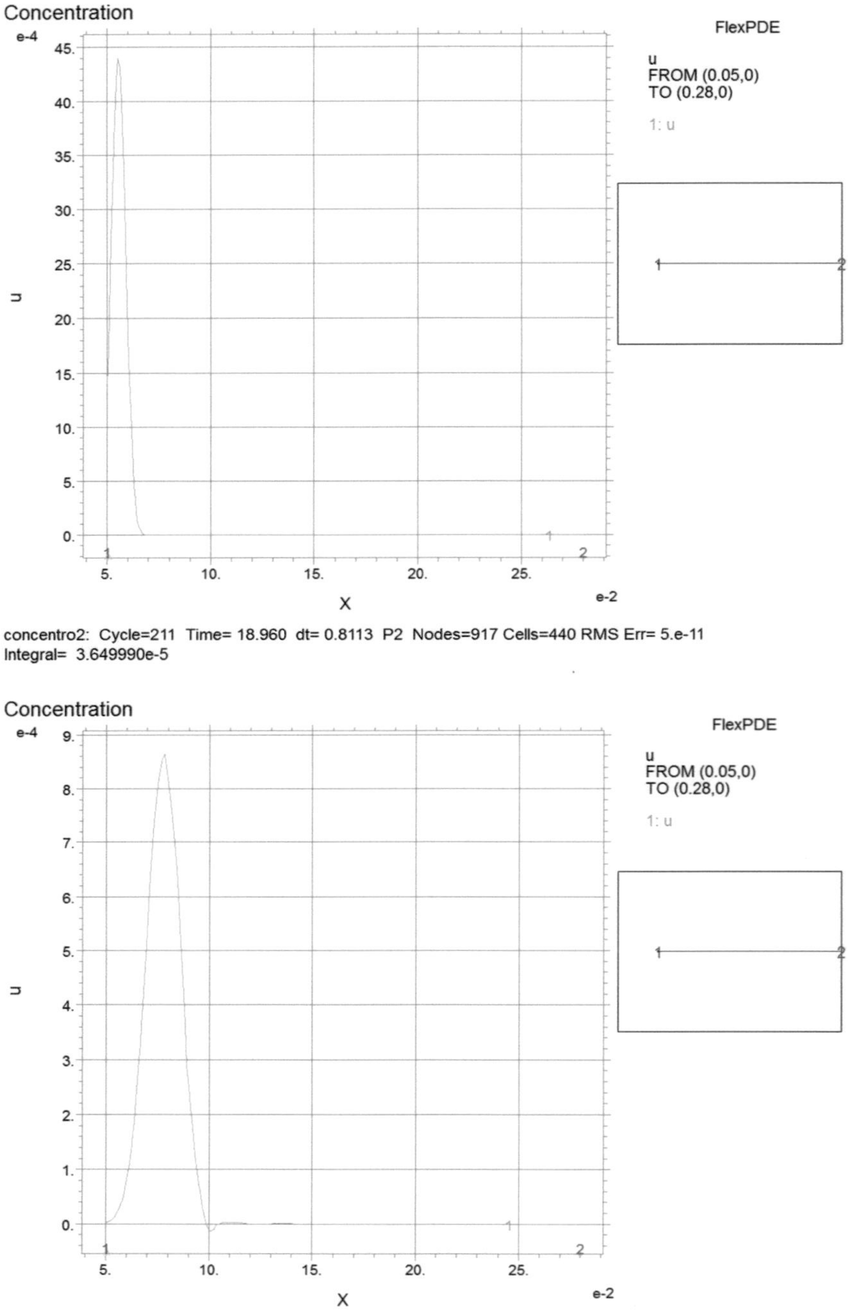

Fig. 13.5 The concentration distributions at 18.96 s (upper) and 94.8 s (lower)

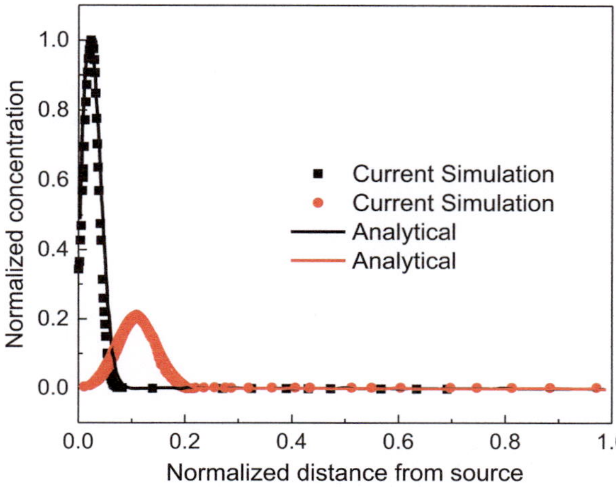

Fig. 13.6 Comparison between simulation and analytical results

Appendix: FlexPDE Script File

```
{ Concentration field with diffusion, advection, and a pulse source }
TITLE 'Concentration'
VARIABLES
  u(1e-12)
DEFINITIONS
  D_x=3.322e-7 { Diffusivity in x/horizontal/longitudinal direction }
  D_y=2.672e-7 { Diffusivity in y/vertical direction }
  vx=2.9e-4
  M=0* UPULSE(t-0, t-01) { Rectangular pulse between t=0 and t=1 }
INITIAL VALUES
  u = 0.0
EQUATIONS
  Dx(D_x*Dx(u)) + Dy(D_y*Dy(u)) - vx*Dx(u)+M= dt(u)
BOUNDARIES
  REGION 1
    M=0
    START(0,-0.1)
      NATURAL(u) = 0
    LINE TO (0.28,-0.1)
      NATURAL(u) = 0
    LINE TO (0.28,0.1)
      NATURAL(u) = 0
    LINE TO (0,0.1)
      NATURAL(u) = 0
    LINE TO CLOSE
  REGION 2
    M=0.1*UPULSE(t-0, t-1) { Dye was injected into Region 2 betweent=
0~1 as a pulse }
    START(0.051,0)
    ARC (CENTER=0.05,0)
```

```
    ANGLE=360
    TO CLOSE
TIME 0 TO 425 BY 1
PLOTS
  FOR t=0 BY 0.001 TO 0.01 BY 0.1 TO 3 BY 1 TO ENDTIME
    CONTOUR(u)
    ELEVATION(u) FROM (0.05,0) TO (0.28,0)
  FOR t=94.8/5
    ELEVATION(u) FROM (0.05,0) TO (0.28,0)
  FOR t=94.8
    ELEVATION(u) FROM (0.05,0) TO (0.28,0)
END
```

Chapter 14
Mechano: Stress and Strain

Introduction

This chapter discusses the mechanical response. In the context of multiphysics, the word "mechanical field" is usually used to describe physical processes involving this aspect of the material behavior, in which quantities such as stress, strain, and displacement change with reference to a geometrical description of an object. A more widely used term for the topic is stress-strain analysis or stress analysis. Stress analysis is an engineering practice to determine the stresses and strains in materials and structures subjected to forces or loads. Stress analysis is a primary interest of civil, mechanical, and aerospace engineers. The practice lays down a theoretical basis for the design of structures and components, such as buildings, bridges, aircrafts, mechanical parts, and earth structures. Stress analysis is also used in the maintenance and retrofit of such objects for the investigations into the causes of their failures.

The input data for stress analysis are a geometrical description of the object, material properties, and a loading condition. The output data are typically a quantitative description of the stress as well as strain and deformation across the computational domain under the given loading condition. Stress analysis for porous materials is also built on the assumption of REV and the basic theories of continuum mechanics. Therefore, all material properties are homogeneous at scales smaller than that of the REV. The assumption of REV needs to be ensured to make sure that the mathematical description extended from continuum mechanics can surrender satisfactory results for problems in porous materials.

The scope of mechanical response needs to be more clearly defined to differentiate the mechanical field from other relevant topics and also to clarify what will be discussed in the rest of this chapter.

First, stress and strain analysis is restricted to solids. The study of stresses in liquids and gasses is the subject of fluid mechanics, which is a multiphysical

Z. Liu, *Multiphysics in Porous Materials*,
https://doi.org/10.1007/978-3-319-93028-2_14

phenomenon and thus will not be counted as the mechanical field which is a monolithic physical phenomenon.

Second, the introduction to the mechanical field in this book will be limited to homogeneous linear elastic materials and the small strain theory. The relationships between several common quantities, i.e., stress, strain, and displacement, and the choice of the dependent variable(s) from these terms make the mechanical field appear much more complicated than other physical fields such as heat transfer. The explanation of these relationships is thus the major content of this chapter. An intuitive goal of stress analysis is to determine the distribution of the internal stresses throughout the system under given external forces. In principle, this implies the determination of, either implicitly or explicitly, the Cauchy stress tensor at every point. However, the stress is usually not the dependent variable. The stress needs to be connected to the displacement, which is preferred as the dependent variable. For the purpose, the stress needs to be linked to the strain and then to the displacement. For the purpose, the relationship between stress and strain, i.e., constitutive relationship, is required as predetermined material properties. Among different types of constitutive relationships, e.g., elastic, elastoplastic, and viscoplastic, the linear elastic relationship is one of the simplest and most common ones and thus will be introduced in this chapter. The second step for the purpose is to relate the strain to the displacement via strain theories, i.e., small strain theory (infinitesimal strain theory) and large strain theory (finite strain theory). The former will be discussed in this book considering its predominant use in numerical simulation.

Third, the external forces (loading) may be body forces such as gravity or magnetic attraction, which act throughout the volume of a material, or concentrated loads such as friction between parts. These forces are usually given or can be easily identified for real problems. Therefore, they will not be extensively discussed. Furthermore, in stress analysis, the physical causes of external forces and the precise nature of the materials, i.e., other aspects of the materials apart from the mechanical behavior, are usually not of interest.

In the following sections, we will first introduce the stress concept including its physical meaning and mathematical formulations. Then the strain and its relationship to the displacement including the small strain theory will be explained. Followed is the introduction to the constitutive relationship for linear elasticity. The stress, strain, and constitutive relationship will be integrated into the mechanical balance equation to obtain the governing equation for the mechanical field. Finally, an example will be given to demonstrate the implementation of the mechanical field.

Stress

The most common mathematical description of stress is the Cauchy stress tensor. In continuum mechanics, the Cauchy stress tensor $\boldsymbol{\sigma}$, or called the true stress tensor or simply the stress tensor, is a second-order tensor. The Cauchy stress tensor can be

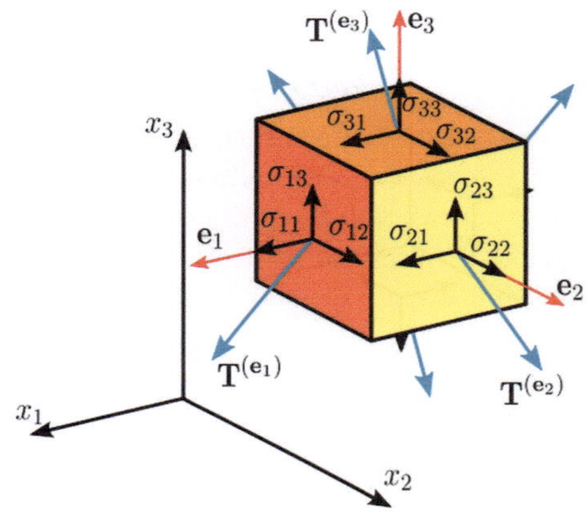

Fig. 14.1 Stress tensor and components

used to formulate the stress state at any point inside a material. In a general 3D space, the Cauchy stress tensor has nine components σ_{ij}:

$$\boldsymbol{\sigma} = \begin{bmatrix} \sigma_{11} & \sigma_{12} & \sigma_{13} \\ \sigma_{21} & \sigma_{22} & \sigma_{23} \\ \sigma_{31} & \sigma_{32} & \sigma_{33} \end{bmatrix} = \begin{bmatrix} \sigma_{xx} & \sigma_{xy} & \sigma_{xz} \\ \sigma_{yx} & \sigma_{yy} & \sigma_{yz} \\ \sigma_{zx} & \sigma_{zy} & \sigma_{zz} \end{bmatrix} = \begin{bmatrix} \sigma_{x} & \tau_{xy} & \tau_{xz} \\ \tau_{yx} & \sigma_{y} & \tau_{yz} \\ \tau_{zx} & \tau_{zy} & \sigma_{z} \end{bmatrix}. \tag{14.1}$$

The three stress components on the diagonal of the matrix are normal stresses, which act perpendicular to the plane, while those of the diagonal are called shear stresses as they are parallel to the plane. Equation 14.1 shows three common math representations of the stress tensor. The rightmost one is frequently used in soil mechanics to better differentiate normal stresses from shear stresses.

The use of the Cauchy stress tensor for formulating the stress state at a point can be understood in the following way. If we "zoom in," any infinitesimal point can be viewed as a cube as shown in Fig. 14.1. The stress state at the point can be represented by the stresses on the surfaces of the cube. Among the six surfaces of the cube, three surfaces have identical stresses to the other three. Therefore, the stress tensor can be made with the nine stresses on these three surfaces. This is what was obtained in Eq. 14.1.

We can further define the stress vector $\mathbf{T}^{(\mathbf{n})}$ to represent the stresses on a plane whose outward normal vector is $\mathbf{n}$. With the stress vector, the Cauchy stress tensor can be written in the following form:

$$\boldsymbol{\sigma} = \sigma_{ij} = \begin{bmatrix} \mathbf{T}^{(1)} \\ \mathbf{T}^{(2)} \\ \mathbf{T}^{(3)} \end{bmatrix} = \begin{bmatrix} \sigma_{11} & \sigma_{12} & \sigma_{13} \\ \sigma_{21} & \sigma_{22} & \sigma_{23} \\ \sigma_{31} & \sigma_{32} & \sigma_{33} \end{bmatrix}. \tag{14.2}$$

However, there is an infinite number of ways to align the cube. If the cube is aligned along different directions, the stresses on its surfaces, i.e., the components of the Cauchy stress tensor, will also be different. This leads to a contradiction: the stress state for one point is the same regardless of the alignment of the cube; however, the components of the stress tensor are different for different alignments. This contradiction can be compromised by the Cauchy stress theorem (or called the Cauchy fundamental theorem). This theorem states that the stress vector on any other plane passing through a point can be found through coordinate transformation equations. Equivalently, we can understand the three stress vectors as the projections of the stress tensor onto three directions such as three axes in the Cartesian coordinate system. If we rotate the axes, the projections will be different; however, the tensor stays the same. This understanding leads to the following equation:

$$\mathbf{T}^{(\mathbf{n})} = \mathbf{n} \cdot \boldsymbol{\sigma}, \tag{14.3}$$

or

$$T_j^{(\mathrm{n})} = \sigma_{ij} n_i. \tag{14.4}$$

This equation implies that the stress vector $\mathbf{T}^{(\mathbf{n})}$ on a plane with a normal unit vector $\mathbf{n}$ is the projection of the stress tensor $\boldsymbol{\sigma}$ on that plane. In a 3D space, there are three independent stress vectors or projections for each stress tensor. Likewise, a stress tensor for a 2D case consists of two independent stress vectors.

In this chapter and, in fact, in most of stress analyses, the Cauchy stress tensor is usually introduced and adopted acquiescently. However, please note that the Cauchy stress tensor is used for stress analysis of material bodies experiencing small deformations. It is thus a central concept in the linear theory of elasticity. For large deformations, which are also called finite deformations, other measures of stress are required, such as the Piola-Kirchhoff stress tensor, Biot stress tensor, and Kirchhoff stress tensor (Hackett 2016).

Strain

The definition of strain usually requires the introduction of two other closely related terms: displacement and deformation. Displacement can be defined as the movement of individual material points in an object due to external loads. This movement of the point with respect to a reference system of axes can be described using a vector: displacement vector. Displacements can be classified into the following four types as shown in Fig. 14.2:

1. Translation of points
2. Rotation of lines
3. Change in length, i.e., elongation and contraction
4. Distortion, i.e., angle change between lines

Fig. 14.2 Schematic of translation, rotation, length change, and distortion

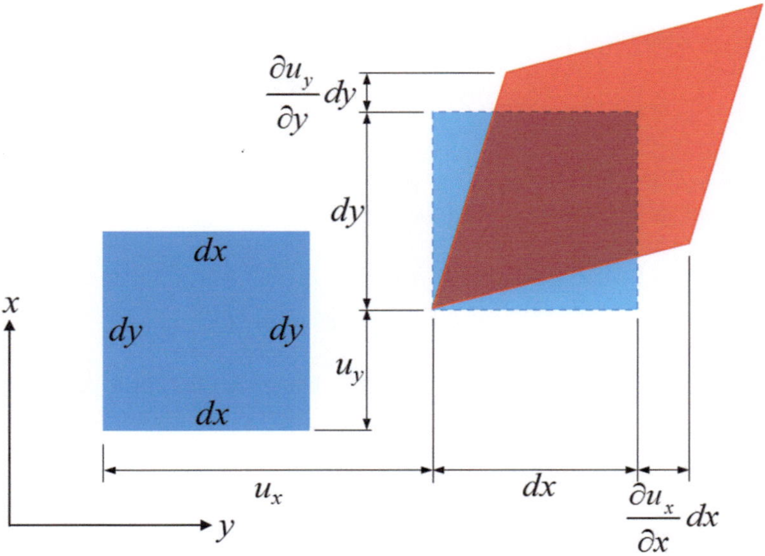

Fig. 14.3 Displacement and deformation

The first two types of displacements occur because of the movement of the entire body, i.e., rigid body movement, while the latter two are associated with the local deformation of the body. In the latter two, individual points of the body move relative to each another as displacements induced by applied loads alter the size and/or the shape of the body. The change in any dimension associated with these two types of relative displacements is defined as deformation. Deformation provides a global description of the object. However, the understanding of the internal of the object requires local descriptions for all the points inside the object. This calls for stresses and strains. Strain can be roughly understood as deformation normalized with respect to the original or current geometry of the object or other things, depending on the choice of strain theories. The strain theory that we usually use is the small strain theory, in which the original geometry is adopted for the normalization. For example, the strain of the rod subjected to axial loading (1D) can be defined as the deformation per unit length of a body.

Similar to stresses, strains may also be classified into the normal strain and shear strain, i.e., acting perpendicular to and parallel to the plane (surface of a material element), respectively. In order to derive the mathematical description of the normal strain, let us consider a 2D infinitesimal rectangular material element with dimensions $dx \times dy$ as illustrated in Fig. 14.3. After deformation, the rectangular becomes a rhombus. Taking the side with the length of dx, for example, the length of this side in the deformed rectangular, i.e., the rhombus, will be:

$$\text{Deformed length} = \sqrt{\left(dx + \tfrac{\partial u_x}{\partial x} dx\right)^2 + \left(\tfrac{\partial u_y}{\partial x} dx\right)^2}$$
$$= dx\sqrt{1 + 2\frac{\partial u_x}{\partial x} + \left(\frac{\partial u_x}{\partial x}\right)^2 + \left(\frac{\partial u_y}{\partial x}\right)^2}$$

As mentioned, this book restricts the introduction to the small strain theory. In the case of small strains, the squares of the derivatives are negligible as the derivatives are small, thus we have

$$\text{Deformed length} \approx dx + \frac{\partial u_x}{\partial x} dx$$

The normal strain in the x direction of the rectangular element is defined by

$$\varepsilon_x = \frac{\text{deformed length}}{\text{original length}} \approx \frac{\partial u_x}{\partial x}. \tag{14.5}$$

Similarly, the normal strain in the y direction and z direction, becomes

$$\varepsilon_y = \frac{\partial u_y}{\partial y} \tag{14.6}$$
$$\varepsilon_z = \frac{\partial u_z}{\partial z}. \tag{14.7}$$

Likewise, we can obtain the formulation of the shear strain in the small strain theory:

$$\gamma_{xy} = \alpha + \beta = \frac{\partial u_y}{\partial x} + \frac{\partial u_x}{\partial y}. \tag{14.8}$$

Combining the formations of the normal and shear strain, we can obtain the infinitesimal strain tensor as

$$\boldsymbol{\varepsilon} = \frac{1}{2}\left[\nabla \mathbf{u} + (\nabla \mathbf{u})^T\right], \tag{14.9}$$

or in the index notation as

$$\varepsilon_{ij} = \frac{1}{2}\left(u_{i,j} + u_{j,i}\right), \tag{14.10}$$

or in the matrix notation as

$$\begin{bmatrix} \sigma_{xx} & \sigma_{xy} & \sigma_{xz} \\ \sigma_{yx} & \sigma_{yy} & \sigma_{yz} \\ \sigma_{zx} & \sigma_{zy} & \sigma_{zz} \end{bmatrix} = \begin{bmatrix} \frac{\partial u_x}{\partial x} & \frac{1}{2}\left(\frac{\partial u_x}{\partial y} + \frac{\partial u_y}{\partial x}\right) & \frac{1}{2}\left(\frac{\partial u_x}{\partial z} + \frac{\partial u_z}{\partial x}\right) \\ \frac{1}{2}\left(\frac{\partial u_y}{\partial x} + \frac{\partial u_x}{\partial y}\right) & \frac{\partial u_y}{\partial y} & \frac{1}{2}\left(\frac{\partial u_y}{\partial z} + \frac{\partial u_z}{\partial y}\right) \\ \frac{1}{2}\left(\frac{\partial u_z}{\partial x} + \frac{\partial u_x}{\partial z}\right) & \frac{1}{2}\left(\frac{\partial u_z}{\partial y} + \frac{\partial u_y}{\partial z}\right) & \frac{\partial u_z}{\partial z} \end{bmatrix}. \tag{14.11}$$

Constitutive Relationship: Linear Elastic Hooke's Law

The general equation for Hooke's law is

$$\boldsymbol{\sigma} = \mathbf{C} : \boldsymbol{\varepsilon}, \tag{14.12}$$

where $\boldsymbol{\sigma}$ is the Cauchy stress tensor, $\boldsymbol{\varepsilon}$ is the infinitesimal strain tensor, and $\mathbf{C}$ is the fourth-order stiffness tensor. However, in engineering applications, we usually do not deal with the above equation in which the stress and strain are second-order tensors, while the stiffness is a fourth-order tensor. Instead, we usually adopt the Voigt notation which reduces the orders of the above tensors by means of specified rules and various types of symmetry.

Voigt Notation

The Voigt notation reduces the order of the stress and strain tensor from 2 to 1. Accordingly, the fourth-order stiffness tensor $\mathbf{C}$ (C_{ijkl}) degenerates into a second-order tensor $\mathbf{C}(C_{\alpha\beta})$. To achieve the goal, we first take advantage of the symmetry of the stress and strain tensors: $\sigma_{ij} = \sigma_{ji}$ and $\varepsilon_{ij} = \varepsilon_{ji}$. Therefore, there are only six independent components in the following stress tensor:

$$\boldsymbol{\sigma} = \begin{bmatrix} \sigma_{11} & \sigma_{12} & \sigma_{13} \\ \sigma_{12} & \sigma_{22} & \sigma_{23} \\ \sigma_{13} & \sigma_{23} & \sigma_{33} \end{bmatrix}. \tag{14.13}$$

So we can use a vector (or 1D array) including the six independent components to represent the above second-order tensor (or 2D array):

$$\boldsymbol{\sigma} = \begin{bmatrix} \sigma_{11} \\ \sigma_{22} \\ \sigma_{33} \\ \sigma_{12} \\ \sigma_{13} \\ \sigma_{23} \end{bmatrix}. \tag{14.14}$$

Similarly, the strain tensor can be formulated as

$$\boldsymbol{\varepsilon} = \begin{bmatrix} \varepsilon_{11} \\ \varepsilon_{22} \\ \varepsilon_{33} \\ 2\varepsilon_{12} \\ 2\varepsilon_{13} \\ 2\varepsilon_{23} \end{bmatrix} = \begin{bmatrix} \varepsilon_{11} \\ \varepsilon_{22} \\ \varepsilon_{33} \\ \gamma_{12} \\ \gamma_{13} \\ \gamma_{23} \end{bmatrix}. \tag{14.15}$$

To comply with the above changes, the fourth-order stiffness tensor is reduced to a second-order tensor as follows:

$$\mathbf{C} = \begin{bmatrix} C_{1111} & C_{1122} & C_{1133} & C_{1112} & C_{1113} & C_{1123} \\ C_{2211} & C_{2222} & C_{2233} & C_{2212} & C_{2213} & C_{2223} \\ C_{3311} & C_{3322} & C_{3333} & C_{3312} & C_{3313} & C_{3323} \\ C_{1211} & C_{1222} & C_{1233} & C_{1212} & C_{1213} & C_{1223} \\ C_{1311} & C_{1322} & C_{1333} & C_{1312} & C_{1313} & C_{1323} \\ C_{2311} & C_{2322} & C_{2333} & C_{2312} & C_{2313} & C_{2323} \end{bmatrix}. \tag{14.16}$$

Besides symmetries, Voigt notation also adopts the following mapping rules for tensor indices:

$$\begin{array}{ccccccccc} ij \text{ or } kl & = & 11 & 22 & 33 & 12,21 & 13,31 & 23,32 \\ \Downarrow & & \Downarrow & \Downarrow & \Downarrow & \Downarrow & \Downarrow & \Downarrow \\ \alpha \text{ or } \beta & = & 1 & 2 & 3 & 4 & 5 & 6 \end{array}$$

With the mapping, the stress tensor will be simplified into

$$\boldsymbol{\sigma} = \begin{bmatrix} \sigma_{11} \\ \sigma_{22} \\ \sigma_{33} \\ \sigma_{12} \\ \sigma_{13} \\ \sigma_{23} \end{bmatrix} = \begin{bmatrix} \sigma_{1} \\ \sigma_{2} \\ \sigma_{3} \\ \sigma_{4} \\ \sigma_{5} \\ \sigma_{6} \end{bmatrix}; \tag{14.17}$$

and the strain tensor is transformed in a similar way:

$$\boldsymbol{\varepsilon} = \begin{bmatrix} \varepsilon_{11} \\ \varepsilon_{22} \\ \varepsilon_{33} \\ 2\varepsilon_{12} \\ 2\varepsilon_{13} \\ 2\varepsilon_{23} \end{bmatrix} = \begin{bmatrix} \varepsilon_{1} \\ \varepsilon_{2} \\ \varepsilon_{3} \\ 2\varepsilon_{4} \\ 2\varepsilon_{5} \\ 2\varepsilon_{6} \end{bmatrix} = \begin{bmatrix} \varepsilon_{11} \\ \varepsilon_{22} \\ \varepsilon_{33} \\ \gamma_{12} \\ \gamma_{13} \\ \gamma_{23} \end{bmatrix} = \begin{bmatrix} \varepsilon_{1} \\ \varepsilon_{2} \\ \varepsilon_{3} \\ \gamma_{4} \\ \gamma_{5} \\ \gamma_{6} \end{bmatrix}. \tag{14.18}$$

With $C_{ijkl} \Rightarrow C_{\alpha\beta}$, the degenerated stiffness tensor $\mathbf{C}$ can be expressed with the new indices as:

$$\mathbf{C} = \begin{bmatrix} C_{11} & C_{12} & C_{13} & C_{14} & C_{15} & C_{16} \\ C_{22} & C_{22} & C_{23} & C_{24} & C_{25} & C_{26} \\ C_{33} & C_{33} & C_{33} & C_{34} & C_{35} & C_{36} \\ C_{41} & C_{42} & C_{43} & C_{44} & C_{45} & C_{46} \\ C_{51} & C_{52} & C_{53} & C_{54} & C_{55} & C_{56} \\ C_{61} & C_{62} & C_{63} & C_{64} & C_{65} & C_{66} \end{bmatrix}. \tag{14.19}$$

The matrix $C_{\alpha\beta}$ can be proven to be symmetric. This symmetry is attributable to the existence of a strain energy density function which satisfies $\sigma_{ij} = \frac{\partial W}{\partial \varepsilon_{ij}}$. Hence, there are at most 21 independent elements in the matrix of $C_{\alpha\beta}$.

In the case of orthotropic materials, which have material properties that differ along three mutually orthogonal twofold axes of rotational symmetry, the above stiffness matrix further reduces to the following form with nine independent elements:

$$C_{\alpha\beta} = \begin{bmatrix} C_{11} & C_{12} & C_{13} & 0 & 0 & 0 \\ C_{12} & C_{22} & C_{23} & 0 & 0 & 0 \\ C_{13} & C_{23} & C_{33} & 0 & 0 & 0 \\ 0 & 0 & 0 & C_{44} & 0 & 0 \\ 0 & 0 & 0 & 0 & C_{55} & 0 \\ 0 & 0 & 0 & 0 & 0 & C_{66} \end{bmatrix}. \tag{14.20}$$

The most common materials or assumptions that we encounter in applications are homogeneous isotropic linear elastic materials. For such materials, the constitutive relationship can be formulated using an even simpler stiffness matrix. One of the most common forms of the matrix is

$$\begin{bmatrix} \sigma_{11} \\ \sigma_{22} \\ \sigma_{33} \\ \sigma_{12} \\ \sigma_{13} \\ \sigma_{23} \end{bmatrix} = \begin{bmatrix} \lambda+2\mu & \lambda & \lambda & 0 & 0 & 0 \\ \lambda & \lambda+2\mu & \lambda & 0 & 0 & 0 \\ \lambda & \lambda & \lambda+2\mu & 0 & 0 & 0 \\ 0 & 0 & 0 & \mu & 0 & 0 \\ 0 & 0 & 0 & 0 & \mu & 0 \\ 0 & 0 & 0 & 0 & 0 & \mu \end{bmatrix} \begin{bmatrix} \varepsilon_{11} \\ \varepsilon_{22} \\ \varepsilon_{33} \\ 2\varepsilon_{12} \\ 2\varepsilon_{13} \\ 2\varepsilon_{23} \end{bmatrix}, \tag{14.21}$$

where λ and μ are the Lamé constants. μ also has the physical meaning of shear modulus and is frequently represented using G in engineering.

Using the index notation, we can write the above equation as

$$\sigma_{ij} = \lambda \varepsilon_{kk} \delta_{ij} + 2\mu \varepsilon_{ij}. \tag{14.22}$$

Elastic Modulus

The constitutive relationship of any homogeneous isotropic linear elastic material can be formulated by two independent elastic moduli, such as the Lamé constants, i.e., μ and λ. Common elastic moduli also include Young's modulus E, Poisson's ratio ν, and bulk modulus K. As shown in the following table, any two out of the five can be represented by any other two moduli.

	$E\nu$	$E\mu$	$\lambda,\ \mu$	$K,\ \mu$	$K\nu$
K	$\frac{E}{2(1-2\nu)}$	$\frac{E\mu}{3(3\mu-E)}$	$\lambda+\frac{2}{3}\mu$	K	K
λ	$\frac{E\nu}{(1+\nu)(1-2\nu)}$	$\frac{\mu(E-2\mu)}{3\mu-E}$	λ	$K-\frac{2}{3}\mu$	$\frac{3K\nu}{1+\nu}$
μ	$\frac{E}{2(1+\nu)}$	μ	μ	μ	$\frac{3K(1-2\nu)}{2(1+\nu)}$
E	E	E	$\frac{\mu(3\lambda+2\mu)}{\lambda+\mu}$	$\frac{9K\mu}{3K+\mu}$	$3K(1-2\nu)$
ν	ν	$\frac{E}{2\mu}-1$	$\frac{\lambda}{2(\lambda+\mu)}$	$\frac{3K-2\mu}{2(3K+\mu)}$	ν

Among the five quantities, the combination of E and ν is possibly the most popular one in engineering applications. The constitutive relationship for isotropic linear elastic materials is formulated using E and ν as follows:

$$\begin{bmatrix} \sigma_{11} \\ \sigma_{22} \\ \sigma_{33} \\ \sigma_{12} \\ \sigma_{13} \\ \sigma_{23} \end{bmatrix} = \frac{E}{(1+\nu)(1-2\nu)} \begin{bmatrix} 1-\nu & \nu & \nu & 0 & 0 & 0 \\ \nu & 1-\nu & \nu & 0 & 0 & 0 \\ \nu & \nu & 1-\nu & 0 & 0 & 0 \\ 0 & 0 & 0 & \frac{1-2\nu}{2} & 0 & 0 \\ 0 & 0 & 0 & 0 & \frac{1-2\nu}{2} & 0 \\ 0 & 0 & 0 & 0 & 0 & \frac{1-2\nu}{2} \end{bmatrix} \begin{bmatrix} \varepsilon_{11} \\ \varepsilon_{22} \\ \varepsilon_{33} \\ 2\varepsilon_{12} \\ 2\varepsilon_{13} \\ 2\varepsilon_{23} \end{bmatrix}. \tag{14.23}$$

If we inverse the stiffness matrix for the material, then we can obtain the following equation in terms of $\boldsymbol{\varepsilon} = \mathbf{C}^{-1}\boldsymbol{\sigma}$, in which $\mathbf{C}^{-1}$ is the compliance matrix:

$$\begin{bmatrix} \varepsilon_{11} \\ \varepsilon_{22} \\ \varepsilon_{33} \\ 2\varepsilon_{12} \\ 2\varepsilon_{13} \\ 2\varepsilon_{23} \end{bmatrix} = \frac{1}{E} \begin{bmatrix} 1 & -\nu & -\nu & 0 & 0 & 0 \\ -\nu & 1 & -\nu & 0 & 0 & 0 \\ -\nu & -\nu & 1 & 0 & 0 & 0 \\ 0 & 0 & 0 & 2+2\nu & 0 & 0 \\ 0 & 0 & 0 & 0 & 2+2\nu & 0 \\ 0 & 0 & 0 & 0 & 0 & 2+2\nu \end{bmatrix} \begin{bmatrix} \sigma_{11} \\ \sigma_{22} \\ \sigma_{33} \\ \sigma_{12} \\ \sigma_{13} \\ \sigma_{23} \end{bmatrix}. \quad (14.24)$$

Given the formation of $\mathbf{C}$ or $\mathbf{C}^{-1}$, we can obtain the other using any symbolic calculation programs such as Mathcad and MATLAB.

Please note that these stiffness and compliance matrices are the matrices for a material point rather than the whole object. Thus, they are local matrices. In numerical simulation, we also commonly encounter another type of stiffness matrix or compliance matrix, which is the global stiffness or compliance matrix for the object.

Plane Strain and Plane Stress

The degeneration from 3D to 2D in the mechanical field is different from and more complicated than other physical fields: We usually need to make it clear whether it is a plane strain or a plane stress condition.

Plane Strain

In the plane strain condition, the strain component along the direction perpendicular to the plane is zero. Accordingly, we only have non-zero strains along the two directions in that plane. This is usually a good approximation to the mechanical field in thick objects undergoing a distributed load on its side surfaces along the longitudinal direction, for example, an earth dam under a constant surcharge pressure on its top. The constitutive relationship for the plane strain case can be obtained by turning the off-plane strain components into zero in the following equation:

$$\begin{bmatrix} \sigma_{11} \\ \sigma_{22} \\ \sigma_{33} \\ \sigma_{12} \\ \sigma_{13} \\ \sigma_{23} \end{bmatrix} = \frac{E}{(1+\nu)(1-2\nu)} \begin{bmatrix} 1-\nu & \nu & \nu & 0 & 0 & 0 \\ \nu & 1-\nu & \nu & 0 & 0 & 0 \\ \nu & \nu & 1-\nu & 0 & 0 & 0 \\ 0 & 0 & 0 & \frac{1-2\nu}{2} & 0 & 0 \\ 0 & 0 & 0 & 0 & \frac{1-2\nu}{2} & 0 \\ 0 & 0 & 0 & 0 & 0 & \frac{1-2\nu}{2} \end{bmatrix} \begin{bmatrix} \varepsilon_{11} \\ \varepsilon_{22} \\ \varepsilon_{33} \\ 2\varepsilon_{12} \\ 2\varepsilon_{13} \\ 2\varepsilon_{23} \end{bmatrix}. \quad (14.25)$$

For example, if the non-zero strain plane is parallel to the surface consisting of axes 1 and 2 (normal to axis 3), then all the strain terms with the subscript "3" will diminish, leading to

$$\begin{bmatrix}\sigma_{11}\\ \sigma_{22}\\ \sigma_{33}\\ \sigma_{12}\\ \sigma_{13}\\ \sigma_{23}\end{bmatrix}=\frac{E}{(1+\nu)(1-2\nu)}\begin{bmatrix}1-\nu & \nu & \nu & 0 & 0 & 0\\ \nu & 1-\nu & \nu & 0 & 0 & 0\\ \nu & \nu & 1-\nu & 0 & 0 & 0\\ 0 & 0 & 0 & \frac{1-2\nu}{2} & 0 & 0\\ 0 & 0 & 0 & 0 & \frac{1-2\nu}{2} & 0\\ 0 & 0 & 0 & 0 & 0 & \frac{1-2\nu}{2}\end{bmatrix}\begin{bmatrix}\varepsilon_{11}\\ \varepsilon_{22}\\ 0\\ 2\varepsilon_{12}\\ 0\\ 0\end{bmatrix}. \tag{14.26}$$

The above equation can be further reduced into:

$$\begin{bmatrix}\sigma_{11}\\ \sigma_{22}\\ \sigma_{12}\end{bmatrix}=\frac{E}{(1+\nu)(1-2\nu)}\begin{bmatrix}1-\nu & \nu & 0\\ \nu & \nu & 0\\ 0 & 0 & \frac{1-2\nu}{2}\end{bmatrix}\begin{bmatrix}\varepsilon_{11}\\ \varepsilon_{11}\\ 2\varepsilon_{12}\end{bmatrix}. \tag{14.27}$$

Inversing the above equation, we can obtain the compliance matrix for the plane strain condition:

$$\begin{bmatrix}\varepsilon_{11}\\ \varepsilon_{11}\\ 2\varepsilon_{12}\end{bmatrix}=\frac{1+\nu}{E}\begin{bmatrix}1-\nu & -\nu & 0\\ -\nu & 1-\nu & 0\\ 0 & 0 & 1\end{bmatrix}\begin{bmatrix}\sigma_{11}\\ \sigma_{22}\\ \sigma_{12}\end{bmatrix}. \tag{14.28}$$

Plane Stress

The plane stress conditions apply to thin objects such as membranes which only have non-zero stress components along the directions within the membrane but no stresses on the plane whose unit normal vector is perpendicular to the membrane. Following a similar procedure as above, we can start with the following equation:

$$\begin{bmatrix}\varepsilon_{11}\\ \varepsilon_{22}\\ \varepsilon_{33}\\ 2\varepsilon_{12}\\ 2\varepsilon_{13}\\ 2\varepsilon_{23}\end{bmatrix}=\frac{1}{E}\begin{bmatrix}1 & -\nu & -\nu & 0 & 0 & 0\\ -\nu & 1 & -\nu & 0 & 0 & 0\\ -\nu & -\nu & 1 & 0 & 0 & 0\\ 0 & 0 & 0 & 2+2\nu & 0 & 0\\ 0 & 0 & 0 & 0 & 2+2\nu & 0\\ 0 & 0 & 0 & 0 & 0 & 2+2\nu\end{bmatrix}\begin{bmatrix}\sigma_{11}\\ \sigma_{22}\\ \sigma_{33}\\ \sigma_{12}\\ \sigma_{13}\\ \sigma_{23}\end{bmatrix}. \tag{14.29}$$

Likewise, assuming the non-zero strain plane is parallel to the surface consisting of axes 1 and 2 (normal to axis 3), all the strain terms with the subscript "3" will be equal to 0, leading to

$$\begin{bmatrix} \varepsilon_{11} \\ \varepsilon_{22} \\ \varepsilon_{33} \\ 2\varepsilon_{12} \\ 2\varepsilon_{13} \\ 2\varepsilon_{23} \end{bmatrix} = \frac{1}{E} \begin{bmatrix} 1 & -\nu & -\nu & 0 & 0 & 0 \\ -\nu & 1 & -\nu & 0 & 0 & 0 \\ -\nu & -\nu & 1 & 0 & 0 & 0 \\ 0 & 0 & 0 & 2+2\nu & 0 & 0 \\ 0 & 0 & 0 & 0 & 2+2\nu & 0 \\ 0 & 0 & 0 & 0 & 0 & 2+2\nu \end{bmatrix} \begin{bmatrix} \sigma_{11} \\ \sigma_{22} \\ 0 \\ \sigma_{12} \\ 0 \\ 0 \end{bmatrix}. \tag{14.30}$$

The above equation can be further reduced into

$$\begin{bmatrix} \varepsilon_{11} \\ \varepsilon_{11} \\ 2\varepsilon_{12} \end{bmatrix} = \frac{1}{E} \begin{bmatrix} 1 & -\nu & 0 \\ -\nu & 1 & 0 \\ 0 & 0 & 2+2\nu \end{bmatrix} \begin{bmatrix} \sigma_{11} \\ \sigma_{22} \\ \sigma_{12} \end{bmatrix}. \tag{14.31}$$

The matrix between stress and strain is the compliance matrix for the plane strain condition. Inversing the above equation, we can obtain the stiffness matrix:

$$\begin{bmatrix} \sigma_{11} \\ \sigma_{22} \\ \sigma_{12} \end{bmatrix} = \frac{E}{1-\nu^2} \begin{bmatrix} 1 & \nu & 0 \\ \nu & 1 & 0 \\ 0 & 0 & \dfrac{1-\nu}{2} \end{bmatrix} \begin{bmatrix} \varepsilon_{11} \\ \varepsilon_{11} \\ 2\varepsilon_{12} \end{bmatrix}. \tag{14.32}$$

Stiffness and compliance matrices in terms of the other elastic moduli can be derived following the above procedure. Taking the plane strain condition, for example, the stiffness matrix can be obtained from

$$\begin{bmatrix} \sigma_{11} \\ \sigma_{22} \\ \sigma_{33} \\ \sigma_{23} \\ \sigma_{13} \\ \sigma_{12} \end{bmatrix} = \begin{bmatrix} \lambda+2\mu & \lambda & \lambda & 0 & 0 & 0 \\ \lambda & \lambda+2\mu & \lambda & 0 & 0 & 0 \\ \lambda & \lambda & \lambda+2\mu & 0 & 0 & 0 \\ 0 & 0 & 0 & \mu & 0 & 0 \\ 0 & 0 & 0 & 0 & \mu & 0 \\ 0 & 0 & 0 & 0 & 0 & \mu \end{bmatrix} \begin{bmatrix} \varepsilon_{11} \\ \varepsilon_{22} \\ 0 \\ 0 \\ 0 \\ 2\varepsilon_{12} \end{bmatrix}, \tag{14.33}$$

which is reduced to:

$$\begin{bmatrix} \sigma_{11} \\ \sigma_{22} \\ \sigma_{12} \end{bmatrix} = \begin{bmatrix} \lambda+2\mu & \lambda & 0 \\ \lambda & \lambda+2\mu & 0 \\ 0 & 0 & \mu \end{bmatrix} \begin{bmatrix} \varepsilon_{11} \\ \varepsilon_{11} \\ 2\varepsilon_{12} \end{bmatrix}. \tag{14.34}$$

Governing Equation: Navier's Equation

The governing equation for the mechanical field can be obtained based on the balance law: The Cauchy equation of motion. According to the principle of conservation of linear momentum, if the continuum body is at mechanical equilibrium, it can be demonstrated that the components of the Cauchy stress tensor at every material point in the body satisfy the equilibrium equations. Let us consider a continuum body occupying a volume V, having a surface area S, with defined traction or surface forces $T_i^{(\mathrm{n})}$ per unit area acting on every point of the body surface and body forces F_i per unit of volume on every point within the volume V. If the body is at mechanical equilibrium, the resultant force acting on the volume should be zero; mathematically, this is:

$$\int_S T_i^{(\mathrm{n})} dS + \int_V F_i dV = 0. \tag{14.35}$$

Then recall the definition of the stress vector, $T_j^{(\mathrm{n})} = \sigma_{ji} n_j$, yielding:

$$\int_S \sigma_{ji} n_j dS + \int_V F_i dV = 0. \tag{14.36}$$

Using Gauss's divergence theorem to convert a surface integral to a volume integral gives

$$\int_V \sigma_{ji,j} dV + \int_V F_i dV = 0, \tag{14.37}$$

or

$$\int_V \left(\sigma_{ji,j} + F_i\right) dV = 0. \tag{14.38}$$

For an arbitrary volume, the integral vanishes, so we have the following differential form of the equilibrium equation:

$$\sigma_{ji,j} + F_i = 0. \tag{14.39}$$

But as mentioned above, we usually deal with displacement directly, which serves as the dependent variable. This leads to the formation of Navier's equation.

The displacement is also a physical field as it exists everywhere across the domain. To turn the displacement into the dependent variable, we need to connect the current unknown, i.e., stress, to the displacement. This goal will be achieved following two steps. First, the strain-displacement equations are substituted into the constitutive equations (Hooke's law), yielding the following expression of stress in terms of displacement:

$$\sigma_{ij} = \lambda \varepsilon_{kk} \delta_{ij} + 2\mu \varepsilon_{ij} = \lambda \delta_{ij} \varepsilon_{k,k} + \mu \left(u_{i,j} + u_{j,i} \right). \tag{14.40}$$

Differentiating the stress gives:

$$\sigma_{ij,j} = \lambda \varepsilon_{k,ki} + \mu \left(u_{i,jj} + u_{j,ij} \right). \tag{14.41}$$

Second, substituting the above gradient of the stress into the equilibrium equation yields

$$\lambda u_{k,ki} + \mu \left(u_{i,jj} + u_{j,ij} \right) + F_i = 0, \tag{14.42}$$

or equivalently,

$$\mu \varepsilon_{i,jj} + (\lambda + \mu) u_{j,ji} + F_i = 0. \tag{14.43}$$

This governing is also called the Navier-Cauchy equation, Navier's equation, or elastostatic equation. The equation is more frequently formulated using the tensor notation as:

$$\mu \nabla^2 \mathbf{u} + (\lambda + \mu) \nabla (\nabla \cdot \mathbf{u}) + \mathbf{F} = 0. \tag{14.44}$$

Practice Problem

This section shows a 2D example. 3D examples with an analytical solution can be found in the *Physics of Continuous Matter* (Lautrup 2005).

Problem Description

A distributed load of 1 kPa is applied on the right side of a square element. As shown in Fig. 14.4, the element has a side length of 1 m. The material consisting the element has Young's modulus of 10 GPa and Poisson's ratio of 0.2. The left boundary is fixed, while the upper and lower boundary was constrained using "rollers": displacement is allowed along the boundary surface while not allowed along the normal direction. A plane strain condition is assumed for this 2D problem. Please calculate the stress distribution inside of the element.

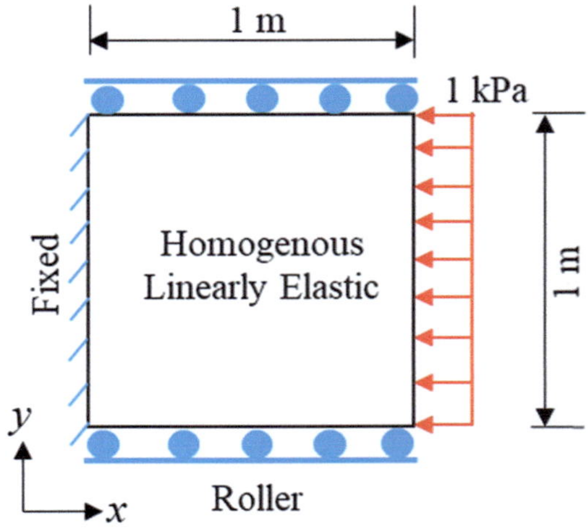

Fig. 14.4 Computation domain and boundary conditions

Governing Equation Natural Boundary Condition

The governing equation can be formulated in several different equivalent ways. In FlexPDE, we can implement using 1. Navier's equation with vector operators (nabla symbol) (Method 1), 2. the equilibrium equation with Cauchy stress (Method 2), and 3. Navier's equation with the derivatives of each stress component (no nabla symbol) (Method3).

Method 1

$$\mathrm{u} : \mathrm{mu}^*\mathrm{DIV}(\mathrm{GRAD}(\mathrm{u})) + (\mathrm{lambda} + \mathrm{mu})*\mathrm{GRAD}(\mathrm{DIV}(\mathrm{u})) = 0$$

Method 2

$$\mathrm{u} : \mathrm{DIV}(\mathrm{sigma\_tensor}) = \mathrm{VECTOR}(0, 0)$$

Method 3

$$\mathrm{ux} : \mathrm{mu}^*(\mathrm{Dxx}(\mathrm{ux}) + \mathrm{Dyy}(\mathrm{ux})) + (\mathrm{lambda} + \mathrm{mu})*(\mathrm{Dxx}(\mathrm{ux}) + \mathrm{Dyx}(\mathrm{uy})) = 0$$
$$\mathrm{uy} : \mathrm{mu}^*(\mathrm{Dyy}(\mathrm{uy}) + \mathrm{Dxx}(\mathrm{uy})) + (\mathrm{lambda} + \mathrm{mu})*(\mathrm{Dxy}(\mathrm{ux}) + \mathrm{Dyy}(\mathrm{uy})) = 0$$

The Dirichlet boundary condition can be implemented by assigning a displacement value to the corresponding boundary segment, while the Neumann boundary condition needs special attention. Details regarding the Neumann boundary condition for FEM packages such as FlexPDE have been given in Chap. 9. According to the previous introduction, the use of Navier's equation defines that the Neumann boundary condition for any boundary segment gives out the value of $\boldsymbol{\sigma} \cdot \mathbf{n}$ on that boundary segment. In the mechanical field, this value corresponds to the magnitude of the traction or pressure on the boundary. Therefore, a distributed load of 1 kPa on the right boundary can be applied using the command "NATURAL (ux) = 1000" or "LOAD (ux) = 1000." The two commands are equivalent in FlexPDE.

Results

The results of the above simulation are obvious. A tensile stress (positive) of 1 kPa is generated along the horizontal direction.

Appendix: FlexPDE Script File

```
{ Basic linear elastic mechanical problems for 2D (plain strain)
homogenous cases}
TITLE '2D Plain Strain'
VARIABLES
 u=VECTOR(ux,uy)    { Methods 1&2 use the displacement vector as
dependent variable }
   !ux,uy           { Methods 3 uses components of the vector as
dependent variable }
DEFINITIONS
 E=10e9
 nu=0.2
 mu=E/2/(1+nu)
 lambda =E*nu/(1+nu)/(1-2*nu)
 C=TENSOR((2*mu+lambda,lambda,0),(lambda,2*mu+lambda,0),(0,0,
mu)) ! Stiffness matrix
 epsilon=VECTOR( dx(ux),dy(uy),0.5*(dx(uy)+dy(ux)) )
 sigma=dot( C,epsilon)
   { The above C and epsilon are defined using the Voigt notation;
sigma_tensor uses the tensor notation }
 sigma_tensor=TENSOR((XCOMP(sigma),ZCOMP(sigma)),(ZCOMP
(sigma),YCOMP(sigma)))
! INITIAL VALUES    { Not time dependent }
EQUATIONS
   { Two equivalent ways to implement the governing equation }
 u: mu*DIV(GRAD(u))+(lambda+mu)*GRAD(DIV(u))=0    { Method 1 }
 !u: DIV(sigma_tensor)=VECTOR(0,0)                     { Method 2 } !
ux: mu*(Dxx(ux)+Dyy(ux))+(lambda+mu)*(Dxx(ux)+Dyx(uy))=0 {
Method 3 }
 !uy: mu*(Dyy(uy)+Dxx(uy))+(lambda+mu)*(Dxy(ux)+Dyy(uy))=0  {
Method 3 } BOUNDARIES
 REGION 1      { For each material region }
START(0,0)
   VALUE(ux)=0
   VALUE(uy)=0
LINE TO (0,1)
   VALUE(uy)=0
   NATURAL(ux)=0
LINE TO (1,1)
   NATURAL(ux)=1000
   VALUE(uy)=0
  LINE TO (1,0)
   NATURAL(ux)=0
   VALUE(uy)=0
```

```
 LINE TO CLOSE
! TIME 0 TO 1  { Not time dependent }
MONITORS
PLOTS
 VECTOR(u) !for Method 3: VECTOR(VECTOR(ux,uy))
  CONTOUR(XCOMP(sigma))
 ELEVATION(uy) FROM (0,0) TO (0,1)
 ELEVATION(ux) FROM (1,0) TO (1,1)
END
```

Chapter 15
Dyno: Dynamics

Introduction: Meaning of Dynamics

Dynamics, which is abbreviated as dyno in this book, is a relatively ambiguous concept in multiphysics. The word or the corresponding root is used in some multiphysics applications such as aerodynamics and elastodynamics. However, it is also used in many topics out of the traditional multiphysics area such as thermodynamics and molecular dynamics. In addition, dynamics has been historically studied with respect to material types, leading to two distinct areas: elastodynamics for solids and aerodynamics (gas) and hydrodynamics (liquid) for fluids. Acoustics consists of a small part from each of the above two areas but is more focused on the vibration of the material points and the energy transfer associated with the vibration. These facts prohibit the wide use of dynamics or dyno as a word for general dynamics phenomena in multiphysics.

Broadly speaking, dynamics is the study of the motion of entities resulting from forces. Traditionally, these entities are generally an assembly of solid inertial elements and rigid particles. Dynamics is usually differentiated from kinematics and statics. Kinematics is concerned with the motion of a material point, which generalizes particles, bodies, conglomerates of such particles or systems, without reference to the underlying cause, i.e., forces. The scope of statics is confined to bodies at rest and forces at equilibrium – a system undergoes no displacements with respect to a specified frame of reference. We also have a quasi-static state if such equilibria can be assumed within a sufficiently long time.

Dynamics and dyno are used in this book as a generic term for the dynamic phenomena in internal systems involving forces or momenta. As shown in Fig. 15.1, internal systems include deformable solids and fluids, whereas external systems refer to particles or rigid bodies, which correspond to multibody dynamics and rigid body dynamics, respectively. Therefore, theories such as molecular dynamics, discrete element method, and Lattice Boltzmann method are out of the scope of dynamics in multiphysics in this book. An interesting example is the smoothed particle

Z. Liu, *Multiphysics in Porous Materials*,
https://doi.org/10.1007/978-3-319-93028-2_15

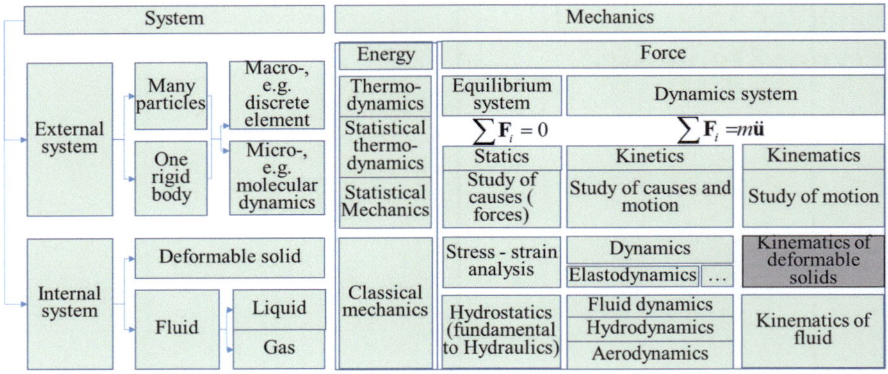

Fig. 15.1 From general mechanics areas to the dynamics in multiphysics

hydrodynamics (SPH). The particles used in the SPH are more like a mesh for discretization than particles to which Newton's second law is applied. Instead, governing equations are still discretized and solved. Therefore, the SPH can still be viewed as a discretization method for implementing dynamics in multiphysics.

The relationship between dynamics and the wave propagation theory needs to be clarified. In deformable solids or fluids, the systems to be considered are not rigid. This turns dynamics into the study of the correlation between forces, either surface forces or body forces, and the motion of material points within the flexible body. Further clarification needs to be made with respect to both the phase type, i.e., solids and fluids, and the scale of deformation.

In solids, the above fact results in a knowledge pool consisting of both dynamics, i.e., Newton's second law, and the understanding of the constitutive relationships of solids, e.g., elastic, plastic, elastoplastic, and viscoelastic. As a result, the dynamics of solids is subdivided into areas such as elastodynamics and plastodynamics. Among these areas, elastodynamics is the predominant one. The scale of the strain amplitudes also determines the nature of dynamic phenomena in solids. If a study is confined to small-amplitude oscillations in strains, then this problem can be regarded as vibration. A close look at the governing equation of the elastodynamics, i.e., Navier's equation with a momentum term, reveals that the equation has the form of a wave equation. Therefore, the dynamics in solids, at least elastodynamics, can be viewed as an application of wave theories in solids. However, the dynamic phenomena in solids, especially those involving large strains, are a broader scope of studies than merely wave propagation which focuses mostly on the oscillation and the associated transfer of energy. In such dynamic phenomena with large strains, the stress, strain, complicated constitutive relationships, and fatigue are also of great interest.

Fluids are much different from solids especially in that the shear stresses in fluids are related to the first-order time derivative of strain, i.e., strain rate or velocity gradient, instead of the strain. Due to the reason, the fluids can have bulk mass movement without altering the stresses, for example, at a constant strain rate.

The above difference leads to two noteworthy outcomes. First, fluids can hardly transmit shearing waves (transverse waves) because a shearing strain does not lead to a shearing stress. Consequently, the shearing motion and energy can be hardly transmitted. The volumetric train is still proportional to the mean stress component (pressure) of the stress tensor. Therefore, pressure (longitudinal) waves can still propagate in fluids. Second, another type of dynamics exists in fluids in addition to pressure waves. This is because, though shear strains do not cause stresses in fluids, their time derivatives can still cause stresses and further contribute to the motion of fluid parcels. As a result, this fluid motion exists as a form of dynamics in addition to mechanical waves (or acoustics in fluids). This type of dynamics, which is the fluid dynamics that we are familiar with, exists as the "background" on which the wave propagation occurs. The Navier-Stokes equations are used to describe the first type of dynamics, i.e., fluid dynamics, while the acoustics can be derived by investigating the disturbance to the volumetric stresses terms and velocity. Acoustics excludes the hydrostatic pressure and the fluid velocity associated with fluid dynamics. Therefore, acoustic phenomena can occur in fluids at rest in the same way as in moving fluids. Surface waves are excluded in the above discussion as we focus on physical fields here.

Therefore, the acoustics in fluids can be viewed as a part or a derivative of the area of fluid dynamics, but it is different from the "fluid dynamics" which is directly described by the Navier-Stokes equations. It is noted that fluid dynamics involves both mass and momentum transport, so it is treated as a multiphysical process and will be discussed in the next part of this book. By contrast, acoustics will be treated as a monolithic physical field in this chapter because only one dependent governing equation is needed for the acoustics in one material.

However, the wave theory is more than acoustics. The wave equations in elastodynamics and fluid dynamics share the same characteristics: a wave equation, oscillation of physical properties (stress and strain), and transfer of energy. The wave equation in this case depicts the process that a wave propagates through media, i.e., solids, gasses, or liquids. This type of wave is pertained to materials and mechanical responses and thus termed the mechanical wave. The mechanical wave is frequently encountered in earth materials in seismic engineering and in exploratory geophysics. The wave is also called acoustic waves in acoustics applications. However, wave phenomena, as oscillation accompanied by a transfer of energy, also widely exist in other forms such as electromagnetic waves. The behavior of particles in quantum mechanics can also be described by waves. In addition, a vibration or movement in gravitational fields can lead to gravitational waves traveling through space.

In summary, acoustics as the study of mechanical waves can be classified as a special case of dynamics, consisting of part of elastodynamics and part of fluid dynamics. In the meanwhile, the wave theory discussed in acoustics is part of the wave theory which also includes other types of waves.

Due to the above consideration, in this chapter, elastodynamics will be introduced first, in which we will show that the governing equation of elastodynamics directly describes the mechanical waves in solids. Then we will use two different ways to show that the wave equation for the acoustics in fluids can be derived as a special case (or a derivative) of fluid dynamics via the Navier-Stokes equation. The wave equation generalized from acoustics will be discussed next. In the end, a practice problem will be provided to show the idea and implementation of elastodynamics. Fluid dynamics, which is treated as a multiphysical field, will be introduced in the next part of the book.

Elastodynamics

Governing Equation of Isotropic, Homogeneous, and Elastic Material

Elastodynamics is the study of waves and time-dependent mechanical responses, e.g., strain and stress, in linearly elastic materials. An elastic wave is a mechanical wave that propagates in elastic or viscoelastic materials. Elastic waves occurring in the Earth as the result of earthquakes or other disturbance are usually called seismic waves. The disturbance, either in terms of stress or strain, produces the wave, while the elasticity of the material provides the restoring force for the wave. Therefore, elastodynamics covers the dynamics in materials with an elastic constitutive relationship.

The wave equation of elastodynamics is simply the equilibrium equation of elastostatics with an additional inertial term:

$$\sigma_{ji,j} + F_i = \rho \ddot{u}_i = \rho \partial_{tt} u_i. \tag{15.1}$$

Substituting the linearly elastic constitutive relationship into the above equation, we can obtain the following governing equation by following a procedure similar to that was introduced in the previous chapter:

$$\mu u_{i,jj} + (\lambda + \mu)\mu u_{j,ji} + F_i = \rho \partial_{tt} u_i. \tag{15.2}$$

Using the tensor notation, the above governing equation of elastodynamics turns into

$$\mu \nabla^2 \mathbf{u} + (\lambda + \mu)\nabla(\nabla \cdot \mathbf{u}) + \mathbf{F} = \rho \frac{\partial^2 \mathbf{u}}{\partial t^2}. \tag{15.3}$$

Acoustics in Solids

To obtain a better understanding of the acoustics in solids, we can separate the two major types of waves: pressure wave (P-wave, primary wave, or longitudinal wave) from the shear wave (S-wave or transverse/transversal wave).

For the pressure wave, there are only volumetric changes but no shearing. Mathematically, this type of physical field can be formulated as:

$$\nabla \cdot \mathbf{u} \neq 0 \text{ and } \nabla \times \mathbf{u} = 0.$$

The physical field with such a dependent variable is called an irrotational field, a curl-free/curl-less field, or a longitudinal field.

For the shear wave, there are no volumetric changes but only shearing. Mathematically, this type of field can be formulated as:

$$\nabla \cdot \mathbf{u} = 0 \text{ and } \nabla \times \mathbf{u} \neq 0.$$

The physical field with such a dependent variable is called a solenoidal field, an incompressible field, a divergence-free field, a transverse field, or an equivoluminal field.

To separate the two types of waves, we can recall the following identity from the tensor analysis:

$$\nabla \times (\nabla \times \mathbf{u}) = \nabla(\nabla \cdot \mathbf{u}) - \nabla^2 \mathbf{u}. \tag{15.4}$$

Then Navier's equation can be reformulated into the following equation by neglecting the body force:

$$(\lambda + 2\mu)\nabla(\nabla \cdot \mathbf{u}) - \mu\nabla \times (\nabla \times \mathbf{u}) = \rho\frac{\partial^2 \mathbf{u}}{\partial t^2}. \tag{15.5}$$

Then in the case of pressure waves, we obtain:

$$(\lambda + 2\mu)\nabla(\nabla \cdot \mathbf{u}) = \rho\frac{\partial^2 \mathbf{u}}{\partial t^2}. \tag{15.6}$$

The wave velocity is the square root of the ratio between the "modulus" and the density, i.e., $\sqrt{\frac{\lambda+2\mu}{\rho}}$.

In the case of shear waves, we obtain

$$-\mu\nabla \times (\nabla \times \mathbf{u}) = \mu\nabla^2 \mathbf{u} = \rho\frac{\partial^2 \mathbf{u}}{\partial t^2}, \tag{15.7}$$

where the shear wave velocity is $\sqrt{\frac{\mu}{\rho}}$.

Fluid Dynamics and Acoustics in Fluids

In fluids such as water, it is more common to derive the propagation of sound waves from the equation of continuity, i.e., conservation of mass and the equation of motion, i.e., conservation of momentum, instead of Navier's equation as for solids. We start the derivation with a few simplifications such as a constant density and negligible viscous and gravitational force, which are reasonable considering the magnitude and time scale of vibration. More importantly, we need to extract the mass and momentum balance from the "background" fluid dynamics (mass and momentum transfer) associated with the viscosity (shearing), gravity, and advection. Then we will obtain the following equations

$$\frac{\partial p}{\partial t} + k\nabla \cdot \mathbf{v} = 0 \quad \text{(mass balance)} \tag{15.8}$$

$$\rho_0 \frac{\partial \mathbf{v}}{\partial t} + \nabla p = 0 \quad \text{(momentum balance)} \tag{15.9}$$

where p is the acoustic pressure, which can be viewed as the deviation from the hydraulic pressure in fluid dynamics due to the existence of the wave, and $\mathbf{v}$ is the flow velocity excluding the background velocity associated with the background fluid dynamics. The use of the p and $\mathbf{v}$ from the fluid dynamics will not impact the following derivation because all the terms in Eqs. 15.8 and 15.9 are about differences instead of absolute values. However, when assigning boundary conditions, p and $\mathbf{v}$ need to use the acoustic values instead of the hydraulic pressure and the real fluid velocity that are used in background fluid dynamics.

A liquid can usually be assumed as elastic under small strains, which is most likely to be true for the case of waves in a liquid. Accordingly, the bulk modulus and the density can be related in the following way

$$k = \rho_0 c_0^2. \tag{15.10}$$

where c_0 can be later proven to be the velocity of the acoustic wave.

As mentioned above, the resistance of fluids such as water to shearing deformation is negligible. Thus, we can infer that the flow velocity field in the fluids is irrotational, $\nabla \times \mathbf{v} = 0$. We then can substitute this irrotational fluid equation and the above constitutive relationship between the bulk modulus and wave velocity, i.e., Eq. 15.10, into the two governing equations (Eqs. 15.8 and 15.9). The following governing equation for the elastic wave in fluids can be obtained

$$\frac{\partial^2 \mathbf{v}}{\partial t^2} - c_0^2 \nabla^2 \mathbf{v} = 0, \tag{15.11}$$

or equivalently in terms of pressure

$$\frac{\partial^2 p}{\partial t^2} - c_0^2 \nabla^2 p = 0. \tag{15.12}$$

In the above general form of the wave equation, c_0 has the physical meaning of wave velocity. The following identity in the tensor analysis was employed in the above deduction:

$$\nabla^2 \mathbf{v} = \nabla(\nabla \cdot \mathbf{v}) - \nabla \times (\nabla \times \mathbf{v}). \tag{15.13}$$

It is also very common to express the acoustic wave equation using a scalar potential ψ where $\mathbf{v} = \nabla\psi$. In that case, the acoustic wave equation is written as:

$$\frac{\partial^2 \psi}{\partial t^2} - c_0^2 \nabla^2 \psi = 0. \tag{15.14}$$

Comprehensive Derivation from Fluid Dynamics

The above derivation of the wave equation for the acoustics in fluids adopts simple governing equations. However, the meanings of p and $\mathbf{v}$ require some effort to comprehend. A more comprehensive version of the above derivation is conducted with the Navier-Stokes equations with step-by-step assumptions. Also, the concept of disturbance is utilized so that the physical meaning of p and $\mathbf{v}$ can be more easily understood. Since the Naiver-Stokes equations have not been introduced yet, this section is provided only as a reference. If necessary, please finish the chapter for fluid dynamics first.

Conservation of Momentum

The equation for the momentum conservation of a fluid with a constant density is

$$\rho\left(\frac{\partial \mathbf{v}}{\partial t} + \mathbf{v} \cdot \nabla \mathbf{v}\right) = -\nabla p + \nabla \cdot \boldsymbol{\tau} + \rho \mathbf{g}, \tag{15.15}$$

where $\mathbf{g}$ is the body force per unit mass, p is the pressure, and $\boldsymbol{\tau}$ is the deviatoric stress. Using the Cauchy stress for $\boldsymbol{\tau}$, then we have

$$p = -\frac{1}{3}\mathrm{trace}(\boldsymbol{\sigma}) \tag{15.16}$$

$$\boldsymbol{\tau} = \boldsymbol{\sigma} + p\mathbf{I} \tag{15.17}$$

where $\mathbf{I}$ is the rank-2 identity tensor.

Several assumptions will be made step by step in the following subsections to derive the momentum balance equation for acoustics in fluids.

Assumption 1: Newtonian Fluid

In acoustics, the fluid material can be assumed to be Newtonian. Accordingly, the deviatoric stress tensor is related to the flow velocity by

$$\boldsymbol{\tau} = \mu\left[\nabla\mathbf{v} + (\nabla\mathbf{v})^T\right] + \lambda(\nabla\cdot\mathbf{v})\mathbf{I}, \tag{15.18}$$

where μ is the shear viscosity and λ is the bulk viscosity. The divergence of $\boldsymbol{\tau}$ is

$$\begin{aligned}\nabla\cdot\boldsymbol{\tau} &= \mu\left[\frac{\partial}{\partial x_i}\left(\frac{\partial u_i}{\partial x_j} + \frac{\partial u_j}{\partial x_i}\right)\right] + \lambda\left[\frac{\partial}{\partial x_i}\left(\frac{\partial u_k}{\partial x_k}\right)\right] = \mu\frac{\partial^2 u_i}{\partial x_i \partial x_j} + \mu\frac{\partial^2 u_i}{\partial x_i \partial x_i} + \lambda\frac{\partial^2 u_k}{\partial x_k \partial x_i} \\ &= (\mu+\lambda)\frac{\partial^2 u_i}{\partial x_i \partial x_j} + \mu\frac{\partial^2 u_i}{\partial x_j \partial x_j} \\ &= (\mu+\lambda)\nabla(\nabla\cdot\mathbf{v}) + \mu\nabla^2\mathbf{v}.\end{aligned} \tag{15.19}$$

Employing the identity $\nabla^2\mathbf{v} = \nabla(\nabla\cdot\mathbf{v}) - \nabla\times\nabla\times\mathbf{v}$, the divergence of the deviatoric stress then becomes

$$\nabla\cdot\boldsymbol{\tau} = (2\mu+\lambda)\nabla(\nabla\cdot\mathbf{v}) - \mu\nabla\times\nabla\times\mathbf{v}. \tag{15.20}$$

Substituting the above formulation of the deviatoric stress into the momentum balance equation yields

$$\rho\left(\frac{\partial\mathbf{v}}{\partial t} + \mathbf{v}\cdot\nabla\mathbf{v}\right) = -\nabla p + (2\mu+\lambda)\nabla(\nabla\cdot\mathbf{v}) - \nabla\times\nabla\times\mathbf{v} + \rho\mathbf{g}. \tag{15.21}$$

Assumption 2: Irrotational Flow

The flow is irrotational in acoustics, hence

$$\nabla\times\mathbf{v} = 0 \tag{15.22}$$

Substituting the above equation into the momentum equation, we obtain

$$\rho\left(\frac{\partial\mathbf{v}}{\partial t} + \mathbf{v}\cdot\nabla\mathbf{v}\right) = -\nabla p + (2\mu+\lambda)\nabla(\nabla\cdot\mathbf{v}) + \rho\mathbf{g}. \tag{15.23}$$

Assumption 3: No Body Forces

The gravitational force as a body force is negligible in acoustic waves in fluids. This further simplifies the momentum equation into

$$\rho\left(\frac{\partial\mathbf{v}}{\partial t} + \mathbf{v}\cdot\nabla\mathbf{v}\right) = -\nabla p + (2\mu+\lambda)\nabla(\nabla\cdot\mathbf{v}). \tag{15.24}$$

Assumption 4: No Viscous Forces

The viscous forces can be excluded as explained in the introduction section, leading to the following momentum equation:

$$\rho\left(\frac{\partial \mathbf{v}}{\partial t}+\mathbf{v}\cdot\nabla\mathbf{v}\right)=-\nabla p. \tag{15.25}$$

Assumption 5: Small Disturbances

Instead of understanding the acoustic field as another field on top of the background fluid dynamics field, we also introduce the concept of disturbance to explain the pressure and fluid velocity associated with the acoustic field. The underlying assumption of the disturbance concept is that the amplitude of the disturbance of the field quantities is small compared with the magnitudes of these quantities. Then we can express the variables as the sum of a time-averaged mean field (–) varying in space and a small fluctuating field (~) varying in space and time:

$$p=\bar{p}+\tilde{p};\rho=\bar{\rho}+\tilde{\rho};\mathbf{v}=\bar{\mathbf{v}}+\tilde{\mathbf{v}}$$

and

$$\frac{\partial\bar{p}}{\partial t}=0;\quad\frac{\partial\bar{\rho}}{\partial t}=0;\quad\frac{\partial\bar{\mathbf{v}}}{\partial t}=0.$$

Then the momentum equation is reformulated into

$$(\bar{\rho}+\tilde{\rho})\left[\frac{\partial\tilde{\mathbf{v}}}{\partial t}+(\bar{\mathbf{v}}+\tilde{\mathbf{v}})\cdot\nabla(\bar{\mathbf{v}}+\tilde{\mathbf{v}})\right]=-\nabla(\bar{p}+\tilde{p}). \tag{15.26}$$

Considering the fluctuations are small, products of the fluctuation terms can be neglected, yielding:

$$\bar{\rho}\frac{\partial\tilde{\mathbf{v}}}{\partial t}+(\bar{\rho}+\tilde{\rho})\bar{\mathbf{v}}\cdot\nabla\bar{\mathbf{v}}+\bar{\rho}(\tilde{\mathbf{v}}\cdot\nabla\bar{\mathbf{v}}+\bar{\mathbf{v}}\cdot\nabla\tilde{\mathbf{v}})=-\nabla(\bar{p}+\tilde{p}). \tag{15.27}$$

Assumption 6: Homogeneous Medium

Assuming the material is homogeneous, the time-averaged variables $\bar{p}$ and $\bar{\rho}$ have zero gradients:

$$\nabla\bar{p}=0; \tag{15.28}$$

and

$$\nabla \overline{p} = 0. \tag{15.29}$$

The momentum equation then reduces into:

$$\overline{\rho}\frac{\partial \tilde{\mathbf{v}}}{\partial t} + (\overline{\rho} + \widetilde{\rho})\bar{\mathbf{v}} \cdot \nabla \bar{\mathbf{v}} + \overline{\rho}(\tilde{\mathbf{v}} \cdot \nabla \bar{\mathbf{v}} + \bar{\mathbf{v}} \cdot \nabla \tilde{\mathbf{v}}) = -\nabla(\tilde{p}). \tag{15.30}$$

Assumption 7: Medium at Rest

The last step is to remove the background fluid dynamics field. For the purpose, we just need to assume that the medium is at rest, i.e., $\bar{\mathbf{v}} = 0$. Then the momentum balance equation becomes:

$$\overline{\rho}\frac{\partial \tilde{\mathbf{v}}}{\partial t} = -\nabla(\tilde{p}). \tag{15.31}$$

Dropping the tildes and using $\overline{\rho} = \rho_0$, we obtain the commonly used form of the acoustic momentum equation:

$$\rho_0 \frac{\partial \mathbf{v}}{\partial t} + \nabla p = 0. \tag{15.32}$$

Conservation of Mass

The mass conservation equation for acoustics in fluids can also be derived in a manner similar to what is used for the conservation of momentum. The mass conservation equation in a fluid volume without any sources/sinks is:

$$\frac{\partial \rho}{\partial t} + \nabla \cdot (\rho \mathbf{v}) = 0. \tag{15.33}$$

Assumption 1: Small Disturbances

The assumption of small disturbances gives

$$\frac{\partial \widetilde{\rho}}{\partial t} + (\overline{\rho} + \widetilde{\rho})\nabla \cdot (\bar{\mathbf{v}} + \tilde{\mathbf{v}}) + \nabla(\overline{\rho} + \widetilde{\rho}) \cdot (\bar{\mathbf{v}} + \tilde{\mathbf{v}}) = 0. \tag{15.34}$$

Expanding the equation and neglecting the products of disturbance terms, the mass balance equation becomes

$$\frac{\partial\tilde{\rho}}{\partial t}+(\overline{\rho}+\tilde{\rho})\nabla\cdot(\overline{\mathbf{v}})+\overline{\rho}\nabla\cdot(\tilde{\mathbf{v}})+\nabla(\overline{\rho}+\tilde{\rho})\cdot\overline{\mathbf{v}}+\nabla\overline{\rho}\cdot\tilde{\mathbf{v}}=0. \tag{15.35}$$

Assumption 2: Homogeneous Medium

Adopting the homogeneous assumption, the mass balance equation becomes

$$\frac{\partial\tilde{\rho}}{\partial t}+(\overline{\rho}+\tilde{\rho})\nabla\cdot(\overline{\mathbf{v}})+\overline{\rho}\nabla\cdot(\tilde{\mathbf{v}})+\nabla\tilde{\rho}\cdot\overline{\mathbf{v}}=0. \tag{15.36}$$

Assumption 3: Medium at Rest

Similar to the momentum balance, remove the background fluid dynamics by assuming the fluid is at rest, i.e., $\overline{\mathbf{v}}=0$. Then the mass balance equation can be simplified as

$$\frac{\partial\tilde{\rho}}{\partial t}+\overline{\rho}\nabla\cdot\tilde{\mathbf{v}}=0. \tag{15.37}$$

Assumption 4: Ideal Gas, Adiabatic, and Reversible

An equation of state for relating the pressure and density is needed to close the equation system. For the purpose, we assume that the fluid is an ideal gas and all acoustic waves compress the medium in an adiabatic and reversible manner. Accordingly, the following equations of the thermodynamic state are valid:

$$\frac{dp}{d\rho}=\frac{\gamma p}{\rho};\gamma=\frac{C_{\mathrm{p}}}{C_{\mathrm{v}}};\text{and } c^2=\frac{\gamma p}{\rho},$$

where C_{p} is the specific heat at a constant pressure, C_{v} is the specific heat at a constant volume, and c is the wave speed. The value of γ is 1.4 for air.

For small disturbances

$$\frac{\frac{\frac{dp}{d\rho}\approx\bar{p}}{\tilde{\rho}};\frac{\frac{p}{\rho}\approx\bar{p}}{\bar{\rho}};\text{and } c^2\approx c_0^2=\gamma\bar{p}}{\bar{\rho},}$$

where c_0 is the speed of sound in the fluid.Therefore, we have

$$\frac{\tilde{p}}{\widetilde{\frac{\rho=\gamma\bar{p}}{\bar{\rho}=c_0^2}}} \tag{15.38}$$

$$\frac{\partial \tilde{p}}{\partial t} + c_0^2 \frac{\partial \widetilde{\rho}}{\partial t} = 0. \tag{15.39}$$

The balance of mass can then be formulated as

$$\frac{1}{c_0^2} \frac{\partial \tilde{p}}{\partial t} + \bar{\rho} \nabla \cdot \tilde{\mathbf{v}} = 0. \tag{15.40}$$

Dropping the tildes and defining $\bar{\rho} = \rho_0$, we obtain the following mass balance equation for acoustic waves in fluids:

$$\frac{\partial p}{\partial t} + \rho_0 c_0^2 \nabla \cdot \mathbf{v} = 0. \tag{15.41}$$

Combining the momentum and mass balance equation. We can obtain a wave equation identical to what we obtained in the simple derivation.

Acoustic Field

Wave Equation

The acoustic field here is used to represent the propagation of mechanical waves. Therefore, acoustics is an interdisciplinary area that deals with all mechanical waves in gasses, liquids, and solids. It covers topics such as vibration, sound, ultrasound, and infrasound. Acoustic applications widely exist in modern society, while the most obvious examples are the audio and noise control industries.

It was demonstrated in previous sections that the governing equation of the acoustic field can be directly obtained from the Navier's equation in elastodynamics for solids or derived from the Navier-Stokes equations for fluids. However, the wave equation has a much more general sense than just a description of mechanical waves. It is also the governing equation of electromagnetic waves, which can be derived from Maxwell's equations. In addition, the wave equation has applications out of the scope of multiphysics: quantum mechanics, plasma physics, and general relativity.

The wave equation is a hyperbolic partial differential equation. It typically concerns a time variable t, one or more spatial variables $x_1, x_2, \ldots, x_n$, and a scalar function $u = u\ (x_1, x_2, \ldots, x_n; t)$, which can be any of these oscillating quantities associated with the wave, e.g., the mechanical displacement of a wave. The general form of the wave equation is

$$\frac{\partial^2 u}{\partial t^2} = c^2 \nabla^2 u, \tag{15.42}$$

where c is the velocity of wave propagation. As shown in the previous sections, this velocity depends on materials for mechanical waves. However, it could also be independent of material properties in other types of waves.

The wave equation alone does not specify a physical solution. Further constraints or conditions on time and boundaries are needed to ensure a unique solution. These constraints could be initial conditions, which prescribe the amplitude and phase of the wave. Another important class of problems occurs in enclosed spaces specified by boundary conditions. In these problems, the solutions represent standing waves or harmonics, which are analogous to the harmonics of musical instruments.

To discuss the boundary conditions for the wave equation, let us use the pressure as the dependent variable:

$$\frac{\partial^2 p}{\partial t^2} = c^2 \nabla^2 p. \tag{15.43}$$

The Dirichlet and Neumann boundary conditions now take the following form

$$\begin{aligned} \text{Dirichlet}: \quad & p = q,, \\ \text{Neumann}: \quad & \mathbf{n} \cdot \nabla p = \rho \ddot{u},, \end{aligned}$$

where $\mathbf{n}$ is the magnitude of traction on the surface (or boundary pressure) and $\ddot{u}$ is the acceleration prescribed on the boundary. Special cases of the above equations could generate geometric or physical boundary conditions that we frequently encounter, for example, $p = 0$ corresponds to the sound soft boundary, and $\mathbf{n} \cdot \nabla p = \mathbf{0}$ corresponds to the symmetry boundary condition.

Speed of Wave

In general, the speed of sound c is given by the Newton-Laplace equation:

$$c = \sqrt{\frac{K_s}{\rho}}, \tag{15.44}$$

where K_s is a coefficient of stiffness, e.g., the isentropic bulk modulus for longitudinal waves, and ρ is the density. Therefore, the speed of sound increases with the stiffness of the material while decreases with the density.

For general equations of state, the speed of sound c is given by

$$c = \sqrt{\left(\frac{\partial p}{\partial \rho}\right)_s}, \tag{15.45}$$

where p is the pressure and ρ is the density. The derivative is taken adiabatically, i.e., at a constant entropy s.

A common practice in acoustics is to decompose a wave into a pressure wave (longitudinal or irrotational) and a shear wave (transverse or equivoluminal). This has been shown in the elastodynamics section. In summary, a longitudinal plane wave propagating in an infinite medium has no equivoluminal components but solely a dilatational component. This wave propagates at a phase speed of $\sqrt{(\lambda + 2\mu)/\rho}$ – longitudinal wave speed. Similarly, a shear plane wave propagating in an infinite medium has no dilatational components but only an equivoluminal component and is thus named the shear wave. This wave propagates at a phase speed of $\sqrt{\mu/\rho}$. For mechanical waves, the wave velocity has a general expression as the square root of the ratio between the "modulus" and the density. The definition of the "modulus" changes with the type of the mechanical wave.

Practice Problem

Problem Description

This problem is modified from the practice problem in the mechanical field chapter. A square element with a side length of 1 m is composed of a material with Young's modulus of 10 GPa, Poisson's ratio of 0.2, and a density of 1000. An electric wave is generated by repeatedly compressing and stretching the upper boundary of the element at a given frequency. Please simulate the wave inside the element to tell what you observe at a frequency of 500 Hz and 5000 Hz and explain the reason for the difference.

Please note that many numerical simulation programs including FlexPDE do not handle the second-order temporal derivatives directly. Therefore, we need to split the original governing into two by defining a new dependent variable for the first derivative of the displacement. The boundary condition for this new dependent variable will also need to be set up accordingly.

Results

Fig. 15.2 shows the displacement field at $t = 0.125$ ms. The figure clearly shows that the wave is starting to propagate into the element from the top.

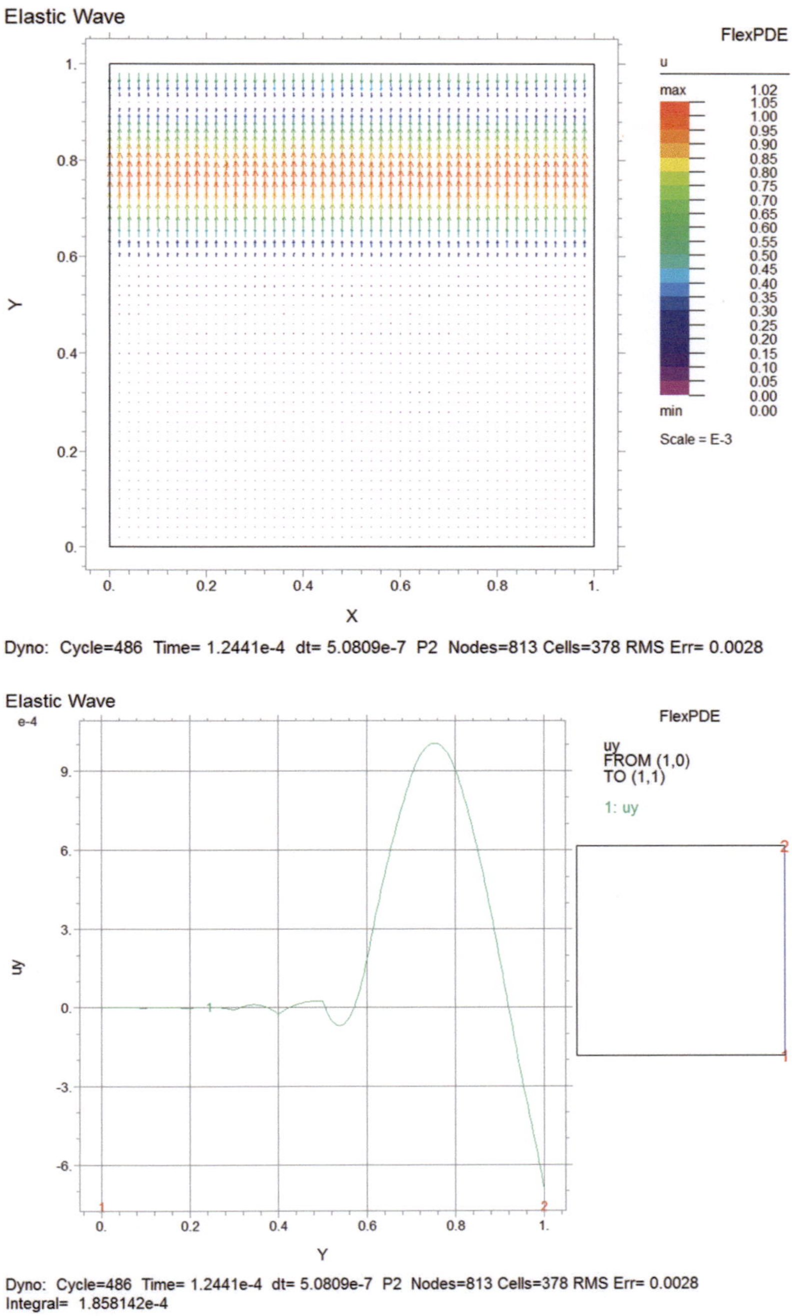

Fig. 15.2 Displacement in the domain (upper) and the horizontal displacement along $x = 0.5$ m (lower) at $t = 0.125$ s

Appendix: FlexPDE Script File

```
{ Elastodynamics in 2D (Plain Strain): Stretch a membrane }
TITLE 'Elastic Wave'
VARIABLES
  u(1e-6)=VECTOR(ux,uy)
  v(1e-6)=VECTOR(vx,vy)
DEFINITIONS
  E=1e10
  nu=0.2
  mu=E/2/(1+nu)
  lambda=E*nu/(1+nu)/(1-2*nu)
  rho=1000
INITIAL VALUES
  v=VECTOR(0,0)
  u=VECTOR(0,0)
EQUATIONS
  u: v=Dt(u)
  v: mu*DIV(GRAD(u))+(lambda+mu)*GRAD(DIV(u))=rho*dt(v)
BOUNDARIES
  REGION 1
    START(0,0)
      VALUE(ux)=0
      VALUE(vx)=0
      LOAD(uy)=0
      LOAD(vy)=0
    LINE TO (0,1)
      LOAD(ux)=0
      LOAD(vx)=0
      VALUE(uy)=1e-3*sin(2*pi*5000*t)
      VALUE(vy)=1e-3*2*pi*5000*cos(2*pi*5000*t)
    LINE TO (1,1)
      VALUE(ux)=0
      VALUE(vx)=0
      LOAD(uy)=0
      LOAD(vy)=0
    LINE TO (1,0)
      VALUE(uy)=0
      VALUE(vy)=0
      VALUE(ux)=0
      VALUE(vx)=0
    LINE TO CLOSE
TIME 0 TO 10 BY 0.01 { Time dependent }
MONITORS
PLOTS
  FOR cycle=1
    SURFACE(uy)
    VECTOR(u)
    ELEVATION(uy) FROM (1,0) TO (1,1)
END
```

Chapter 16
Chemico: Chemical Reaction

Introduction

In this chapter, basics of chemical reactions will be presented first to provide background knowledge for the chemical field. Then chemical kinetics will be introduced as it defines the governing equation of the chemical field. An example will be given to demonstrate the use of chemical kinetics for establishing the governing equation. The reaction rate coefficient is the key auxiliary relationship in the governing equation. The formulation of this coefficient and its dependence on various environmental factors will be discussed at last.

Uniqueness of Chemical Field

The chemical field is unique in multiphysics. As described in Part I, it is still counted as a physical field in multiphysics though chemical reactions are not counted as a physical process. This treatment generates an aftereffect in its mathematical descriptions, differentiating the chemical field from the others. One major difference is that the chemical field adopts the concentration or quantities depending on it such as reaction quotients as the state variables. However, the concentration is the dependent variable of the concentration field. An understanding of this contradiction is that the chemical field is more like an auxiliary relationship in mass transfer. This relationship defines the source/sink term and thus describes how the mass of a phase/species is consumed or produced. The relationship usually affects the mass balance of several phases/species simultaneously since a chemical reaction usually involves multiple materials. However, the chemical field still owns the characteristics of a field, that is, a spatial distribution of the state variable which can change with time in transient processes or stay unchanged at equilibrium.

Z. Liu, *Multiphysics in Porous Materials*,
https://doi.org/10.1007/978-3-319-93028-2_16

The uniqueness of the chemical field in multiphysics also endows it with a special treatment in numerical analysis. Since it is similar to auxiliary relationships, it is possible to treat it as a normal auxiliary relationship during the discretization and solution processes. Alternatively, to improve the accuracy, its treatment can be similar to that of governing equations, for example, more synchronized with the iteration of the governing equations. However, because it is a time-dependent ordinary differential equation associated with every point or element instead of the whole region, it cannot be discretized and placed into the matrix like other governing equations. To reduce the computational load, it is also possible to divide the computation regions into connected regions so that only one governing equation of the chemical field is needed for a region rather than each element or grid point within the region. But this simplification should comply with the real situation of the chemical reactions because one pre-assumption of the chemical reaction is that the reactants are perfectly mixed within each reactor, i.e., "regions" in the above treatment.

Understanding of Chemical Reactions

A chemical reaction is a process within which one set of chemical substances is transformed into another. This transformation only involves the relocations of electrons in the forming and breaking of chemical bonds between atoms without any change to the nuclei that determine the types of the elements. In much less common cases, both electronic and nuclear changes exist in the chemical reactions of unstable and radioactive elements, leading to nuclear chemistry, which is excluded in this book.

The substances initially involved in a chemical reaction are reactants or reagents. These reactants undergo the changes involving the forming and breaking of bonds and produce other substance(s), i.e., product(s), which usually have properties different from that of the reactants. Such a reaction usually occurs as a sequence of individual sub-steps, i.e., elementary reactions, instead of one step. Many reactions require no input of extra energy other than the thermal energy and thus termed spontaneous. Non-spontaneous reactions usually happen too slowly at human time scales. As a result, such reactions demand the input of some types of additional energy such as heat, light, and electricity to proceed to completion, i.e., chemical equilibrium, at a rate that we can observe.

Chemical reactions are usually characterized by a characteristic reaction rate which is dependent on the given temperature and chemical concentration. The rate at which a reaction takes place is studied in the scope of reaction kinetics. As a branch of chemistry, chemical kinetics studies both the rate of a chemical reaction and factors that affect the rate. The rate depends on various parameters such as temperature, pressure, reactant concentrations, surface area, activation energy, and the presence of a catalyst. Temperature facilitates chemical reactions because a higher temperature provides more energy to the molecules, resulting in more

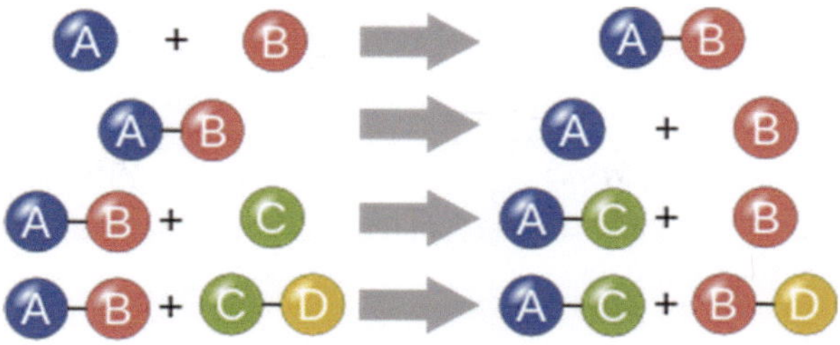

Fig. 16.1 Schematic of basics chemical reaction types

collisions per unit time. Pressure increases lead to decreases in the volume between molecules and consequently increase in the frequency of collisions between the molecules and the reaction rates. Reactant concentrations usually affect the reaction rate as higher concentrations promote the collisions between particles. However, there are also reactions whose rates are independent of reactant concentrations, i.e., zero order reactions. The surface area determines the area of sites which are available for the reactions. Increasing the surface area generally increases the reaction rates. This is particularly important in porous materials with complicated internal structures. The activation energy is defined as the amount of energy required to make the reaction start and carry on spontaneously. Therefore, a higher activation energy corresponds to a lower reaction rate. Catalysts are additives which remain unchanged but are capable of changing the pathway or mechanism of a reaction. For example, a catalyst can expedite a reaction by lowering the activation energy of the reaction.

The types of chemical reactions need to be briefly introduced to lay down a basis for the following introduction to the mathematical description of the chemical reactions. There are four basic types of chemical reactions shown in Fig. 16.1: synthesis, decomposition, single replacement, and double replacement.

In a synthesis reaction, two or more simple substances combine to yield a more complex substance. As shown in the above figure, these reactions have a general form of $A + B \rightarrow AB$. One example is the burning of hydrogen gas, which combines hydrogen and oxygen into water:

$$2H_2 + O_2 \rightarrow 2H_2O. \tag{16.1}$$

In a decomposition reaction, a relatively complex substance breaks down into multiple simple substances in the form of $AB \rightarrow A + B$. One example is the electrolysis of water which produces oxygen and hydrogen gas:

$$2H_2O \rightarrow 2H_2 + O_2. \tag{16.2}$$

In a single replacement reaction, a single element replaces an element in another compound. These reactions occur in the general form of $A + BC \rightarrow AC + B$. One example is the replacement of magnesium for hydrogen in water to form magnesium hydroxide and hydrogen gas:

$$Mg + 2H_2O \rightarrow Mg(OH)_2 + H_2. \tag{16.3}$$

A double replacement reaction involves the exchange of places for the anions and cations in two compounds, which produces two entirely different compounds. These reactions can be represented in the general form of AB + CD → AD + CB. A typical example is the reaction of lead nitrate with potassium iodide to form lead iodide and potassium nitrate:

$$Pb(NO_3)_2 + 2KI \rightarrow PbI_2 + 2KNO_3. \tag{16.4}$$

Chemical Kinetics and Its Mathematical Description

Chemical kinetics or reaction kinetics investigates the rates of chemical processes and the influence of experimental conditions on the rates. For the chemical field in numerical simulation, the information needed from chemical kinetics is the mathematical models for describing the characteristics of a chemical reaction. The determination of reaction rates including both rate laws and rate constants can be fulfilled by means of experiments, theories, or ab initio simulations.

The mechanisms are usually analyzed and summarized from a macroscopic perspective. This is consistent with multiphysics, which relies on statistical mechanics, molecular dynamics, and quantum mechanics but is established primarily with a language of continuum mechanics. Understanding of reaction rates on the molecular level falls into the realm of chemical dynamics or reaction dynamics. To understand chemical kinetics, let us consider a typical chemical reaction:

$$a\mathrm{A} + b\mathrm{B} \rightarrow c\mathrm{C} + d\mathrm{D}, \tag{16.5}$$

where the lowercase letters (a, b, c, and d) represent the stoichiometric coefficients and the capital letters represent the reactants (A and B) and the products (C and D).

The reaction rate r for a chemical reaction occurring in a closed isochoric system such as in Eq. 16.5 (no reaction intermediates) is defined as [IUPAC‘s Gold Book]

$$r = -\frac{1}{a}\frac{d[\mathrm{A}]}{dt} = -\frac{1}{b}\frac{d[\mathrm{B}]}{dt} = \frac{1}{c}\frac{d[\mathrm{C}]}{dt} = \frac{1}{d}\frac{d[\mathrm{D}]}{dt}, \tag{16.6}$$

where [X] denotes the concentration of the substance X.

The rate of a reaction is always positive. A negative sign indicates the reactant concentration is decreasing. As can be seen, the rates of reaction for different reactants and products are related by the concentrations of the substances and the stoichiometric numbers. The reaction rate usually has the unit of mol L^{-1} s^{-1}. It is noted that the above relationship is only valid for a single reaction in a closed system with a constant volume.

Reaction Rate Equation

The relationship between the reaction rates of different substances is described above. However, extra work is still needed to determine the absolute value of the reaction rate *r*. A rate equation or a rate law is a mathematical expression used in chemical kinetics to link the rate of a reaction to the concentration of each reactant. In gas phase reactions, the rate is often alternatively related to the partial pressures. For the chemical reaction $a\mathrm{A} + b\mathrm{B} \rightarrow c\mathrm{C} + d\mathrm{D}$, the rate equation is of the following form:

$$r = k[\mathrm{A}]^a[\mathrm{B}]^b, \tag{16.7}$$

where *k* is the reaction rate coefficient or rate constant. However, this parameter is not necessarily a constant. The reaction rate coefficient allows for the influences of all the parameters except concentration. Among these factors being considered, temperature is usually the most important one. The exponents *a* and *b* are the reaction orders which can be determined based on the reaction mechanism. For elementary or single-step reactions, the order of each reactant is equal to its stoichiometric coefficient. However, for complex or multistep reactions, the rate equation needs to be determined based on the detailed step-by-step mechanisms. In the above example, the reaction orders are *a* and *b* for A and B, respectively, while the overall reaction order is $a + b$.

Combining Eqs. 16.6 and 16.7 for the reaction rate, we then can obtain the following first-order ordinary differential equation:

$$\frac{d[\mathrm{A}]}{dt} = -ak(T)[\mathrm{A}]^a[\mathrm{B}]^b. \tag{16.8}$$

This can be viewed as the governing equation for the chemical field, though a comprehensive description of the field requires the mass transport as well. A theoretical solution is possible for simple cases such as simple reactions. Taking the reaction $a\mathrm{A} \rightarrow c\mathrm{C} + d\mathrm{D}$, for example, the reaction rate of this first-order reaction can be obtained as

$$r = -\frac{d[\mathrm{A}]}{dt} = k[\mathrm{A}]. \tag{16.9}$$

The integration of the above equation yields

$$[\mathrm{A}] = [\mathrm{A}]_0 \cdot \mathrm{e}^{-kt} \tag{16.10}$$

where $[\mathrm{A}]_0$ is the concentration of A at the beginning of the chemical reaction.

As can be seen, the rate of a first-order reaction depends only on the concentration and the properties of the involved substance. More than one time constant is needed when describing reactions of a higher order. Taking the reaction of hydrogen and nitric oxide, for example:

$$2H_2 + 2NO \rightarrow N_2 + 2H_2O, \tag{16.11}$$

the observed rate equation is

$$r = k[H_2][NO]^2. \tag{16.12}$$

Just like many reactions, the above rate equation does not correspond to the stoichiometric coefficients in the overall reaction – H_2 is the first order while the stoichiometric coefficient in the overall reaction is 2. This is because not all of the elementary steps affect the rate of reaction. We need to investigate the elementary steps to understand the overall reaction rate:

$$\begin{gathered} 2NO \rightarrow N_2O_2 \quad \text{(fast equilibrium)} \\ N_2O_2 + H_2 \rightarrow N_2O + H_2O \quad \text{(slow)} \\ N_2O + H_2 \rightarrow N_2 + H_2O \quad \text{(fast)} \end{gathered}$$

Usually, the slowest elementary step controls the reaction rate. In this case, the second reaction is the slowest one and thus controls the overall rate. This is a bimolecular elementary reaction whose rate is given by the second-order equation: $r = k_2[H_2][N_2O_2]$, where k_2 is the rate constant for the second step. However, N_2O_2 is an unstable intermediate, and its concentration depends on the first action: $[N_2O_2] = k_1[NO]^2$, in which k_1 is the equilibrium constant for this first step. Substituting this equation into the overall reaction rate equation leads to

$$r = k_1 k_2 [H_2][NO]^2. \tag{16.13}$$

As can be seen, only one H_2 appears in the rate equation. This is because the other H_2 reacts in the third step, which is fast and thus does not affect the overall reaction rate.

Integration with Mass Transfer

Although a closed system, or reactor, is assumed to obtain the above reaction rate, the real computational domain consists of a number of such mini "closed" systems involving mass transport. Therefore, these mini systems are not really closed due to the mass transport between them. However, we assume the chemical actions occur in a way similar to that in a closed system, i.e., not obviously impacted by the mass transport, within a short time span or a time step in numerical analysis. Within this step, the system is closed. However, the system is open between steps so that the concentrations are updated by solving the mass transfer equation including the source term, which is calculated with the chemical reaction results. The full mass balance must be taken into account: accumulation = in – out + generation – consumption, or mathematically:

$$\frac{dc}{dt} = q_{\text{in}} + q_{\text{out}} + \int_0^V r dV, \tag{16.14}$$

where c is the concentration, i.e., [A], q_{in} is the inflow rate of a material in molecules per second, q_{out} is the outflow, and r is the instantaneous reaction rate of A. When applied to a closed system with a constant volume, this equation reduces to $r = \frac{d[\text{A}]}{dt}$.

Reaction Rate Coefficient

Temperature Dependence

The temperature dependence of the reaction rate coefficient can be considered using the Arrhenius equation:

$$k = k_0 e^{-\frac{E_a}{RT}}, \tag{16.15}$$

where E_a is the activation energy and R is the gas constant. Since at temperature T, the molecules have energies given by the Boltzmann distribution, one can expect the number of collisions with energy greater than E_a to be proportional to $e^{-\frac{E_a}{RT}}$. k_0 is a pre-exponential factor or frequency factor.

The values of k_0 and E_a are dependent on the reaction. There are also more complex equations, which describe the temperature dependence of other rate constants that do not follow this pattern.

A chemical reaction takes place only when the reacting particles collide. However, not all collisions are effective in causing the reaction. Products are formed only when the colliding particles possess a certain minimum energy called the threshold energy. Basically, the number of activated particles nearly doubles for a temperature of T + 10 K. For convenience, the temperature coefficient, which is the ratio of its rate constant at a higher temperature to its rate constant at a lower temperature, is usually adopted. For example, Q_{10} is commonly used as the ratio of rate constants that are 10 °C apart.

Pressure Dependence

The pressure dependence varies with different materials and with the ranges of volume. For condensed-phase reactions, i.e., solids and liquids, the dependence is usually sufficiently weak in the range of pressures that we normally encounter. Therefore, the pressure dependence of the reaction constant in such materials is negligible. For gas reactions, the pressure dependence of the rate constant is associated with the activation volume. For the following reaction,

$$A + B \rightleftharpoons |A\cdots B|^{\dagger} \Rightarrow AB,$$

the activation volume, $\Delta V^{\dagger}$, is

$$\Delta V^{\dagger} = V^{\dagger} - V_{A} - V_{B}, \tag{16.16}$$

where V is the partial molar volumes of the reactants and products and † indicates the activation-state complex.

The change of the reaction rate constant in the above equation with pressure at a constant temperature can be formulated as

$$-RT\left(\frac{\partial \ln k_x}{\partial P}\right)_T = \Delta V^{\dagger}. \tag{16.17}$$

The conditions in real reactions can be complicated because the partial molar volumes and the activation volume can also vary with the pressure.

The reaction rate can increase or decrease with the pressure, depending on the value of $\Delta V^{\dagger}$. As an example of the possible magnitude of the pressure effect, the reaction rate in some organic reactions can be doubled when the pressure is increased from one atmosphere, i.e., 0.1 MPa, to 50 MPa.

Chapter 17
Electro: Electrostatics

Introduction

Electrostatics is the study of the processes and properties of stationary or low-velocity electric charges with negligible acceleration. Traditionally, electrostatics includes both the electric field and the forces on the objects in the field. This definition and scope are fixed in the above way attributed to the way we discovered and understood the electric field. From the perspective of multiphysics, the above definition of electrostatics involves both the electric field and forces. From this viewpoint, electrostatics can be categorized as a multiphysics topic. However, we still discuss it as monolithic physics considering the fact that the force is just one way we understand the electric field and that our interest is more about the origin and nature of the force rather than the mechanical behavior of the objects under the influence of the forces.

As mentioned above, our interest in and understanding of the electric field arose from forces that electrically charged objects exert on each other. This type of phenomenon is especially common in tribomaterials. For example, plastic wrap can be easily attracted to our hands. It also can be observed that the attractive force between the hands and the plastic wrap will become stronger as they move closer to each other. For many materials, such attractive forces only become apparent after a material gets into contact with and separated from another material that is usually different from the first material. This contact-and-separation process enables electron transfer between the materials, and as a result, the materials are oppositely charged. That is, one material loses electrons, while the other receives electrons. Then we can understand that one of the two materials, say material 1, generates an electric field, and the other material, say material 2, as an ensemble of charged particles, suffers from the forces due to the existence of the electric field generated by material 1. But to be more accurate, this electric field is just a superposition of the numerous electric fields generated by the material points comprising material 1. On the contrary, we can also understand the process as that the material points in

Z. Liu, *Multiphysics in Porous Materials*,
https://doi.org/10.1007/978-3-319-93028-2_17

material 2 also generate a total electric field, which exerts forces on the material points in material 1. From this description, we can infer three things:

1. The occurrence of such forces requires the presence of both charged particles and an external electric field.
2. The forces are mutual, and thus the description is also mutual when there are two charged objects.
3. The forces on macroscopic objects stem from the forces on the material points. Thus an explanation of electrostatics using such point charges is more fundamental.

Without involving magnetics and heating, the electric field is mostly noticeable via the forces it generates. In many cases, electrostatically induced forces could appear to be rather weak, such as the force between the plastic wrap and hands. However, when compared with the gravitational force, the electrostatic force between an electron and a proton, which together make up a hydrogen atom, is about 36 orders stronger than gravity in magnitude. Another factor which makes the effects of charge exchanges during contacts less noticeable is that the electrons gaining or losing in the contact-and-separation process remain at site only in materials with a high resistance to electric currents. This is because only such materials can prevent the electrons or cavities from moving away or being filled within a time which is long enough to be observed. These electrons have a tendency to move away also because of electric fields: each electron generates an electric field and pushes away other nearby electrons. In some cases, quick neutralization could happen if two oppositely charged objects with high resistance meet, leading to a dramatic discharging process such as the static 'shock' or lightning.

In this chapter, we will first introduce the forces between electrically charged particles. Coulomb's law for describing these forces will be discussed. Next, we will move to the forces that a charged particle undergoes in an electric field. Then, Gauss's law will be introduced to link the amount of the electric charge to the electric field generated by the charged particles. Followed will be the derivation of the governing equation of electrostatics in terms of the electric potential. We will also show the energy associated with the electric field. Electric displacement field and dielectric materials will be explained to guide the application of electrostatics in materials. A practice problem will be given in the final section to demonstrate the numerical simulation of electrostatics.

Coulomb's Law

Let us first take a look at the electrostatic force between two objects. Charles-Augustin de Coulomb reported that the magnitude of the electric force between two point charges is directly proportional to the product of the charges and inversely proportional to the square of the distance between them based on his test for studying the repulsion and attraction forces of charged particles using a torsion balance in 1785.

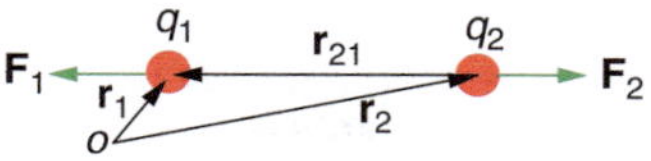

Fig. 17.1 Forces on two close point charges

For two charges q_1 and q_2, Coulomb's law can be written out using the following equation:

$$F = \frac{1}{4\pi\varepsilon_0}\frac{q_1 q_2}{r^2} = k_e\frac{q_1 q_2}{r^2}, \tag{17.1}$$

where ε_0 is the vacuum permittivity,r is the distance between two charges, and k_e is Coulomb's constant. The vacuum permittivity and Coulomb's constant have the following values:

$$\varepsilon_0 = \frac{10^{-9}}{36\pi}\mathrm{C^2N^{-1}m^{-2}} \approx 8.854187817 \times 10^{-12}\mathrm{C^2N^{-1}m^{-2}} \tag{17.2}$$

$$k_e = \frac{1}{4\pi\varepsilon_0} \approx 8.987551787 \times 10^9\mathrm{C^{-2}N\,m^2}. \tag{17.3}$$

This equation also can be written in terms of vectors considering that forces are vectors as shown in Fig. 17.1

$$\mathbf{F}_1 = k_e\frac{q_1 q_2}{|\mathbf{r}_{21}|^3}\mathbf{r}_{21}, \tag{17.4}$$

where $\mathbf{F}_1$ is the force applied on q_1 by q_2 and $\mathbf{r}_{21}$ is the vectorial distance pointing from q_2 to q_1:

$$\mathbf{r}_{21} = \mathbf{r}_1 - \mathbf{r}_2, \tag{17.5}$$

where $\mathbf{r}_1$ and $\mathbf{r}_2$ are the vectors representing the position of the two point charges in a given frame of axes. Therefore, if q_2 to q_1 are positive, the direction of the force on q_1 will be the direction of the vector pointing from q_2 to q_1.

The rule of superposition can be applied to the above electrostatic forces. For example, if there exist other point charges, $q_3, \ldots, q_n$, the electrostatic force on the point q_1 is

$$\mathbf{F}_1 = \sum_{i=2}^{n} k_e\frac{q_1 q_i}{|\mathbf{r}_{i1}|^3}\mathbf{r}_{i1}. \tag{17.6}$$

Electric Field

The electric field can be defined based on the above understanding of the electrostatic force. In the above example, if the point charge q_1 changes its position, then its position and $\mathbf{r}_{i1}$'s will also change. As a result, the force on q_1 will change. But if the point charge remains at the same location, q_1 will undergo the same force no matter how the charge moves to the location. Therefore, each point in the space has a unique quantity which determines the magnitude and direction of the force acting on q_1 and this quantity will change with locations. This quantity thus demonstrates the characteristic of a physical field. Therefore, we can define a physical field for this quantity in the following way:

$$\mathbf{E} = \sum_{i=2}^{n} k_e \frac{q_i}{|\mathbf{r}_{i1}|^3} \mathbf{r}_{i1}. \tag{17.7}$$

This is, in fact, the definition of the electric field caused by all the point charges except q_1. We can see that this electric field is independent of q_1. Therefore, an electric field is independent of the charge that we place in the field if we overlook the electric field generated by this charge. To reach a more general definition, let us take a look at the electric field generated by n charges: point charges $q_1,\ldots,q_n$. The total electric field generated by these charges is

$$\mathbf{E} = \sum_{i=1}^{n} k_e \frac{q_i}{|\mathbf{r}_i|^3} \mathbf{r}_i, \tag{17.8}$$

where $\mathbf{r}_i$ is the unit vector pointing from point q_i to the point of interest. The above $\mathbf{E}$ is the total electric field: the superposition of the electric fields generated by all the point charges. In a similar way, we can define the electric field generated by the point charge q_i as

$$\mathbf{E}_i = k_e \frac{q_i}{|\mathbf{r}_i|^3} \mathbf{r}_i. \tag{17.9}$$

If q_i is a positive source point charge, then the direction of the electric field points along lines directed radially outward from the charge. On the contrary, a negative q_i generates an electric field with a radially inward direction.

With the above definition of the electric field $\mathbf{E}$, we then can easily write down the force acts on any given point charge q as

$$\mathbf{F} = \mathbf{E}q. \tag{17.10}$$

The above definition of the electric field can be graphically represented using electric field lines as shown in Fig. 17.2. Electric field lines are lines with arrows pointing from positive and/or pointing to negative charges. Electric field lines are parallel to the direction of the electric field, and the density of the lines at any point indicates the magnitude of the electric field therein. We can further define the electric flux as a measure of the flow of the electric field through a given area. Graphically, the electric flux is proportional to the number of electric field lines going through a normally perpendicular surface. Another field can be developed by combining this proportionality and the electric field: electric displacement field $\mathbf{D}$, which will be introduced later. $\mathbf{D}$ quantifies the direction and density of the electric field lines

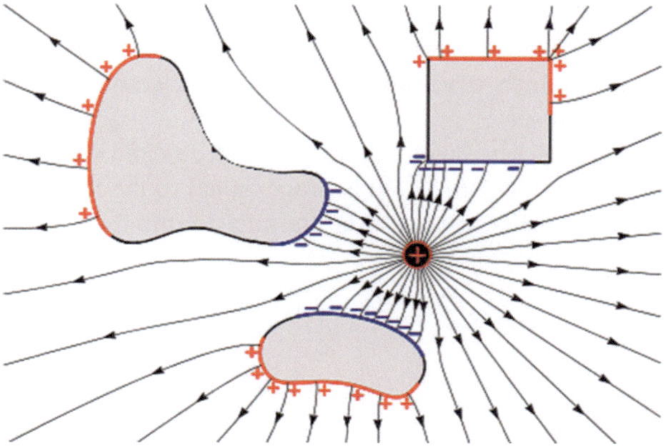

Fig. 17.2 Conductors in the electric field generated by a point charge

directly without a proportionality. As a result, **D** is also called the electric flux density with a unit of C/m^2. Since the electric field, **E**, is a vector field defined everywhere except at the location of point charges, electric lines theoretically should not get into these points.

One thing that needs to be pointed out is that there are no electric field lines inside conductors even if they are placed in an electric field. This reason is that perfect conductors have no resistance to the movement of electrons. Therefore, when it is placed in an electric field, the electrons in the conductors will feel the electric force and move immediately. Based on the direction of the force, the positive charges will move along the electric field lines while negative charges move against the direction of the lines. As shown in the above figure, they will move to the peripheries of the conductors. The movement of charges will cease when the charges on the periphery of a conductor are able to generate an electric field which is strong enough to cancel out the external electric field within the conductor. As a result, there is no electric field or electric field lines in the conductor.

The above electric field is defined with discrete charges. In continuous materials, it is more common to use the charge density ρ, which could change with position **r**. In this case, we can define the electric field generated by a continuous charge "cloud" with a charge density ρ at a point with a position vector $\mathbf{r}_0$ as

$$\mathbf{E}(\mathbf{r}) = \frac{1}{4\pi\varepsilon_0} \int \frac{1}{\|\mathbf{r}_0 - \mathbf{r}\|^2} \rho(\mathbf{r}) d\mathbf{r}. \tag{17.11}$$

Gauss's Law

In the above section, we showed how to define the electric field based on the forces between point charges and extended this definition to continuous materials. In fact, the above definition can be further generalized into a relationship between charges and the electric field: Gauss's law. Gauss's law can serve as a key relationship in establishing the governing equation for the electric field. The law relates the electric charge, which serves as the source in the electric field, to the electric field, either **E** or **D**, which serves as the flux.

Gauss's law states that the total electric flux through any closed surface of any shape in an electric field is proportional to the total electric charge enclosed within the surface. This law can be formulated into the following form:

$$\int_{\Omega} \mathbf{E} \cdot \mathbf{n} d\Omega = \frac{q}{\varepsilon_0} = \int_{\Omega} \frac{\rho}{\varepsilon_0} dV. \tag{17.12}$$

Gauss's law applies to both 2D and 3D, so Ω can be a 2D domain (area) or a 3D one (volume). Accordingly, ρ can be both a surface charge density and a volume charge density. The above equation in the integral form can also be reformatted into an equivalent equation in the differential form:

$$\nabla \cdot \mathbf{E} = \frac{\rho}{\varepsilon_0}. \tag{17.13}$$

This theorem was proposed by Gauss based on his study of the electric flux. The Gauss theorem states a physical and mathematical law which has more general implications in physical fields: the outward flux of a vector field through a closed surface is equal to the volume integral of the divergence over the region inside the surface.

Governing Equation and Electric Potential

The above equation for Gauss's law cannot be used as the governing equation for the electric field. From the physical perspective, as mentioned in the previous section, Gauss's law serves more as a constitutive relationship for relating the source to the electric field flux. From the mathematical perspective, the equation relates a vector, i.e., **E** or **D**, to a scalar, i.e., ρ. Gauss's law provides one equation for multiple dependent variables. Each component of the vector-dependent variable counts as one dependent variable, and the order of quantity of either the left- or right-hand side of the equation determines the number of governing equations (although in many cases, we treat the number of PDEs as the number of governing equations).

The governing equation of the electric field stems from the observation that the electric field is irrotational or equivoluminal, mathematically, that is,

$$\nabla \times \mathbf{E} = \mathbf{0}, \tag{17.14}$$

where "**0**" is a vector. So in a 3D case, the above PDE actually represents three dependent variables, i.e., three components of **E**, and three governing equations for the three components. The above equation is a special case of Faraday's law of induction in electrostatics. We can simulate the electric field with the above governing equation (here we still count it as one following the convention used in other places of this book) and the Gauss's law. However, the practice may be not easy because we are dealing with two separate equations containing the same vector. We usually need to calculate **E** caused by a separate charge or charge cluster at special locations and then use these **E** values as boundary conditions together with the governing equation to calculate **E** throughout the domain.

A more convenient and common way for the use of the governing equation in the electric field is to combine the above two first-order PDEs into one second-order PDE with a new scalar-dependent variable. This is, in fact, a very common practice in both mathematics and physics. First, for any irrotational physical field, we can relate the vector field, i.e.,**E** in this example, to a scalar variable, which usually has the meaning of a potential, in the following way:

$$\mathbf{E} = -\nabla \psi, \tag{17.15}$$

where ψ has a physical meaning electric potential with a unit of volt (V) in this case. The electric potential difference is the "voltage" that we use widely in daily life. Therefore, the electric field has a unit of V/m. The magnitude of the electric field at a particular location is usually called the electric field strength. The negative sign is added considering that the electric field lines point from locations with a high potential to those with a low potential.

If we write the above equation in terms of a line integral, then we can obtain the conclusion that the voltage is the amount of work required to move a unit charge from the start point to the end point of the line:

$$-\int_a^b \mathbf{E} \cdot d\mathbf{l} = \psi(\mathbf{b}) - \psi(\mathbf{a}), \tag{17.16}$$

where a and b are the start and end points.

In fact, the above assumption that the electric field is proportional to the gradient of another scalar field is not just a postulate. It can be deducted based on the observation that the electric field is irrotational, mathematically, that is,

$$\nabla \cdot \mathbf{E} = 0. \tag{17.17}$$

This can also be seen from a key feature of the electric field: work is path independent, or the choice of any path between two points does not change the value of the line integral.

The definition of the electric potential, combined with the differential form of Gauss's law, provides the following governing equation for the electric field:

$$\nabla^2 \psi = -\frac{\rho}{\varepsilon_0}. \tag{17.18}$$

This equation is in the form of Poisson's equation. In the absence of electric charges, the equation degenerates into Laplace's equation:

$$\nabla^2 \psi = 0. \tag{17.19}$$

This PDE can be solved with appropriate boundary conditions. A Dirichlet boundary condition for an electric field just gives out the electric potential on the boundary. For example, when the boundary is far enough from charges, we can assume the boundary is infinitely far and thus has an electric potential of zero. By contrast, a Neumann boundary condition places a constraint on the gradient of the electric potential, i.e., electric field. For example, for any conductor in an electric field, the direction of the electric field needs to be parallel to the outward normal of the surface. Therefore, the component of the electric field parallel to the surface (or perpendicular to the outward normal vector) will be zero. More complicated boundary conditions will be discussed in the chapter for electromagnetics in the next part of the book.

Electrostatic Approximation

From another perspective, electrostatics can be viewed as a special case of electromagnetics which is more common in reality. The validity of the electrostatic approximation rests on the assumption that the electric field is irrotational. From Faraday's law, this assumption implies the absence or near absence of time-varying magnetic fields:

$$\frac{\partial \mathbf{B}}{\partial t} = 0, \tag{17.20}$$

where **B** is the magnetic field.

Therefore, electrostatics does not require the magnetic field or electric current to be totally absent. Rather, the magnetic field or electric current, if exists, must not change with time. When the magnetic field or electric current changes with time very slowly, we may also be able to simplify the problem into electrostatics with an acceptable loss in accuracy.

Electrostatic Energy

Energy is an important aspect of a physical field. The energy of the previously introduced physical fields is mostly straightforward, either in terms of mechanical work, potential energy, strain energy, or chemical potential. However, the energy of the electric field is slightly different and needs discussions. This energy is usually needed when the electric field is coupled to other fields such as the thermal field.

To obtain a better understanding, let us still start with point charges. The simplest scenario is to consider the energy of a point charge q in an external electric field. The total energy or, more accurately, the electric potential energy, owned by this point charge is equal to the energy that is needed to move this charge from an infinitely far location (with zero electric potential) to the current location:

$$U = q\psi. \tag{17.21}$$

If ψ is positive, i.e., higher than the potential of the start point, then extra work is needed to make the charge move against the electric field line. Accordingly, the potential energy owned by the point charge will be positive.

In the above example, if the external field is generated by another point charge q_2 at $\mathbf{r}_2$, then the energy needed to move q, which is marked as q_1 for convenience, from an infinitely far location to a location $\mathbf{r}_1$ will be

$$U = q_1\phi_1 = q_1 k_e \frac{q_2}{|\mathbf{r}_{12}|}, \tag{17.22}$$

where $\mathbf{r}_{12} = \mathbf{r}_1 - \mathbf{r}_2$. In fact, this is not merely the energy of charge q_1 but that of the system consisting of $q_1$and q_2. In the above example, we would do the same amount of work for moving q_2 to the same relative position of q_1. In short, we need to do this amount of work to the system consisting of the two charges by moving them to positions with a distance of $|\mathbf{r}_{12}|$. Therefore, U in the above equation is the potential energy of the whole system. In the above equation, we can also write the system's electric potential energy as

$$U = \frac{1}{2}\sum_{i=1}^{2} q_i\psi_i = \frac{1}{2}\sum_{j=1}^{2}\left(q_i \sum_{j=1(j\neq i)}^{2} \frac{q_j}{r_{ij}}\right). \tag{17.23}$$

In order to extend the above equation to a more general condition, then let us consider a collection of n charged particles, i.e., $q_1, q_2, \ldots, q_i, \ldots q_n$. q_i is located at the point with a coordinate vector of $\mathbf{r}_i$. Then the electric potential energy owned by the whole system is

$$U = \frac{1}{2}\sum_{i=1}^{n} q_i\psi_i = \frac{1}{2}\sum_{j=1}^{n}\left(q_i \sum_{j=1(j\neq i)}^{n} \frac{q_j}{r_{ij}}\right). \tag{17.24}$$

Then we can extend the above energy calculation from discrete systems to continuous systems. In a continuous system, the charges are distributed in a domain consisting of elementary volumes. Then for each elementary volume dV, the energy can be calculated as

$$dU = \frac{1}{2}(\rho dV)\psi. \tag{17.25}$$

The total energy of the total domain can be obtained via integration

$$U = \int_{\Omega} dU = \int_{V} \left(\frac{1}{2}\rho\psi\right) dV. \tag{17.26}$$

The electric potential energy of a continuum can also be formulated using the electric field and permittivity. To derive the equation, we first substitute Poisson's equation into the above equation:

$$U = \int_{\Omega} dU = \int_{V} \left[\frac{1}{2}(\varepsilon\nabla \cdot \mathbf{E})\psi\right] dV. \tag{17.27}$$

Then we can take advantage of the following divergence vector identity:

$$\nabla(\mathbf{E}\psi) = (\nabla \cdot \mathbf{E})\psi + \mathbf{E}\nabla\psi. \tag{17.28}$$

Substituting Eq. 17.28 into Eq. 17.27, we obtain

$$U = \int_{\Omega} dU = \frac{\varepsilon}{2}\int_{V} \nabla \cdot (\psi\mathbf{E})dV - \frac{\varepsilon}{2}\int_{V} \nabla\psi \cdot \mathbf{E}dV. \tag{17.29}$$

Then, we apply the divergence theory:

$$U = \int_{\Omega} dU = \frac{\varepsilon}{2}\int_{V} (\psi\mathbf{E}) \cdot \mathbf{n}dS - \frac{\varepsilon}{2}\int_{V} (-\mathbf{E}) \cdot \mathbf{E}dV. \tag{17.30}$$

If we set the electric potential at the boundary of the domain to zero, for example, by assuming the boundary is at infinity, then the first term on the right-hand side of the above equation will be zero. The energy then has the following formulation:

$$U = \frac{\varepsilon}{2}\int_{V} \mathbf{E} \cdot \mathbf{E}dV. \tag{17.31}$$

Based on the above introduction, the electric potential energy density of the electric field can be formulated using either $\rho\psi/2$ or $\varepsilon E^2/2$.

Electric Displacement Field and Dielectric Materials

As introduced above, the electric field **E** multiplied by another scalar, e.g., ε, will become a new field: the electric displacement field. This field is not just randomly created with two existing physical quantities but rather a very useful quantity with a physical meaning in dielectric materials. Nevertheless, the physical meaning is not as clear as that of the measurable quantity **E**. In vacuum, **E** is caused by free charges only. However, when we consider the electric field in a normal dielectric (material), we will find the dielectric material will resist the external electric field by moving its electric dipoles. As a result, the external electric field will be weakened.

To better understand this, let us introduce polarization which quantifies this resistance. When an external electric field is applied to a dielectric material, bound charged elements within a molecule will move in response to the external electric field: positively charged elements move in the direction of the field while negatively charged elements move opposite to the direction of the field. These movements are confined in a limited distance as these elements are bound to a molecule and are thus not free to move around in the material. Therefore, the molecule remains neutral as a whole even with the movement of the charged elements within it. This movement can be quantified with a vector, polarization density, which is defined as

$$\mathbf{P} = \frac{dq_\mathrm{b}}{dV}\mathbf{d} = \rho_\mathrm{b}\mathbf{d}, \tag{17.32}$$

where the amount of the bound charge q_b within an infinitesimal volume gives out the bound charge density ρ_b and the displaced distance vector $\mathbf{d}$ gives out the direction of the polarization vector. $q_\mathrm{b}\mathbf{d}$ (unit: $\mathrm{C} \cdot \mathrm{m}$) is also called the electric dipole moment, and accordingly $\mathbf{P}$ has the physical meaning of the density of the electric dipole moment (unit: $\mathrm{C} \cdot \mathrm{m/m^3}$). The displaced charges will also generate a secondary "bound electric field," which is opposite to the original external field. We can consider this field as one generated by the bound charges, so we have

$$\nabla \cdot \mathbf{P} = -\rho_\mathrm{b}. \tag{17.33}$$

The negative sign is given to indicate the opposite direction (resisting to the external electric field). This is Gauss's law for bound charges. We can obtain the real electric field by combining the original external field and the resistance field caused by the bound charges. In vacuum ($\rho_\mathrm{b} = 0$), there is no bound charge, and thus Gauss's law gives

$$\nabla \cdot (\varepsilon_0 \mathbf{E}_0) = \rho_0 = \rho_\mathrm{f}, \tag{17.34}$$

where $\mathbf{E}_0$ and ρ_0 are the electric field and total charge density in vacuum. If we replace the vacuum with a dielectric without altering the charges, then we have

$$\nabla \cdot (\varepsilon_0 \mathbf{E}) = \nabla \cdot (\varepsilon_0 \mathbf{E}_0) - \nabla \cdot \mathbf{P} = \rho_\mathrm{f} + \rho_\mathrm{b} = \rho, \tag{17.35}$$

where $\mathbf{E}$ and ρ are the electric field and total charge density, respectively, in the new domain with the dielectric. As $\mathbf{P}$ is usually in the direction of $\mathbf{E}_0$, the electric field is weakened compared to that in the vacuum:

$$\nabla \cdot (\varepsilon_0 \mathbf{E}) + \nabla \cdot \mathbf{P} = \rho_\mathrm{f}. \tag{17.36}$$

A more common way to write the equation is to develop a new quantity, the electric displacement field:

$$\mathbf{D} = \varepsilon_0 \mathbf{E} + \mathbf{P} = \varepsilon \mathbf{E}, \tag{17.37}$$

so that

$$\nabla \cdot \mathbf{D} = \rho_{\mathrm{f}}. \tag{17.38}$$

Therefore, **D** does not change when an external field is placed on different dielectrics. We can further define the dielectric constant ε_r to quantify the permittivity of any material with reference to the permittivity of vacuum:

$$\mathbf{D} = \varepsilon \mathbf{E} = \varepsilon_0 \varepsilon_r \mathbf{E} \tag{17.39}$$

Most minerals in geomaterials have ε_r values ranging from 1 to 5, while water has a ε_r value of 80. Permittivity is the amount of charge needed to generate one unit of electric flux in a particular medium. Thus, permittivity measures a material's ability to resist an electric field, not its ability to "permit" the field, as the name "permittivity" might suggest.

As ε_r is usually larger than 1, while **D** does not change in different materials, the electric field **E** will be smaller (more weakened) in materials with a higher dielectric constant.

Practice Problem

Please simulate the electric field generated by a metal cylinder whose electric potential is 10 V. The metal cylinder has a radius of 0.2 m, and the length is much larger than the radius.

Results

Based on the above description, the problem can be simplified into a 2D case, in which the metal piece is a circular area. The permittivity of the metal can be represented using a very large number such as 1×10^{12}. Theoretically, the electric potential is zero at an infinitely far location. However, we can use a computational area that is much larger than the size of the metal area to approximate the theoretical condition. The following results (Fig. 17.3) were obtained by using a square computation area with a side length of 1 m. Considerable error can be generated due to the size and shape of the domain. Please try to conduct a sensitivity analysis to investigate the influence of the size and shape of the computational domain on the results. The area outside of the metal area is assumed to be vacuum.

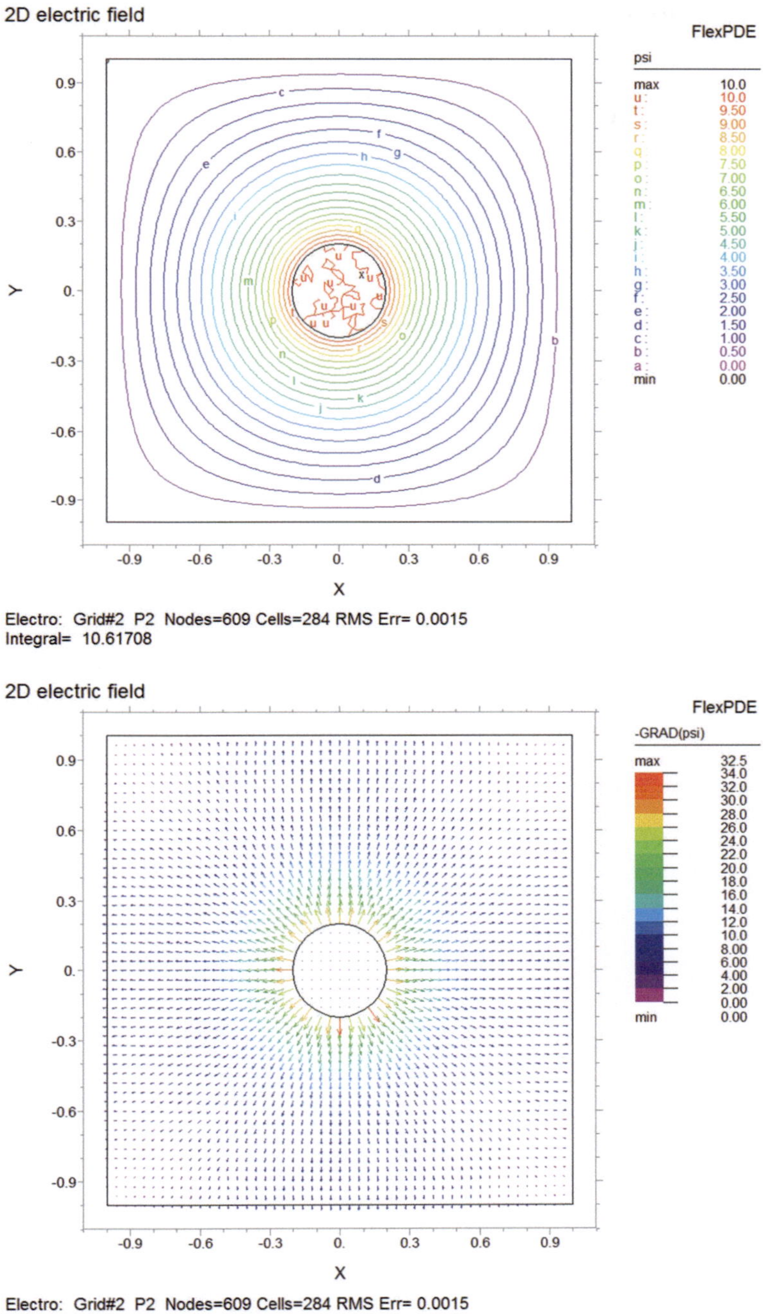

Fig. 17.3 Electric potential (upper) and electric field (lower)

Appendix: FlexPDE Script File

```
{ Electric field generated by a ring with 10 V}
TITLE '2D electric field'
VARIABLES
  psi             { Electric potential }
! SELECT
DEFINITIONS
  E=-GRAD(psi)    { Electric field }
  rho=0
  epsilon=8.654e-12
! INITIAL VALUES  { Not time-dependent }
EQUATIONS
  DIV(E)=rho/epsilon
BOUNDARIES
  REGION 1          { Vacuum }
    START(-1,-1) VALUE(psi)=0
    LINE TO (1,-1) VALUE(psi)=0
    LINE TO (1,1) VALUE(psi)=0
    LINE TO (-1,1) VALUE(psi)=0
    LINE TO CLOSE
REGION 2        { Charge area: a metal cylinder - a circular plate in 2D }
  epsilon=1e12 { Use a large permittivity value to simulate
conductors }
  START(0.2,0) VALUE(psi)=10 ARC(CENTER=0,0) ANGLE=360 TO CLOSE
! TIME 0 TO 1 { Not time-dependent }
MONITORS
PLOTS
CONTOUR(psi)
  VECTOR(-GRAD(psi))
END
```

Chapter 18
Magneto: Magnetostatics

Introduction: Magnetostatics and Magnetic Field

The study of the magnetic field originated from the observations on the magnets and the forces they place on objects made of iron and other metals. The study of these phenomena, i.e., magnetism, has extended far beyond magnets and has been entangled with the study of the electric field in electromagnetics. Similar to the electrostatics, we here try to separate the magnetic field from electromagnetics by confining the systems to a stationary state (no change with time) or changes with low velocities, i.e., currents do not alternate rapidly. With the inclusion of forces, magnetostatics can then be viewed as the magnetic analogy of electrostatics.

Similar to the electric field, there are also two physical fields defined to describe the magnetic field, **H** field and **B** field, which correspond to the **E** field and **D** field in the electrostatics, respectively. However, the use of **H** field and **B** field is possibly more confusing than their compartments in the electric field. Both **H** field and **B** field can be called the magnetic field, depending on the disciplines. It is more common to call **H** the magnetic field (or magnetic field strength/intensity) because the term "magnetic field" was historically reserved for the early proposed **H**. However, many researchers, especially in physics, believe **B** is more fundamental than **H**, so **B** is called the magnetic field. **H** has a unit of A/m, which can be compared to the unit of **E**: V/m. Similar to **D**, which is the electric flux density with the unit of C/m^2, **B** is the magnetic flux density with the unit of Wb/m^2 (webers per square meter, or tesla). The magnetic field lines are a tool to visualize the magnetic field, and their density is measured in terms of Wb.

The magnetic field is generally more difficult to understand than the electric field. In the special relativity of Albert Einstein, the electric field and the magnetic field are just different aspects of one thing, i.e., electromagnetics, and thus one may turn into the other depending on the observational frame of reference. But from a simpler way, we can understand the magnetic field as the result of two things: moving electrons and spin magnetic moments of elementary particles (particles whose substructure is

Z. Liu, *Multiphysics in Porous Materials*,
https://doi.org/10.1007/978-3-319-93028-2_18

unknown, e.g., quarks and leptons). The latter is in general much less obvious than the former except for applications such as nuclear magnetic resonance. For electrons, a rotating electron creates a magnetic dipole with two magnetic poles of equal magnitude but opposite polarity. Assuming an electron rotates about an axis with an angular momentum of **L**, the generated magnetic dipole moment **m** is

$$\mathbf{m} = \frac{-e}{2m_e}\mathbf{L}, \tag{18.1}$$

where e and m_e are the charge (elementary charge) and rest mass of the electron. The magnetic dipole moment can also be equivalently defined with a sufficiently small loop of current, I, and loop area, **A**:

$$\mathbf{m} = I\mathbf{A}. \tag{18.2}$$

The above two definitions are equivalent if the loop of current is caused by the charge: $I = \frac{e}{T} = \frac{2\pi e}{\omega}$, in which ω is the angular velocity. The direction of the vectoral area **A** needs to be the same as the direction determined using the right-hand rule with the loop current, as shown in Fig. 18.1.

Usually, electrons in a material are arranged in a way that their magnetic moments, both orbital and intrinsic, cancel out when the ensemble of the electrons are viewed as a whole. But in some conditions, such as in the mineralization of magnetite or under an applied external magnetic field, the electron magnetic moments will be lined up toward one direction to produce a strong net magnetic field. The induced magnetic field could remain to some extent even after the removal of the external magnetic field, leading to permanent magnets.

Magnetic dipoles are commonly introduced when discussing magnetism. This concept is analogous to yet much different from the positive and negative charges in the electric field. One major difference is that the existence of the magnetic monopole has still not been validated. A common observation is that every magnet comes with a dipole: a "south pole" and a "north pole." If we cut a magnet into two pieces, then each piece will also own its "south pole" and a "north pole." This can be explained with the origin of the magnetic dipoles from the rotating charges (Fig. 18.1). Each rotating charge develops a dipole on two sides of the plane of rotation. We thus cannot separate the two poles as a plane always has two sides. A macroscopic magnetic material consists of numerous dipoles like this. If we cut the

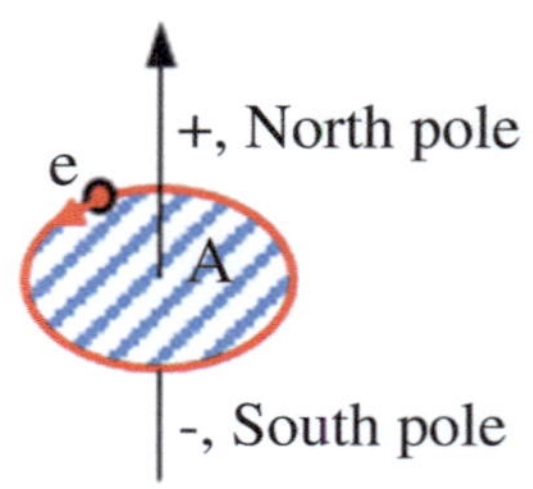

Fig. 18.1 Move electric charges (current) and resultant magnetic dipole

material into two pieces, the dipole ensembles in each piece will still own a magnetic field with the same direction.

In this chapter, we will first discuss the forces associated with a charge in a magnetic field to introduce the concept of the magnetic field (**B** field) just like what we did in the previous chapter for the electric field. Then, we will present a mathematical description of the magnetic field using two different methods relying on the **H** field and **B** field. Gauss's law for magnetism will be given to reach the governing equation for the magnetic field. A vector potential will be introduced to turn the governing equation into a second-order partial differential equation, which is useful and common in numerical analysis. The energy associated with the magnetic field will be introduced afterward. A practice problem will be provided in the final section for simulating the magnetic field generated by a permanent magnet.

Magnetic Field

The magnetic field can be defined in many equivalent ways. One common way is via the study of the force on a moving charged particle in the vicinity of a current-carrying wire. Based on the previous chapter, we know that a particle with charge q in an electric field $\mathbf{E}$ experiences a force $\mathbf{F} = q\mathbf{E}$. However, we can also easily discover via simple experiments that the force also depends on the velocity of the particle. If we incorporate this additional part in the force, we would obtain the Lorentz force law:

$$\mathbf{F} = q\mathbf{E} + q\mathbf{v} \times \mathbf{B}, \tag{18.3}$$

where $\mathbf{v}$ is the particle's velocity. We can understand the first term as the forces between static charges while the second term as the forces between non-stationary charges. For the second term, the electric current in the electric wire generates a magnetic field, which exerts a force on the moving charge. If the field is defined using the second term in the above Lorentz law, then this field, i.e., **B** field, is the way to quantify the magnetic field.

Another common way to define the magnetic field is to study the torque that a magnetic field applies on a magnetic dipole.

As shown in the above figure, we conceptualize the magnetic dipole as two connected magnets with a distance $\mathbf{d}$ and an equal pole strength of q_{m}, within which one represents the south pole and the other represents the north pole. As shown in Fig. 18.2, these two monopoles can be understood as charges with different signs in an electric field: north corresponds to positive, whereas south corresponds to negative. Therefore, the positive (north) pole will move along the direction of the magnetic field lines, while the south will move against the direction of the lines. As a result, a torque develops on the dipole. This torque, in fact, explains why a compass rotates to align itself with Earth's magnetic field. In this example, we find that the torque is proportional to the pole strength q_{m} (unit: $\mathrm{A} \cdot \mathrm{m}$), the distance between the

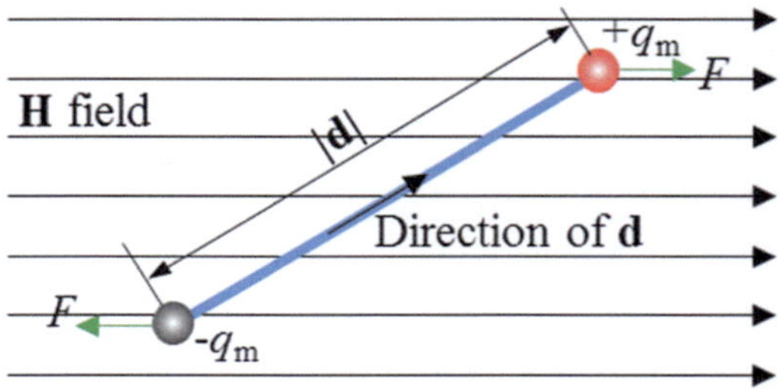

Fig. 18.2 Torque on a magnetic dipole in a magnetic field

poles |**d**|, and the external magnetic field. The magnetic field defined using the following equation is also the **B** field

$$\boldsymbol{\tau} = q_{\mathrm{m}}\mathbf{d} \times \mathbf{B}, \tag{18.4}$$

where **d** is a vectoral distance pointing from the south (negative) pole to the north pole. Accordingly, based on the right-hand rule for the cross product, the direction of the torque is pointing inward.

We can observe a major difference between the electrostatics and magnetostatics in the above two ways of defining the magnetic field: **E**is the field directly relating an electric charge to a force in electrostatics, and **B** is the field directly relating a magnetic dipole to a force in the magnetostatics. However, as mentioned, **E** is analogous to **H** instead of **B**. As the counterpart of **D**, **B** is more fundamental and does not change in different materials. Therefore, for a magnet in vacuum, **B** field lines start from the north pole, extend into the vacuum, and return to the south pole; the lines will continue without any change inside the magnet (a material) until they get back to the north pole. A more accurate and common statement is that "**B**field lines neither start nor end." In electrostatics, charges of different signs can exist separately, i.e., no requirement of electric dipoles, so **D** field lines, i.e., electric field lines, do not have the above feature. However, just like the **E** field, the **H** field will alter in materials due to the resistance of the material to the external magnetic field. Inside the above magnetic, the magnetic materials will raise resistance to alter the **B** field. The part of the **B** field excluding the resistance raised by the magnetic materials is the **H** field. Outside of the magnetic field, there is no magnetic material, neither is resistance; therefore, the **H** field is identical to the **B** field outside of the magnet. The resistance depends on the type and geometry of the magnetic materials and is thus very difficult to calculate.

This resistance varies from location to location, so it is also a physical field: magnetization. Magnetization **M** is also called magnetic polarization. **M** is analogous to the electric polarization**P**: **P** is the density of the electric dipole moment (unit: $\mathrm{C} \cdot \mathrm{m/m^3}$), while **M** is the density of the magnetic dipole moment (unit:$\mathrm{A} \cdot \mathrm{m^2/m^3}$). The relationship between the two types of magnetic fields and magnetization can be mathematically formulated in the following way:

$$\mathbf{B} = \mu_0(\mathbf{H} + \mathbf{M}), \tag{18.5}$$

where μ_0 is the permeability in vacuum, in which $\mu_0 = 4\pi \times 10^{-7}$.

In nonmagnetic materials, in which paramagnetism or diamagnetism dominates, **M** is often proportional to the applied magnetic field. Accordingly, we define

$$\mathbf{M} = \chi\mathbf{H}, \tag{18.6}$$

where χ is the magnetic susceptibility, which is a dimensionless quantity characterizing a magnetic material by telling whether a material is attracted into or repelled to the magnetic field. The magnetic susceptibility can be defined in many equivalent forms; while the definition in the above equation is the most common form: volume magnetic susceptibility. A positive value of χ means the magnetic field will be strengthened in the material and the negative is verse visa. With the magnetic susceptibility, we have the following relationship between **B**, **H**, and **M**:

$$\mathbf{B} = \mu_0(\mathbf{H} + \chi\mathbf{H}) = \mu_0(1 + \chi)\mathbf{H} = \mu_0\mu_r\mathbf{H}, \tag{18.7}$$

where μ_r is the relative permeability.

In a simple way, we can just write

$$\mathbf{B} = \mu\mathbf{H}, \tag{18.8}$$

where μ is the absolute permeability of the material. The above two equations are introduced and referenced widely without specifying the underlying assumption: a relationship, mostly linear, between **B**(**H**) and **M**. It is needed to point out this assumption works for most cases. However, for permanent magnets, who owes **M** without external magnetic field, we need to start from Eq. 18.5.

Magnetic materials are materials that are attracted to a magnet strongly enough that the attraction can be felt. Magnetic materials can be classified as hard and soft magnetic materials – the former can gain a permanent magnetic field after being placed in an external magnetic field, while the latter cannot. A permanent magnet is made by placing a hard magnetic material such as alnico and ferrite into a strong magnetic field, whose magnetic dipoles will be aligned to a direction (on average) and maintain part of the aligned dipoles after the external magnetic field is removed. Soft magnetic materials like annealed iron do not tend to stay magnetized after the external magnetic field is removed. Magnetic materials can also be categorized into ferromagnetic, ferrimagnetic, and antiferromagnetic according to their electron configuration and consequently the way how the magnetic moments of atoms or molecules (usually related to the spins of electrons) response to an external magnetic field. Magnetic materials like iron can have a volume magnetic susceptibility of 200,000, whereas common nonmagnetic materials have absolute values of χ ranging from 10^{-10} to 10^{-4}. In nonmagnetic materials, a positive χ indicates a paramagnetic material, while a negative χ indicates a diamagnetic material, though χ is very small in both cases.

Gauss's Law for Magnetism

Gauss's law can also be applied to the magnetic field. The Gauss's law in the electric field states that the electric density flux is caused by the charges which serve as sources/sinks emitting/adsorbing the flux. When we move to the magnetic field, one major difference is that there is no magnetic monopole like the positive or negative charges which can exist independently. As a result, the **B** field lines never end. In other words, the **B** field is a sourceless or a solenoidal field. For a sourceless field such as **B**, the following equation is valid:

$$\nabla \cdot \mathbf{B} = 0. \tag{18.9}$$

In vacuum, since the **H** field is related to the **B** field via a constant proportionality, the above equation is also valid for the **H** field as follows:

$$\nabla \cdot \mathbf{H} = 0. \tag{18.10}$$

However, when getting into materials, the proportionality μ may vary from point to point depending on the materials and geometry. It is also noted that μ is not necessarily a constant depending on the type of the magnetic material and the range of **B**. Consequently, in the presence of magnetic materials, Eq. 18.10 is not valid anymore, and we can only rely on Eq. 18.9 for **B**.

As explained in the previous chapter, Eq. 18.9 (or Eq. 18.10 in vacuum) is more like a constitutive equation and cannot be used as the governing equation. To describe the magnetic field without any considerable changes in the electric field, i.e., magnetostatics, we need to refer to the following equation as the governing equation:

$$\nabla \times \mathbf{H} = \mathbf{J}, \tag{18.11}$$

where **J** is the current density. This equation presents a special case of Ampère's law, in which any changing external electric field is excluded. In the case of permanent magnets, if no electric currents exist, then the following governing equation can be used:

$$\nabla \times \left(\frac{\mathbf{B}}{\mu_0} - \mathbf{M} \right) = \mathbf{0}. \tag{18.12}$$

This is because magnetization (magnetic polarization) exists without external electric currents or a magnetic field. This permanent magnetization is quantified with **M**, which usually has different magnitudes along different directions as magnets are polarized. A good bar magnet may have an average magnetization magnitude of 100,000 A/m along one direction. However, magnetic materials such as ion can have a magnetization of around a million A/m under external magnetic fields.

However, it is also very common to convert the above equation into a second-order PDE by means of introducing a potential, just like the electric potential for the electric field. This will be described in the following section.

Governing Equation and Auxiliary Equations in Terms of Vector Potential

The magnetic field can be formulated using the magnetic vector potential. Different from the electric field, which is irrotational, the magnetic field is equivoluminal (sourceless or solenoidal). That is, the magnetic flux density satisfies a zero-divergence condition; therefore, **B** can be represented as the curl of another vector such that

$$\mathbf{B} = \nabla \times \mathbf{A}, \tag{18.13}$$

where **A** is the magnetic vector potential. Substituting the above equation into Gauss's law for the magnetic field, we obtain

$$\nabla \cdot (\nabla \times \mathbf{A}) = 0. \tag{18.14}$$

As can be seen, this equation is automatically satisfied, which confirms the effectiveness of the use of the magnetic vector potential.

The above equation needs to be substituted into Ampère's law to yield the governing equation in terms of the vector potential:

$$\nabla \times \left(\frac{\nabla \times \mathbf{A}}{\mu_0} + \mathbf{M} \right) = \mathbf{J}. \tag{18.15}$$

If the linear relationship between **B**and **H** can be assumed, we can obtain

$$\nabla \times \left(\frac{\nabla \times \mathbf{A}}{\mu_0 \mu_\mathrm{r}} \right) = \mathbf{J}. \tag{18.16}$$

As we know, except in metals, electric currents can be ignored in materials. Then the above equation is reduced into

$$\nabla \times \left(\frac{\nabla \times \mathbf{A}}{\mu_0 \mu_\mathrm{r}} \right) = \mathbf{0}. \tag{18.17}$$

The above standard governing equation can be further simplified by introducing extra constraints on **A**. First, let us use the absolute permeability:

$$\nabla \times \left(\frac{1}{\mu} \nabla \times \mathbf{A} \right) = 0. \tag{18.18}$$

Then we apply Lagrange's formula to transform the above equation into

$$\nabla \times \left(\frac{1}{\mu}\nabla \times \mathbf{A}\right) = \nabla\left(\frac{1}{\mu}\nabla \cdot \mathbf{A}\right) - \nabla \cdot \left(\frac{1}{\mu}\nabla \mathbf{A}\right) = 0. \tag{18.19}$$

As **A** is a quantity that we construct, we can impose any further constraints when constructing **A** as long as the above conditions are not violated. For example, we can set $\nabla \cdot \mathbf{A} = 0$, which is called the Coulomb gauge condition. Then, we have

$$\nabla \cdot \left(\frac{1}{\mu}\nabla \mathbf{A}\right) = 0. \tag{18.20}$$

For permanent magnets, permanent magnetization exists, while electric currents are absent, so we have

$$\nabla \times \left(\frac{\nabla \times \mathbf{A}}{\mu_0} - \mathbf{M}\right) = \mathbf{0}, \tag{18.21}$$

or with the Coulomb gauge condition as

$$\nabla \cdot \left(\frac{1}{\mu}\nabla \mathbf{A}\right) - \nabla \times \mathbf{M} = \mathbf{0}. \tag{18.22}$$

The introduction of this vector potential does not appear as useful and common as the electric potential in the electric field. However, the vector potential will help a lot when we move from magnetostatics to electromagnetics.

Magnetostatic Energy

The energy of the magnetic field can be obtained via the analogy between the electric and magnetic fields. The magnetic field also has q_{m} and **A**. However, these two quantities are not as widely used as q and ϕ in the electric field, so here we do not discuss the counterpart of $\rho\phi/2$ in the magnetic field. Instead, we can still easily derive the energy density function in Eq. 18.23 based on the analogy between **E/D** and **H/B**. Corresponding to the electric field, the magnetic potential energy of a continuum can be formulated as

$$U = \frac{1}{2}\int_V \mathbf{H} \cdot \mathbf{B} dV = \frac{\mu}{2}\int_V \mathbf{H} \cdot \mathbf{H} dV = \frac{1}{2\mu}\int_V \mathbf{B} \cdot \mathbf{B} dV. \tag{18.23}$$

The electric potential energy density of the magnetic field can be formulated using either $\mu H^2/2$ or $B^2/(2\mu)$, in which H and B are the magnitudes of the **H** and **B** field, respectively.

Practice Problem

Problem Description

Please simulate the magnetic field generated by a bar magnet that is 0.4 m long and 0.2 m wide (2D) in vacuum. The magnet has magnetic polarization along the length direction whose magnitude is 10,000 A/m. The relative permeability of the magnet material is 5000.

Results

The magn etic field generated by the magnet will be symmetric. So we can analyze only one half or one fourth of the magnet to obtain the magnetic field. Here, if we analyze one half of the magnet, we will obtain the following results for the magnitude of the vector magnetic potential **A** and magnetic flux density **B** (Fig. 18.3).

Appendix: FlexPDE Script File

```
{ Magnetic field generated by a permanent magnet with magnetization
(M) along one direction }
TITLE '2D Magnetic Field'
VARIABLES
  A             { Use magnetic potential }
DEFINITIONS
  mu0=4*pi*1e-7
  mu=1*mu0
  px=0
  py=0
  P=VECTOR(Px,Py)
  B=CURL(A)
EQUATIONS
  CURL(B/mu-P)=0
BOUNDARIES
  REGION 1      { Vacuum }
    mu=1*mu0
    START(0,0) NATURAL(A)=0
    LINE TO (1,0) VALUE(A)=0
    LINE TO (1,1) NATURAL(A)=0
    LINE TO (0,1) VALUE(A)=0
    LINE TO CLOSE
```

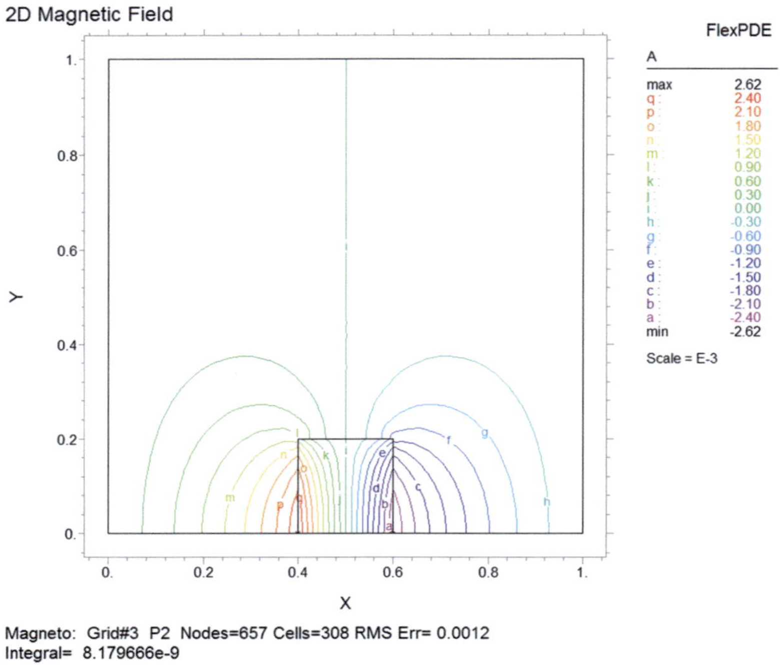

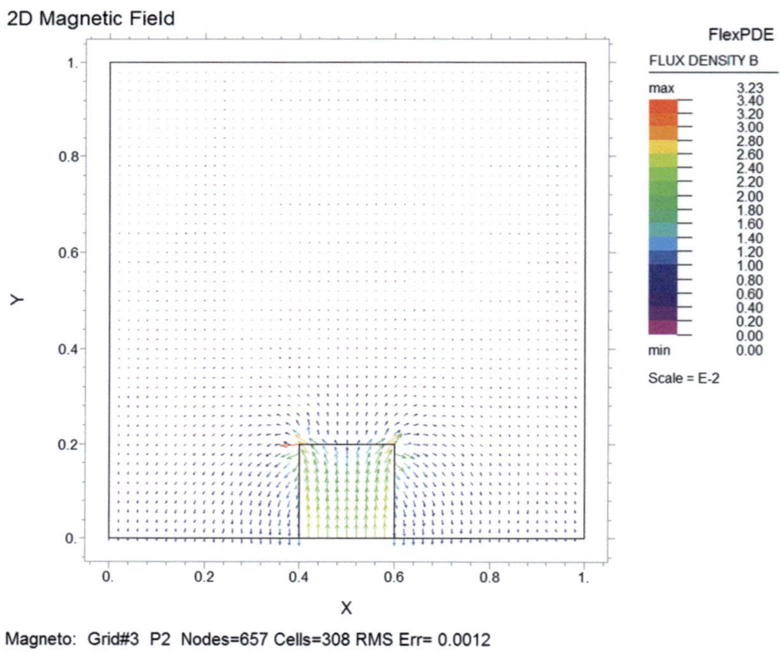

Fig. 18.3 Magnitude of vectoral potential (upper) and B field (lower)

```
  REGION 2     { Magnet }
    mu=5000*mu0
    px=0
    py=10000
    START(0.4,0)
    LINE TO (0.6,0) TO (0.6,0.2) TO (0.4,0.2) TO CLOSE
MONITORS
PLOTS
  CONTOUR(A)
  VECTOR(CURL(A)) AS 'FLUX DENSITY B'
END
```

Part IV
Multiphysics

Chapter 19
Thermomechanics: Non-isothermal Mechanics

Introduction

The coupling between the thermal and mechanical fields leads to one of the most common types of multiphysics: thermomechanics. The coupling could be two-way. However, analysis with one-way coupling is more common. If a thermomechanical process is simply analyzed using a one-way coupled field, then it is most likely that it is about the changes in the mechanical field caused by heat transfer. These changes could be either thermal stresses or thermal strains, depending on the constraints in the mechanical field. This is because most thermomechanical processes of concern are caused by the changes in the thermal field, e.g., heat flux, heat sources, and uneven temperature distributions. Such excitations originating in the thermal field will change the way that the molecules or atoms constituting the material interact with each other and consequently the stresses and/or strains at the macroscopic scale.

Two-way couplings between thermal and mechanical fields are also possible but much less common. This two-way coupling requires to additionally consider the fact that mechanical energy can convert into heat via friction or plastic deformation. As for friction, we may all have the experience that our palms become warmer when being rubbed against each other. This can also occur in fluids as the viscosity of the fluids could turn the kinetic energy into thermal energy. This process is usually termed dissipation. As for plastic deformation, one typical example is that noticeable heat can be generated by repeatedly bending and restoring a thin aluminum plate.

In this chapter, basic concepts in thermomechanics will be introduced first according to the direction of the coupling. The mathematical descriptions of two different types of couplings in thermomechanics will be laid out: thermal expansion/ stress as the coupling from the thermal field to the mechanical field and heat generated by the mechanical processes as the coupling from the mechanical field to the thermal field. After that, a key material property in the thermomechanical field,

Z. Liu, *Multiphysics in Porous Materials*,
https://doi.org/10.1007/978-3-319-93028-2_19

i.e., thermal expansion coefficient, will be introduced. A practice problem modified from the two relevant monolithic physical fields, i.e., thermal and mechanical fields, will be provided in the end.

Thermal Expansion and Thermal Stress

Thermal expansion is the tendency of a material to change its shape, area, and volume in response to a change in the temperature. It is known from the statistical mechanics that temperature is a macroscopic indicator of the average kinetic energy of the molecules/atoms of a material (Fig. 19.1, left). The kinetic energy of these particles increases as the temperature increases. Accordingly, the particles move more dramatically and consequently tend to maintain a greater average separation. As a result, the material will expand if there are no mechanical constraints (Fig. 19.1, right). The magnitude of this expansion is usually proportional to the changes in the temperature. This proportionality is a significant material property to characterize the material's ability to mechanically response to temperature changes: thermal expansion coefficient. Though unusual, some materials may contract as the temperature increases. However, such phenomena are usually limited to specific temperature ranges due to other less common mechanisms.

In reality, a non-constrained mechanical field is usually an idealization as constraints almost always exist. Therefore, compressive (or tensile) stresses will develop if the thermal expansion (or contraction) caused by temperature change is constrained. This can be explained by treating stresses as a density of the energy of the particles (Fig. 19.1, middle). In fact, the equivalence of the virial stress from energy density (Marc and McMillan 1985; Swenson 1983; Tsai 1979) and the Cauchy stress has been investigated using both theoretical arguments and numerical simulations (Subramaniyan and Sun 2008). In the above thermal expansion example, a temperature increase leads to an increase in the average kinetic energy of the particles. However, if the volume does not change, then the energy density will increase, leading to an increase in the Cauchy stress. In fact, depending on the material type, the temperature change may result in changes in other parts of the

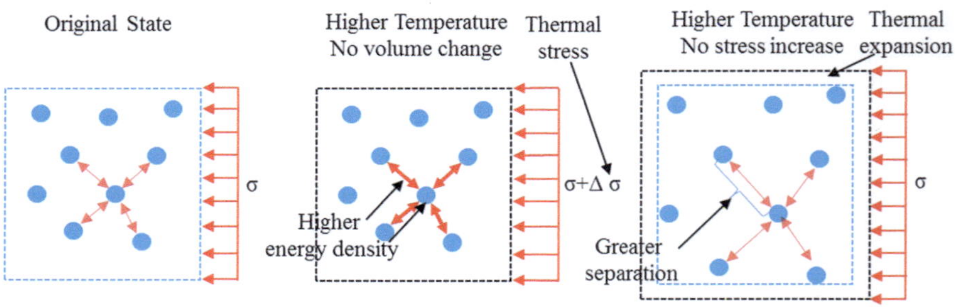

Fig. 19.1 Microscale mechanisms for thermal stresses and strains

particle's energy such as the bond energy and angle energy. The thermal stress is an overall indication of the above changes in the energy density.

Therefore, the thermal stresses and thermal strains are two outcomes produced via the same microscopic mechanisms. Both thermal stresses and thermal strains can cause detrimental or beneficial results, depending on the purpose of the application. Thermal stresses can significantly impact structure's strength and stability and even lead to cracks or breaks. Two typical examples are the residual stress in welding and the longitudinal thermal cracking in the paved roads. Thermal strains are not desired as they can lead to excessive deformation, which will compromise the serviceability of a structure or a component. By contrast, the beneficial use of the thermal expansion process has also led to many significant innovations in the history of science and engineering such as mercury thermometers. Both thermal stresses and strains can be well considered with extensions to the governing equation of the mechanical field, which will be introduced in the following section.

Heat Generated by Mechanical Processes

Friction and plastic deformation are two predominant mechanisms in the conversion of mechanical energy into thermal energy. In both cases, the work done on the material(s) will be transferred to the kinetic energy of the molecules/atoms, which becomes part of the internal energy and exhibits as a temperature increase.

When explaining this conversion process via friction, we usually need to differentiate two different conditions: the friction between two solid objects or that within a fluid object. As for the former, the surfaces between two solid objects are rough like canyons at very small scales. During contacts, some atoms could be brought very close together, and as a result, the kinetic energy of the macroscopic movement will be stored as potential energy of the particles. This potential energy will then be converted to the kinetic energy of the particles when the particles move apart. Real friction also involves a lot of plastic deformation. In a fluid, the resistance of the material to gradual deformations stems from the interparticle forces. When such deformations such as shearing flow happen, the interparticle forces will transfer energy from the motion of the fluid, i.e., the kinetic energy of fluid parcels, into the kinetic energy of the molecules, which is part of the internal energy of the fluid. As the result, the fluid is heated up. This process is partially irreversible and is thus referred to as dissipation or viscous dissipation. Turbulent flows can significantly expedite or even dominate this energy conversion process. Dissipation is high in regions with large gradients, e.g., boundary layers and shear layers, and in regions with very high turbulence levels, e.g., wakes.

The mechanisms associated with plasticity are complicated and vary from material to material. The plasticity in metals is much different from that in amorphous materials, and the plasticity in porous materials could be even more complicated. However, despite those complicated and distinct intermediate steps and mechanisms involved in the process, the energy conversion still can be simplified into the process described above: from external work to the kinetic energy of the material/object and then to the kinetic energy of the atoms/molecules.

As indicated above, the heat generated in the mechanical processes is less common than the thermal stresses and thermal strains in the context of multiphysics. When moving to the porous material, applications with heat generated by a mechanical process are even rarer. Friction between solid objects is the interest of multibody mechanics instead of continuous mechanics, based on which multiphysics is built. The heat caused by plasticity in porous materials is usually negligible. For fluids, the viscous dissipation is an indispensable part of fluid dynamics. However, we mostly focus on the energy loss rather than its conversion to heat and the consequent effect on the heat transfer in the fluids. Due to these considerations, the mathematical formulation of the energy conversion with viscous dissipation and plasticity will be only briefly introduced in the following section.

Mathematical Formulation

Coupling from Thermal Field to Mechanical Field

Both thermal stresses and thermal strains can be considered as the sequence of the coupling from the thermal field to the mechanical field. The coupling can be formulated with an additional term to Navier's equation. Let us first take a look on how to formulate thermal expansion before we introduce how to add in this term.

The coefficient of thermal expansion quantifies a material's ability to change its size with a change in temperature. This ability can be defined according to the dimensions of the deformation, i.e., 1D, 2D, and 3D. In 1D, such as a rod, the linear coefficient of thermal expansion is defined as

$$\alpha = \frac{1}{L}\frac{dL}{dT}, \tag{19.1}$$

where L is the length and dL/dT is the rate of the change of the dimension L per a unit change in temperature. Based on the definition, the change in the dimension can be measured in terms of

$$\frac{\Delta L}{L} = \alpha \Delta T, \tag{19.2}$$

where α is the linear thermal expansion coefficient. The word "linear" here stands for 1D.

This equation works well as long as the linear expansion coefficient does not change too much over the temperature change range ΔT and the fractional change in length is small $\Delta L/L << 1$. When moving to the small strain theory, the above equation can be written as

$$\varepsilon = \alpha \cdot dT. \tag{19.3}$$

In a 3D case with homogeneous and isotropic materials, we will also have strains along the other two perpendicular directions. Therefore, the strains caused by the temperature change can be expressed as follows using the matrix notation:

$$\boldsymbol{\varepsilon}_{\mathrm{T}} = \begin{bmatrix} \alpha \cdot dT & 0 & 0 \\ 0 & \alpha \cdot dT & 0 \\ 0 & 0 & \alpha \cdot dT \end{bmatrix}. \tag{19.4}$$

There are only normal strain components because the temperature change only causes changes along normal directions in the absence of mechanical constraints.

If the above object is completely constrained, the following stress increment will be caused:

$$\boldsymbol{\sigma}_{\mathrm{T}} = \begin{bmatrix} -3K\alpha \cdot dT & 0 & 0 \\ 0 & -3K\alpha \cdot dT & 0 \\ 0 & 0 & -3K\alpha \cdot dT \end{bmatrix}. \tag{19.5}$$

where K is the bulk modulus. Accordingly, we also only have normal stress components. The negative sign implies compressive stresses.

In order to derive the governing equation, let us get back to the equation of equilibrium:

$$\nabla \cdot \boldsymbol{\sigma} + \mathbf{f} = 0. \tag{19.6}$$

The stress $\boldsymbol{\sigma}$ in the above equation is the elastic stress. However, the strain obtained from the strain-displacement relationship is the total strain $\boldsymbol{\varepsilon}$ instead of the elastic strain $\boldsymbol{\varepsilon}_{\mathrm{e}}$. Therefore, the above equation with the thermal strain term can be rewritten considering

$$\boldsymbol{\sigma} = D : \boldsymbol{\varepsilon}_{\mathrm{e}} = D : (\boldsymbol{\varepsilon} - \boldsymbol{\varepsilon}_{\mathrm{T}}):$$

$$\nabla \cdot [\mathbf{D} : (\boldsymbol{\varepsilon} - \boldsymbol{\varepsilon}_{\mathrm{T}})] + \mathbf{f} = 0. \tag{19.7}$$

Substituting the formulation of the stiffness matrix for linear elasticity into the above equilibrium equation (Eq. 19.7), we can obtain the following non-isothermal Navier's equation by following the derivation procedure used in the chapter for the mechanical field:

$$\mu \nabla^2 \mathbf{u} + (\mu + \lambda) \nabla (\nabla \cdot \mathbf{u}) + \mathbf{f} - 3\alpha K \nabla [(T - T_0)\mathbf{I}] = 0, \tag{19.8}$$

or equivalently

$$\nabla \cdot \left[\mathbf{D} : \frac{1}{2} \left(\nabla \mathbf{u} - (\nabla \mathbf{u})^{\mathrm{T}} \right) - 3\alpha K (T - T_0) \mathbf{I} \right] + \mathbf{f} = 0, \tag{19.9}$$

where $\mathbf{D}$ is the stiffness tensor (fourth-order tensor) and $\mathbf{I}$ is

$$\mathbf{I} = \begin{bmatrix} 1 & 0 & 0 \\ 0 & 1 & 0 \\ 0 & 0 & 1 \end{bmatrix}. \tag{19.10}$$

Once the displacement is obtained by solving the above governing equation, the stress will need to be calculated with the following equation:

$$\boldsymbol{\sigma} = \mathbf{D} : \frac{1}{2}\left(\nabla\mathbf{u} - (\nabla\mathbf{u})^{\mathrm{T}}\right) - 3\alpha K(T - T_0)\mathbf{I}. \tag{19.11}$$

Please note that the stress and displacement in the above equation are written using the tensor notation instead of the Voigt notation.

Coupling from Mechanical Field to Thermal Field

The heat generated by the mechanical field serves as the major coupling from the mechanical field to the thermal field. The governing equation of the thermal field, i.e., heat equation, is first recalled to mathematically formulate this coupling:

$$\frac{\partial(\rho CT)}{\partial t} = \nabla \cdot (\lambda \nabla T) - \nabla \cdot (\rho CT\mathbf{v}) + g, \tag{19.12}$$

where ρ is the density, C is the gravimetric heat capacity, T is temperature, λ is the thermal conductivity, $\mathbf{v}$ is the velocity of material particles, and g is the heat source.

For both friction and plasticity deformation, the heat source can be related to the rate of the work done on the material point of interest:

$$g = \eta \frac{dW}{dt}, \tag{19.13}$$

where η is the ratio between the generated heat and the total mechanical energy input or energy dissipation, W. For plasticity, the above total mechanical energy dissipation in the plastic deformation can be formulated as

$$W = -\boldsymbol{\sigma} : \boldsymbol{\varepsilon}_{\mathrm{p}}, \tag{19.14}$$

where $\dot{\boldsymbol{\varepsilon}}_{\mathrm{p}}$ is the plastic strain rate. The heat source in the thermal field then becomes

$$g = \eta\boldsymbol{\sigma} : \dot{\boldsymbol{\varepsilon}}_{\mathrm{p}}. \tag{19.15}$$

The value of η is usually low in friction. However, it could be much higher in the plastic deformation. For example, experiments with an alloy made of Ti, Al, and Cu showed that η could be as high as 70% during high strain rate plastic deformation (Kapoor and Nemat-Nasser 1998). In a liquid, it is common to assume that most of the energy dissipation goes to the internal energy of the liquid, especially the kinetic energy of the liquid molecules. Therefore, η is approximately 1. The heat source attributed to the viscous energy dissipation in the fluid can be formulated as

$$g = \boldsymbol{\sigma} : \nabla \dot{\mathbf{u}}, \tag{19.16}$$

where $\dot{\mathbf{u}}$ is the velocity vector. For compressible Newtonian fluids, a detailed formulation of the viscous dissipation is

$$g = \mu \begin{bmatrix} 2\left(\frac{\partial u}{\partial x}\right)^2 + 2\left(\frac{\partial v}{\partial y}\right)^2 + 2\left(\frac{\partial w}{\partial z}\right)^2 \\ +\left(\frac{\partial v}{\partial x} + \frac{\partial u}{\partial y}\right)^2 + \left(\frac{\partial w}{\partial y} + \frac{\partial v}{\partial z}\right)^2 + \left(\frac{\partial u}{\partial z} + \frac{\partial w}{\partial x}\right)^2 \end{bmatrix} + \lambda(\nabla \cdot \mathbf{u})^2, \tag{19.17}$$

where μ and λ are the viscosities associated with the shearing deformation and volume change, respectively, and u, v, and w are the components of the flow velocity along three mutually perpendicular directions in the Cartesian coordinate system.

Coefficient of Thermal Expansion

Two other types of thermal expansion coefficients have been developed in addition to the linear one: area and volumetric thermal expansion coefficients. The choice of the coefficient type depends on the application and dimensions of interest. Mathematical definitions of these coefficients are defined below for solids, liquids, and gasses. Substances that expand at the same rate in every direction are called isotropic. For isotropic materials, the area and volumetric thermal expansion coefficients are, respectively, approximately twice and three times of the linear thermal expansion coefficient.

Area Expansion Coefficient

The area thermal expansion coefficient is defined as the ratio between the fractional change in the area and the change in the temperature:

$$\alpha_A = \frac{1}{A}\frac{dA}{dT}, \tag{19.18}$$

where A is the area of interest in the object and dA/dT is the rate of the area change per unit change in the temperature. The fractional change in the area can be estimated as

$$\frac{\Delta A}{A} = \alpha_A \Delta T. \tag{19.19}$$

This equation is valid when the area expansion coefficient does not change significantly over the range of ΔT and the fractional change in the area is small, i.e., $\Delta A/A << 1$.

Volume Expansion Coefficient

The volumetric thermal expansion coefficient is the most basic thermal expansion coefficient and the one most relevant to fluids. In general, substances expand or contract when their temperature changes, with expansion or contraction occurring in all directions. The volume thermal expansion coefficient is defined as the ratio between the fractional change in volume and the change in temperature:

$$\alpha_V = \frac{1}{V}\frac{dV}{dT}, \tag{19.20}$$

where V is the volume of the material and dV/dT is the rate of volume change with the temperature. If we already know the expansion coefficient, then we can calculate the change in volume as

$$\frac{\Delta V}{V} = \alpha_V \Delta T, \tag{19.21}$$

where $\Delta V/V$ is the fractional change in volume and ΔT is the change in the temperature.

Practice Problem

Problem Description

A 2D (plane strain) square object with a side length of 1 m has an initial temperature of 273.15 K. The top boundary gets into contact with a water bath with a temperature of 293.15 K, while all the other boundaries are connected to a water bath with a temperature of 273.15 K. The material of the object has a thermal diffusivity of 1×10^{-7}m^2/s. The two sides of the object are fixed, while the top and bottom are constrained using rollers. Young's modulus is 10 GPa and Poisson's ratio is 0.2. The linear thermal expansion coefficient of the material is 1×10^{-5}1/K. Please compute σ_y in the domain at $t = 10$ (Figs. 19.2 and 19.3).

Results

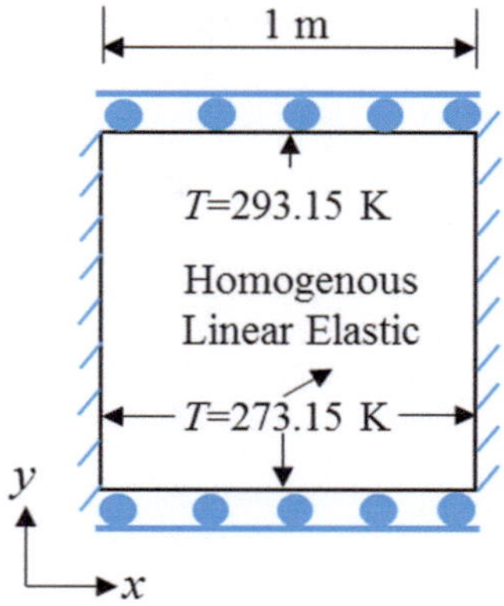

Fig. 19.2 Computational domain and boundary conditions

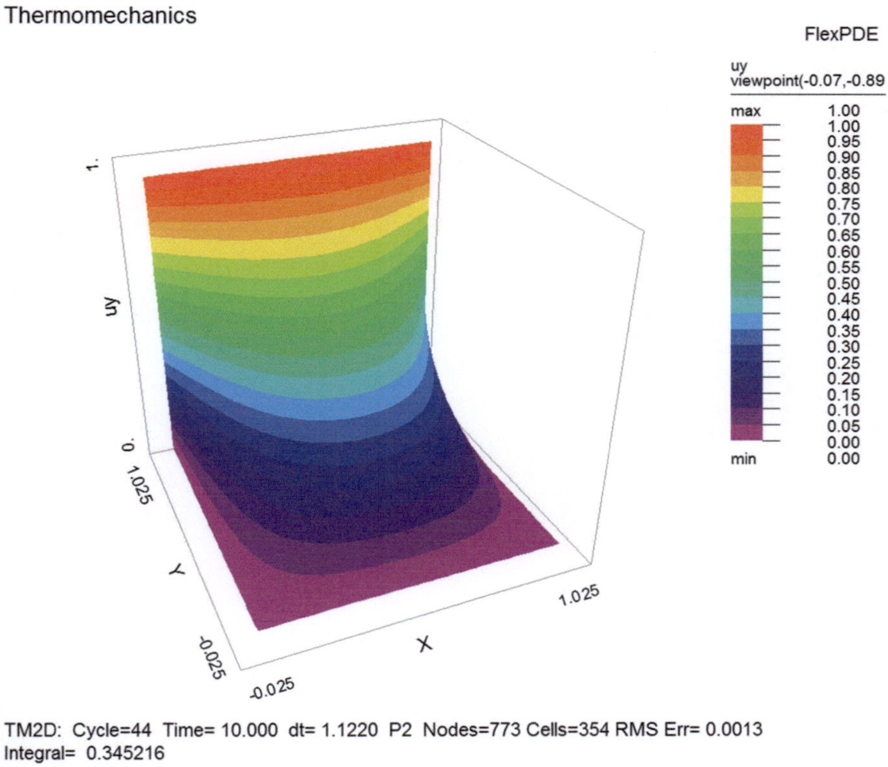

Fig. 19.3 Distribution of vertical stresses

Appendix: FlexPDE Script File

```
{ Thermomechanical process in a plane strain case }
TITLE 'Thermomechanics'
VARIABLES
  u=VECTOR(ux,uy)
  Temp
DEFINITIONS
  E=10e9
  nu=0.2
  mu=E/2/(1+nu)
  lambda=E*nu/(1+nu)/(1-2*nu)
  alpha=10e-6 C=1e7
  delta=VECTOR(1,1)
INITIAL VALUES
  u=VECTOR(0,0)
  Temp=273.15
EQUATIONS
  u: mu*DIV(GRAD(u))+(lambda+mu)*GRAD(DIV(u))-3*alpha*GRAD(T-273)=0
  temp: C*dt(temp)=DIV(GRAD(temp))
BOUNDARIES
  REGION 1
    START(0,0)
      NATURAL(ux)=0
      VALUE(uy)=0
      NATURAL(temp)=0
    LINE TO (1,0)
      VALUE(ux)=0
      VALUE(uy)=0
      NATURAL(temp)=0
    LINE TO (1,1)
      NATURAL(ux)=0
      VALUE(uy)=1
      VALUE (temp)=293.15
    LINE TO (0,1)
      VALUE(ux)=0
      VALUE(uy)=0
      NATURAL(temp)=0
    LINE TO CLOSE
TIME 0 TO 10 by 0.01
MONITORS
PLOTS
  FOR CYCLE=1
    SURFACE(lambda*(dx(ux)+dy(uy))+2*mu*dx(uy)) ! sigma_y
    SURFACE(uy)
END
```

Chapter 20
Hydromechanics: Poroelasticity as a Simple Case

Introduction

Poroelasticity is traditionally used to describe the coupling between the solid and the interpenetrating fluid in their mixture, in which the solid is believed to be saturated by the fluid. In other words, it is an area of studying the movement/flow of material points in both the liquid phase and the solid skeleton. This is distinct from both the groundwater movement and mechanical/dynamic responses of the skeleton. For example, the hydraulic field based on Darcy's law does not take into account the fact that the solid skeleton can be altered. A correction is needed even in the case of steady flow of the pore water under the influence of external forces, let alone problems such as the propagation of elastic vibrations in the porous materials, in which the problems could become untenable (Frenkel 2005).

There are two major parts for poroelasticity: static (or quasi-static) and dynamic theories. The static poroelasticity accounts for a process in which water movement and solid skeleton deformation occur simultaneously and affect each other. Therefore, it can be termed as hydromechanics according to the convention adopted in this book. This static poroelasticity theory is a generalization of the consolidation theory in soil mechanics. In fact, this static theory has been more extensively studied. As a result, poroelasticity is frequently used interchangeably with the consolidation theory or to refer to the static poroelasticity in a narrow sense (Detournay and Cheng 1993; Verruijt 2013). The dynamic poroelasticity is developed for the wave propagation in both the liquid and solid phases of saturated porous materials. So, it is only about dynamics, but the monolithic field involves two different phases. The dynamics here requires using the comprehensive momentum conservation equation for each phase to describe the elastic waves in the representative elementary volume as a mixture of the solid and liquid. It is actually an extension of the mechanical field for mixtures by including the momentum of each phase.

According to Wang (2000), the term "poroelasticity" was first coined by Geertsma as a footnote in his 1966 paper, entitled "Problems of Rock Mechanics

Z. Liu, *Multiphysics in Porous Materials*,
https://doi.org/10.1007/978-3-319-93028-2_20

in Petroleum Production Engineering." In the footnote, Geertsma referred to Biot's work on "the theory of the elasticity and viscoelasticity of fluid-saturated porous solids." Geertsma also pointed out the analogy between poroelasticity and thermoelasticity. The development of the static poroelasticity is primarily attributed to the relevant advances in geomechanics, hydrogeology, and petroleum engineering, for practical problems involving the coupled fluid and mechanical behavior. Typical problems include soil consolidation, water level change in wells due to tides or external loading, oil fields that sink into the sea, and stress around a borehole (Wang 2000). The dynamic poroelasticity is advanced in virtue of the needs from ultrasonic and seismic applications in porous materials such as geological formations (Carcione et al. 2010) and biomaterials (Cowin 1999).

The modern framework of poroelasticity was built on Biot's work on static poroelasticity (1941) and dynamic poroelasticity (1956a, b), though it could trace back to earlier work such as the 1D soil consolidation theory of von Terzaghi (1925). Later work usually focused on reformatting the equation system, validating relationships or parameters with experiments, adding extra terms to account for special effects, relating parameters with microscopic mechanics, and applications to more materials and real problems (Gassmann 1951; Geertsma 1957; Biot and Willis 1957; Detournay and Cheng 1993; De Boer 2000; Wang 2000; Rudnicki 2001; Coussy 2004; Verruijt 2013). The theory of linear poroelasticity (or consolidation) has now reached a stage where there is a consensus on the basic equations. However, there is still no general agreement on the definitions and the notations of the basic physical parameters (Verruijt 2013). In this book, the notation used by Wang (2000) for static poroelasticity and those adopted by van Dalen (2013) will be employed.

As mentioned, poroelasticity has gained a significant development attributed to needs from different disciplines. Due to the same reason, it has widespread applications in various areas. Typical topics include the seepage of liquid waste disposed of underground, oil and gas recovery, soil consolidation, and glacier dynamics (Penta et al. 2014). In recent years, research on the transport, mass exchange, and solid stress in biological tissues has also gained momentum. It is noted that both the static and dynamic poroelasticity theories have been attracting increasing interest in biomaterials (Moeendarbary et al. 2013; Penta et al. 2014). The historical development and future directions of poroelasticity will continue to be interwoven with the advances in both engineering approaches following the work of von Terzaghi and the mixture theory following the work of Fillunger (de Boer 2000; Penta et al. 2014).

It is also necessary to mention that most of the work in poroelasticity, in fact, is devoted to linear poroelasticity (Schanz and Diebels 2003). As commented by Rudnicki (2001), although departures from linear behavior are common in real cases, linear poroelasticity still presents significant improvements to overcome over-idealizations in monolithic physics studies in which either the liquid or the solid is neglected. Moreover, parameters describing the nonlinear material behavior are usually difficult to obtain with accuracy. Therefore, linear poroelasticity often offers a more practical approach than complex nonlinear models. In fact, the insights gained from linear poroelasticity provide valuable assistance to interpret and build more complex models.

In this chapter, the definitions of the basic terms in poroelasticity, including phase compositions, stresses, and strains, will be discussed first. Then static poroelasticity and dynamic poroelasticity will be introduced consequently. A practice problem derived from a classic poroelasticity problem for static poroelasticity will be presented for the numerical implementation of the theory.

Definitions of Phase Composition, Stress, and Strain

The derivation of the mathematical equation system includes the definitions of phase compositions, stresses, and strains, identification of constitutive relationships, establishment of governing equations, and acquisition of relationships between proposed parameters and well-defined properties of porous materials. For the definitions of phase compositions, stresses, and strains, there are usually two ways due to the historical development of poroelasticity: the classic way in soil mechanics and the way in the mixture theory. Most recent poroelasticity studies are formulated using the mixture theory notation, possibly considering its rigorous rationality in physics and potential to be extended to more complicated multiphase and multicomponent systems. Here, we will start the derivation with the mixture theory and then explain its relationship to the classic way in soil mechanics.

Phase Composition

The mixture theory relies heavily on the assumption of the REV and the concept of volume fraction. Every point in a domain of interest is considered as a well-mixed REV, within which any phase i takes a volume fraction of θ_i. In a saturated porous material, there are only two phases, i.e., a solid (skeleton) and a liquid (water). Because the sum of the volume fractions should be 1, in saturated porous materials, only 1 parameter is needed to describe the volume fractions. We can use the porosity which is widely used in most disciplines related to soils and rocks:

$$\phi = \frac{V_\mathrm{f}}{V}, \tag{20.1}$$

where ϕ is the porosity, V_f is the fluid (water) volume, and V is the total volume. Therefore, the volume fractions can be calculated using the porosity:

$$\theta_\mathrm{f} = \phi, \tag{20.2}$$

$$\theta_\mathrm{s} = 1 - \phi, \tag{20.3}$$

where θ_w and θ_f are the volume fractions (or volumetric water content) of the fluid and solid, respectively.

Stress

The total stress is defined for the REV, which is a pseudo bulk material obtained by mixing the solid particle and pore liquid. As a result, the total stress is the sum of the solid stress and fluid stress. Fluid instead of water is used here for a more general framework. To obtain the total stress, let us first obtain the solid and fluid stresses. The fluid stress is the stress caused by the pore fluid pressure, which only has normal components but no shear components:

$$\boldsymbol{\sigma}_{\mathrm{f}} = -\phi p \mathbf{I} = \begin{bmatrix} -\phi p & 0 & 0 \\ 0 & -\phi p & 0 \\ 0 & 0 & -\phi p \end{bmatrix}, \tag{20.4}$$

where $\mathbf{I}$ means only the normal stresses are non-zero. The pore fluid pressure is a compressive stress, whereas a stress is positive when it is a tensile stress. Therefore, there is a negative sign. As can be seen, the pore pressure in the fluid volume fraction of ϕ is averaged with respect to the whole domain to obtain the fluid stress. The solid stress has two parts: the intergranular interaction and the pressure coming from the liquid. Therefore, it is equal to

$$\boldsymbol{\sigma}_{\mathrm{s}} = \boldsymbol{\sigma}_{*} + (1-\phi)(-p)\mathbf{I}, \tag{20.5}$$

where $\boldsymbol{\sigma}_{*}$ is the intergranular stress and can be defined as

$$\boldsymbol{\sigma}_{*} = \begin{bmatrix} \sigma_{11} & \sigma_{12} & \sigma_{13} \\ \sigma_{22} & \sigma_{22} & \sigma_{23} \\ \sigma_{31} & \sigma_{32} & \sigma_{33} \end{bmatrix}. \tag{20.6}$$

The total stress $\boldsymbol{\sigma}^{\mathrm{t}}$ is then obtained as

$$\boldsymbol{\sigma}^{\mathrm{t}} = \boldsymbol{\sigma}_{\mathrm{f}} + \boldsymbol{\sigma}_{\mathrm{s}} = \begin{bmatrix} \sigma_{11}-p & \sigma_{12} & \sigma_{13} \\ \sigma_{22} & \sigma_{22}-p & \sigma_{23} \\ \sigma_{31} & \sigma_{32} & \sigma_{33}-p \end{bmatrix}, \tag{20.7}$$

that is,

$$\boldsymbol{\sigma}^{\mathrm{t}} = \boldsymbol{\sigma}^{*} + \mathbf{p}. \tag{20.8}$$

The above equation is very similar to the principle of effective stress in the binary system (Fig. 20.1) as follows:

$$\boldsymbol{\sigma}^{\mathrm{t}} = \boldsymbol{\sigma}^{'} + \mathbf{p}. \tag{20.9}$$

In fact, the effective stress $\boldsymbol{\sigma}^{'}$ is assumed to be approximately equal to $\boldsymbol{\sigma}_{*}$ in traditional soil mechanics. However, it has been shown by both theoretical and experimental studies that a more accurate definition is

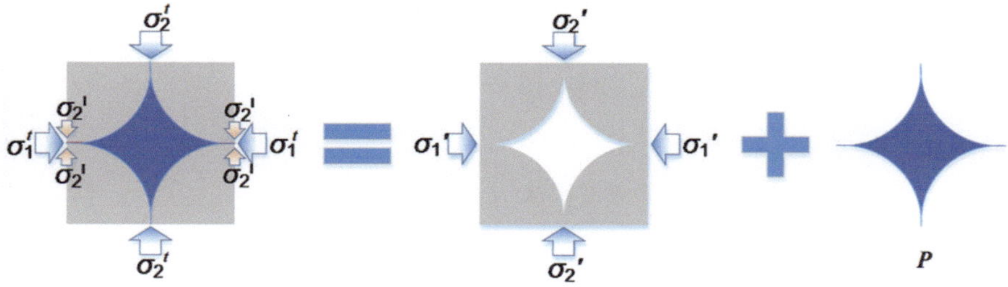

Fig. 20.1 Stresses in binary (two-phase) porous materials

$$\boldsymbol{\sigma}^{\mathrm{t}} = \boldsymbol{\sigma}^{'} + \mathbf{p} - \frac{K}{K_{\mathrm{s}}}\mathbf{p} = \boldsymbol{\sigma}^{'} + \alpha\mathbf{p}, \tag{20.10}$$

where α is the Biot-Willis coefficient. Other formulations which include volume fractions are also available. In soft soils, the bulk modulus of the solid-fluid mixture, K, is much smaller than that of the solid grains K_s, so α is close to 1. That is why the stress principle is applicable to most geotechnical applications involving soft soils. But in stiff soils or rocks, the effective stress may deviate from the value calculated with the effective stress principle significantly. In later constitutive relationships, σ will be used to represent an isotropic stress field applied to the mixture, which is equal to the average of the normal terms in the total stress:

$$\sigma = \frac{1}{3}\mathrm{trace}(\boldsymbol{\sigma}^{\mathrm{t}}) = \frac{1}{3}\sigma_{ii}, \tag{20.11}$$

where trace($\boldsymbol{\sigma}^{\mathrm{t}}$) is the trace of the total stress tensor. The above mean normal stress or hydrostatic stress is used because the stress interaction between the solid and fluid phases resides mostly between the hydrostatic stress terms (i.e., pressure) rather than deviatoric stress terms.

Strain and Displacement

For strains, we will define the strain of the solid, $\boldsymbol{\varepsilon}$; the strain of the fluid, $\mathbf{e}$; and the increment of the fluid content, ζ. The solid constitutes the skeleton of the porous material; therefore the solid deformation reflects what we observe on the solid skeleton. The volume of the solid is assumed to be evenly distributed across the whole volume. The fluid strain is the strain of a pseudo liquid which is evenly distributed over the whole REV. This definition of the fluid strain is similar to Darcy's velocity, which is defined over the whole sectional area including both the liquid and solid, instead of the true velocity (analogous to the true fluid deformation), which is defined over the liquid sectional area (real flow passageway). If we define the displacements of liquid and solid on these pseudo phases, then the solid and fluid strains can be related to the displacement as

$$\boldsymbol{\varepsilon} = \frac{1}{2}\left[\nabla \mathbf{u} + (\nabla \mathbf{u})^{\mathrm{T}}\right], \tag{20.12}$$

and

$$\mathbf{e} = \frac{1}{2}\left[\nabla \mathbf{U} + (\nabla \mathbf{U})^{\mathrm{T}}\right], \tag{20.13}$$

where **u** and **U** are the displacements of the solid and liquid, respectively.

That is to say, we treat a saturated porous material, which is a two-phase composite, as a continuum. Each REV is a well-mixed region which is large enough compared to the solid grains and interconnected pores but small enough compared to the domain of interest. The increment of fluid content, ζ, can be accurately defined as (Wang 2000)

$$\zeta = \frac{\delta m_{\mathrm{f}}}{\rho_{\mathrm{f0}}}, \tag{20.14}$$

where δm_{f} is the change in fluid mass content and ρ_{f0} is the fluid density in the reference state. Or it can be equivalently defined as

$$\zeta = \frac{\delta V_{\mathrm{p}} - \delta V_{\mathrm{f}}}{V}, \tag{20.15}$$

where V_{p} is the pore volume. This increment can be related to the deformations of the solid and liquid via (Biot and Willis 1957)

$$\zeta = \varepsilon_{jj} - e_{ii} = \nabla \cdot \mathbf{u} - \nabla \cdot \mathbf{U}, \tag{20.16}$$

where e and ε are the scalar terms of the liquid and solid strains, respectively. In many cases, the volume strain of the solid skeleton is represented using another symbol:

$$\varepsilon_{v} = \varepsilon_{ii}. \tag{20.17}$$

The theories of static and dynamic poroelasticity usually adopt different forms of constitutive relationships. Therefore, the constitutive relationships will be introduced in the following sections for static and dynamic poroelasticity.

Static Poroelasticity

Constitutive Relationships

The establishment of the constitutive relationships for saturated porous materials needs to consider both the solid and fluid. It is not difficult to imagine that these two phases will interact with each other in the stress-strain relationship. Since the fluid stress/strain only has normal components, the couplings between the solid and fluid only occur via these normal terms. As a result, we usually identify the couplings between normal and bulk/volume stress/strain first and then add in the shear terms in the formulation. The following constitutive relationship can be obtained:

$$\varepsilon_v = \frac{\sigma}{K} + \frac{p}{H}, \tag{20.18}$$

$$\zeta = \frac{\sigma}{H} + \frac{p}{R}, \tag{20.19}$$

where K is the bulk modulus, H is the poroelastic expansion coefficient, and R is the reciprocal of the unconstrained storage coefficient. Their physical meanings of these parameters can be described using the following equations:

$$K = \frac{\delta\sigma}{\delta\varepsilon_v}|_{p=0}, \tag{20.20}$$

$$H = \frac{\delta p}{\delta\varepsilon_v}|_{\sigma=0} = \frac{\delta\sigma}{\delta\zeta}|_{p=0}, \tag{20.21}$$

$$R = \frac{\delta\zeta}{\delta p}|_{\sigma=0}. \tag{20.22}$$

The above constitutive relationships are sometimes written in the following form:

$$\sigma = K\varepsilon_v - \alpha p, \tag{20.23}$$

$$\zeta = \alpha\varepsilon_v + \frac{1}{M}p, \tag{20.24}$$

where $\alpha = \frac{K}{H}$ is the Biot-Willis coefficient introduced above and $\frac{1}{M} = \frac{1}{R} - \frac{K}{H^2}$ is the constrained specific storage coefficient. It is seen that, in either form of the constitutive relationship, only three independent mechanical properties are needed. In order to use the above relationships to derive the governing equation, the shear stress/strain needs to be included. The complete constitutive relationship for the static poroelasticity is

$$\sigma_{ij} = \lambda\varepsilon_{kk}\delta_{ij} + 2G\varepsilon_{ij} - \alpha p\delta_{ij} = \lambda \cdot trace(\boldsymbol{\varepsilon}) + 2G\boldsymbol{\varepsilon} - \alpha p\mathbf{I}, \tag{20.25}$$

$$\zeta = \alpha u_{i,i} + \frac{1}{M}p = \alpha\nabla \cdot \mathbf{u} + \frac{1}{M}p. \tag{20.26}$$

Governing Equation

Then let us recall the Navier's equation for the momentum equilibrium of the solid and continuity equation of the pore fluid:

$$(\lambda + \mu)\nabla(\nabla \cdot \mathbf{u}) + \mu\nabla^2\mathbf{u} + \mathbf{g} = 0, \tag{20.27}$$

$$\frac{1}{M}\frac{\partial p}{\partial t} + \nabla \cdot \left[-\frac{k}{\eta}\nabla(p + \rho g z) \right] = 0, \tag{20.28}$$

where k is the permeability.

Substituting the constitutive equations (Eqs. 20.25 and 20.26) into the governing equations (Eqs. 20.27 and 20.28), we obtain the final governing equations:

$$(\lambda + \mu)\nabla(\nabla \cdot \mathbf{u}) + \mu\nabla^2\mathbf{u} + \mathbf{f} = \alpha\nabla p, \tag{20.29}$$

$$\frac{1}{M}\frac{\partial p}{\partial t} + \nabla \cdot \left[-\frac{k}{\eta}\nabla(p + z) \right] = -\alpha\frac{\partial}{\partial t}(\nabla \cdot \mathbf{u}). \tag{20.30}$$

In poroelasticity, we may also see some other material properties, which can be formulated using K, H, R, and so on. These include the undrained modulus

$$K_u = \frac{\delta\sigma}{\delta\varepsilon_v}\Big|_{\zeta=0} = \frac{1}{\frac{1}{K} - \frac{R}{H^2}}, \tag{20.31}$$

undrained Poisson's ratio

$$\nu_u = \frac{3\nu + \alpha B(1 - 2\nu)}{3 - \alpha B(1 - 2\nu)}, \tag{20.32}$$

and first Skempton's coefficient

$$B = \frac{\delta p}{\delta\sigma}\Big|_{\zeta=0} = -\frac{R}{H} = \alpha M / K_u. \tag{20.33}$$

Further micromechanical relationships are sometimes needed to relate the material properties of the porous material to its phases (Merxhani 2016):

$$\alpha = 1 - \frac{K}{K_s} \text{ (Biot and Willis 1957)} \tag{20.34}$$

$$\frac{1}{M} = \frac{1}{R} - \frac{\alpha}{H} = \left(1 - \frac{K}{K_s} - \phi\right)\frac{1}{K_s} + \phi\frac{1}{K_f} \text{ (Verruijt 1969)} \tag{20.35}$$

This will leads to

$$\frac{1}{H} = \frac{1}{K} - \frac{1}{K_s}, \tag{20.36}$$

$$\frac{1}{R} = \frac{1}{K} + \phi\left(\frac{1}{K_f} - \frac{1}{K_s}\right) - \frac{1}{K_s}, \tag{20.37}$$

where K_s and K_f are the bulk moduli of the solid and fluid phases, respectively. In many documents, compressibilities (reciprocals of bulk moduli) instead of bulk moduli are used in the above equations.

Dynamic Poroelasticity

Constitutive Relationships

In the previous section for static poroelasticity, the inertial and associated kinetic energy was not considered. However, when the speed of the movement of the phases in the porous material is appreciable, e.g., when there is vibration or stress waves, the theory in the previous section cannot well describe the behavior of the porous materials. To include these dynamic effects, constitutive relationships are slightly modified from those adopted in the previous section:

$$\boldsymbol{\sigma}_s = A\varepsilon_v\boldsymbol{\delta} + Q\mathbf{e} + 2G\boldsymbol{\varepsilon} \tag{20.38}$$

$$\mathbf{p} = Q\varepsilon_v\boldsymbol{\delta} + R\mathbf{e} \tag{20.39}$$

$$E_p = \frac{1}{2}(\boldsymbol{\sigma}_s \cdot \boldsymbol{\varepsilon} + \mathbf{p} \cdot \mathbf{e}), \tag{20.40}$$

where G, A, Q, and R are another four parameters quantifying the elastic properties of the porous material. They are not conventional moduli but can be obtained using the moduli of the phases and the bulk mixture.

Governing Equation

In order to derive the governing equation, it is better to resort to the Lagrangian equation. The Lagrangian mechanics is just another description of the classic mechanics which works better for complicated systems involving a lot of constraints. The Lagrange equation is

$$\frac{d}{dt}\frac{\partial}{\partial \dot{x}}(E_{\mathrm{k}} - E_{\mathrm{d}}) - \frac{\partial}{\partial x}(E_{\mathrm{k}} - E_{\mathrm{d}}) = F, \tag{20.41}$$

where x is the generalized coordinate(s), E_{k} is the kinetic energy, E_{d} is the energy dissipation, and F is the total force. The kinetic energy is

$$E_{\mathrm{k}} = \frac{1}{2}\left(\rho_{11}\dot{\mathbf{u}} \cdot \dot{\mathbf{u}} + \rho_{11}\dot{\mathbf{U}} \cdot \dot{\mathbf{U}} + 2\rho_{12}\dot{\mathbf{u}} \cdot \dot{\mathbf{U}}\right), \tag{20.42}$$

where ρ_{11}, ρ_{22}, and ρ_{12} are the effective density of the solid, liquid, and coupling between the solid and the liquid, respectively. These density terms can be related to the densities of the solid and liquid via the following equations (Biot 1956a, b):

$$\rho_{11} = (1 - \phi)\rho_{\mathrm{s}} - \rho_{12}, \tag{20.43}$$

$$\rho_{22} = \phi\rho_{\mathrm{f}} - \rho_{12}, \tag{20.44}$$

$$\rho_{12} = -(\tau - 1)\phi\rho_{\mathrm{f}}. \tag{20.45}$$

The energy dissipation function can be expressed in terms of the velocities of the solid and fluid:

$$E_{\mathrm{d}} = \frac{1}{2}b_0\left(\dot{\mathbf{u}} - \dot{\mathbf{U}}\right) \cdot \left(\dot{\mathbf{u}} - \dot{\mathbf{U}}\right), \tag{20.46}$$

where the coefficient b_0 can be related to the permeability, dynamic viscosity, and porosity via the following equation:

$$b_0 = \frac{\eta\phi^2}{k}. \tag{20.47}$$

Substituting the above constitutive relationships and energy functions into the Lagrange equation, we can obtain the following governing equations:

$$\begin{aligned}\rho_{11}\ddot{\mathbf{u}} + \rho_{12}\ddot{\mathbf{U}} + b_0\left(\dot{\mathbf{u}} - \dot{\mathbf{U}}\right) = {} & (A + G)\nabla(\nabla \cdot \mathbf{u}) + G\nabla^2\mathbf{u} \\ & + Q\nabla(\nabla \cdot \mathbf{U}) + \mathbf{F}_{\mathrm{s}} - Q\nabla\theta_{\mathrm{s}}\end{aligned} \tag{20.48}$$

$$\rho_{12}\ddot{\mathbf{u}} + \rho_{22}\ddot{\mathbf{U}} - b_0\left(\dot{\mathbf{u}} - \dot{\mathbf{U}}\right) = Q\nabla(\nabla \cdot \mathbf{u}) + R\nabla^2\mathbf{U} + \mathbf{F}_{\mathrm{f}} - R\nabla\theta_{\mathrm{f}}. \tag{20.49}$$

Practice Problem

Problem Description

In linear poroelasticity, Mandel's problem (Mandel 1953; Abousleiman et al. 1996) is usually used to validate linear poroelastic models. Mandel (1953) presented an

Fig. 20.2 Computational domain and boundary conditions

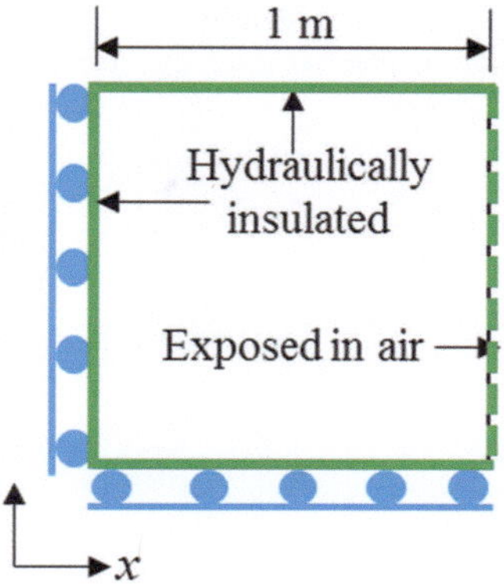

analytical solution to predict the coupled hydromechanical response of porous materials under a set of prescribed boundary conditions (Ferronato et al. 2010). In the Mandel's problem, a 2D rectangular object (plane strain) is subjected to a vertical distributed load of p_0, while the two side boundaries are free to move. The top and bottom boundaries are hydraulically insulated, while the side boundaries are exposed in air (zero pore water pressure). Due to the symmetry in the geometry and loading, we can only analyze one fourth of the object. Accordingly, a 2D domain that has dimensions of 1 m by 1 m ($a \times b$) is adopted here (Fig. 20.2). Let us consider the upper right fourth of the 2D object. Therefore, the left and bottom boundaries will be symmetric. That is, there is no normal displacement and no shear stresses for the mechanical field, while no flux is allowed through the boundaries (hydraulic insulation).

The analytical solution for such a problem is

$$p = 2p_0 \sum_{n=1}^{\infty} \left[\frac{\sin \alpha_n}{\alpha_n - \sin \alpha_n \cos \alpha_n} \left(\cos \frac{\alpha_n x}{a} - \cos \alpha_n \right) \exp \left(\frac{-\alpha_n^2 ct}{a^2} \right) \right], \tag{20.50}$$

where α_n is the nth positive solution to the following equation:

$$\tan \alpha = \frac{\lambda + 2\mu}{\mu} \alpha, \tag{20.51}$$

and c is the consolidation coefficient and can be calculated using poroelastic and other material properties as

$$c = \frac{\kappa}{\rho_w g \left(\frac{1}{M} + \frac{\alpha^2}{\lambda + 2\mu} \right)}, \tag{20.52}$$

in which κ is the hydraulic conductivity.

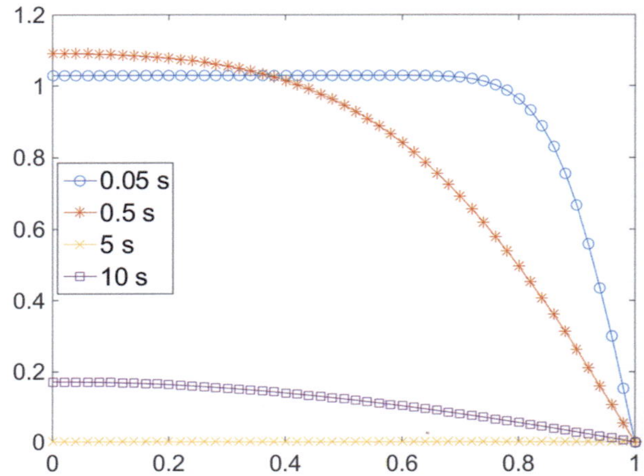

Fig. 20.3 Analytical results for the pore pressure variation

In our example, the two Lamé constants are both 40 MPa, Poisson's ratio is 0.3, the Biot-Willis coefficient is 1.0, the porosity is 0.375, the hydraulic conductivity is 1×10^{-5} m/s, and the compressibilities of the pore water and solid grain are 4.4×10^{-10} 1/Pa and 2.8×10^{-10} 1/Pa, respectively. Please calculate the pore water distributions at 0.05 s, 0.5 s, 5 s, and 10 s and compare them against the analytical solution.

Results

The analytical solution is shown in Fig. 20.3. MATLAB code for the implementation of the solution is attached in the next subsection.

The above figure clearly shows the peculiar non-monotonic pressure behavior, i.e., the so-called Mandel-Cryer effect, in which p rises above the initial p_0 at small time values. This non-monotonic pressure change occurs because the slab contraction at the drained faces induces an additional buildup of pressure in the interior.

The following figures show the numerical solution. As can be seen, the numerical solution reproduces a similar process. However, the default iteration and error control offered in the student version of FlexPDE failed to handle the high nonlinearity residing in the beginning of the process. The Mandel-Cryer effect was not perfectly simulated (Fig. 20.4).

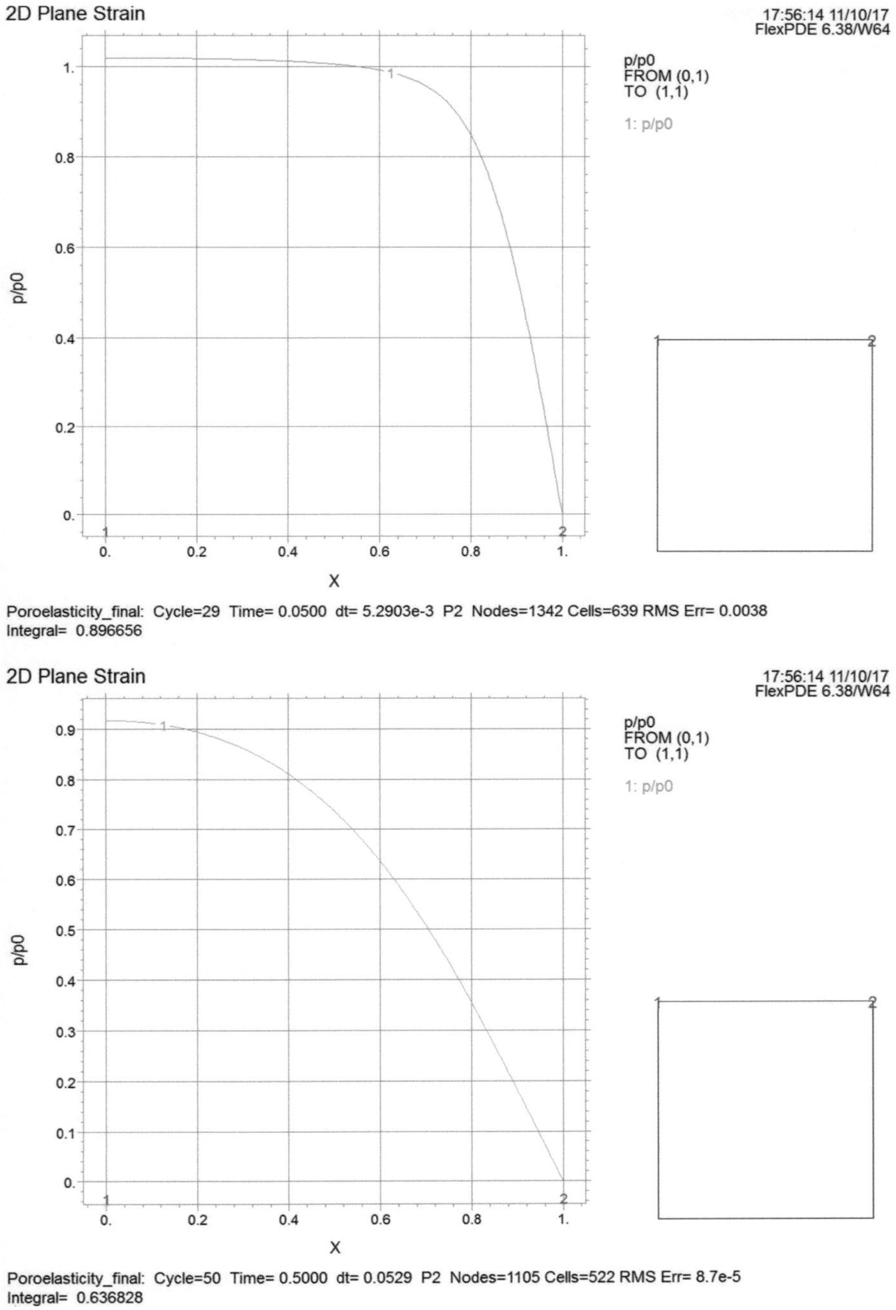

Fig. 20.4 Numerical results for pore pressure variation

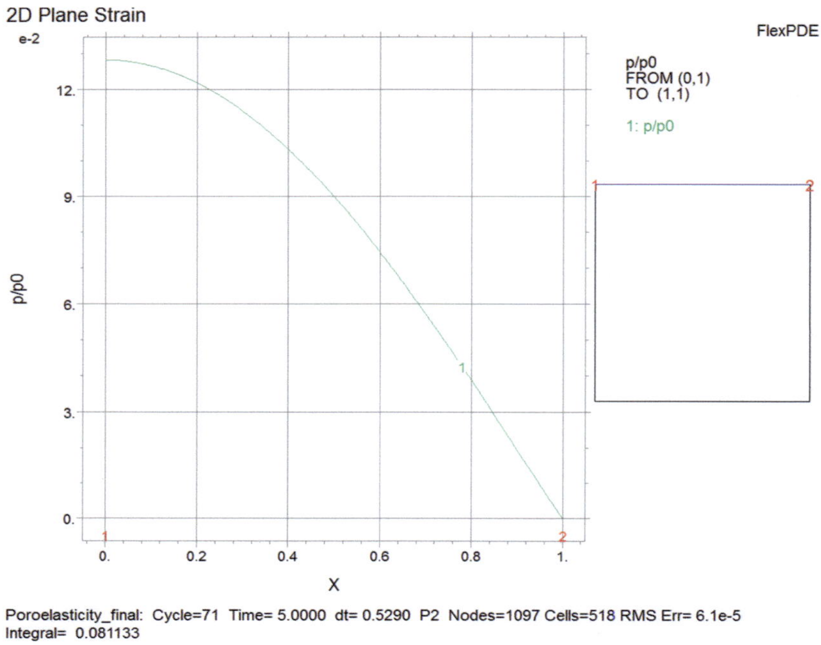

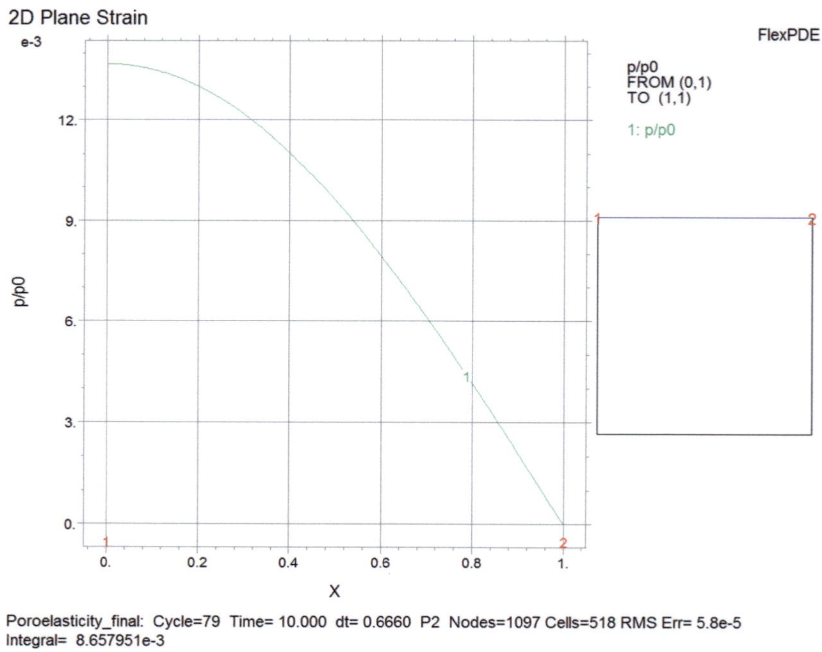

Fig. 20.4 (continued)

Appendix: FlexPDE Script File

Analytical Solution

MATLAB script file 1: main.m

```
%Analytical solution to Mandel's problem
clear;

mu=40e6; %E/2/(1+nu)
lambda=40e6; %E*nu/(1+nu)/(1-2*nu)
alpha=1;
kappa=1e-5;
phi=0.375;
beta=4.8e-10;
c_s=2.8e-10; % Compressibility of solid grain
M=1/(phi*beta+(alpha-phi)*c_s);
p0=1e4;

K=lambda+2/3*mu;
R=1/(1/M+alpha^2/K);
H=K/alpha;
Ku=1/(1/K-R/H^2);
B=alpha*M/Ku;
nu=lambda/2/(lambda+mu);
nu_u=(3*nu+alpha*B*(1-2*nu))/(3-alpha*B*(1-2*nu));
c=kappa/(9800*(M^(-1)+alpha^2/(lambda+2*mu)));

a=1;

time=[0.05,0.500,50,10];
x=0:0.02:1;
n_time=0;
p=zeros(length(time),length(x));
for t=time
    n_time=n_time+1;
for n=1:100
    fun=@f;
    alpha_n=fzero(fun,[0+(n-1)*pi+0.1 pi/2+(n-1)*pi]);
    %We can also use vpasolve to calculate alpha_n, but it is much
slower!
%{ syms xxx
%   alpha_n=vpasolve(-abs(tan(xxx))+(lambda+2*mu)*xxx/mu==0,xxx,
[0+(n-1)*pi+0.1 pi/2+(n-1)*pi]);
%}
    %p(x,z,t), x=0~1, z=1 t=0.5
    p_n=2*p0*(lambda+2*mu)*cos(alpha_n)/(mu-(lambda+2*mu)*(cos
(alpha_n))^2)*(cos(alpha_n*x/a)-cos(alpha_n))*exp
(-alpha_n^2*c*t/a^2); %Original
```

```
    %p_n=2*p0*sin(alpha_n)/(alpha_n-sin(alpha_n)*cos(alpha_n))*
(cos(alpha_n*x/a)-cos(alpha_n))*exp(-alpha_n^2*c*t/a^2); %
Equation in Ferronato et al. (2010)
    p(n_time,:)=p(n_time,:)+p_n;
end
end

plot(x,p(:,:)/p0);
```

MATLAB script file 2: f.m

```
function y=f(x)
mu=40e6; %E/2/(1+nu)
lambda=40e6; %E*nu/(1+nu)/(1-2*nu)
y=-abs(tan(x))+(lambda+2*mu)*x/mu
```

Numerical Solution

```
{ Basic linear elastic mechanical problems for 2D (plane strain)
homogenous cases }
TITLE '2D Plane Strain'
SELECT
  !PREFER_STABILITY
  ERRLIM 0.01
  REMATRIX=ON
  NEWTON=3
VARIABLES
  u=VECTOR(ux,uy)
  p
DEFINITIONS
  mu=40e6
  lambda=40e6
  alpha=1
  kappa=1e-5
  phi=0.375
  beta=4.8e-10 ! Compressibility of fluid
  c_s=2.8e-10 ! Compressibility of solid grain
  M=1/(phi*beta+(alpha-phi)*c_s)
  p0=1e4
INITIAL VALUES
  ux=0
  uy=0
  p=p0
EQUATIONS
  u: mu*DIV(GRAD(u))+(lambda+mu)*GRAD(DIV(u))=alpha*GRAD(p)
  p: dt(1/M*p+alpha*DIV(u))+DIV(-kappa/9800*GRAD(p))=0
BOUNDARIES
  REGION 1
    START(0,0)
      NATURAL(ux)=0
      VALUE(uy)=0
      NATURAL(p)=0
```

```
    LINE TO (1,0)
      NATURAL(uy)=0
      NATURAL(ux)=0
      VALUE(p)=0
    LINE TO (1,1)
      NATURAL(ux)=0
      NATURAL(uy)=-p0
      NATURAL(p)=0
    LINE TO (0,1)
      VALUE(ux)=0
      NATURAL(uy)=0
      NATURAL(p)=0
    LINE TO CLOSE
TIME 0 TO 10
MONITORS
  FOR CYCLE=1
    ELEVATION(p/p0) FROM (0,0) TO (1,0)
    VECTOR(u)
    VECTOR(-GRAD(p))
    CONTOUR(p/p0)
PLOTS
  FOR t=0.05
    ELEVATION(p/p0) FROM (0,1) TO (1,1)
  FOR t=0.5
    ELEVATION(p/p0) FROM (0,1) TO (1,1)
  FOR t=5
    ELEVATION(p/p0) FROM (0,1) TO (1,1)
  FOR t=10
    ELEVATION(p/p0) FROM (0,1) TO (1,1)
HISTORY(p/p0) at (0,0.05)
END
```

Chapter 21
Thermohydromechanics

Introduction

In porous materials, thermohydromechanics is one of the most common types of multiphysical phenomena in addition to the static poroelasticity discussed in the previous chapter. Thermohydromechanics is responsible for many applications related to unsaturated porous materials such as hydro-diffusion and subsidence, drying/hydration and shrinkage, frost heave, spalling, geothermal applications, and hydraulic cracking. This is because, in nature or in labs, most processes occur in non-isothermal conditions. Therefore, the existence of heat transfer is more a rule than an exception. Also, theoretically, the water flow and the accumulation or dissipation of strain energy can alter or trigger a heat transfer process. For the water movement, any type of potential difference within porous materials, e.g., matric, osmotic, and pneumatic, will give rise to water movement. These differences can result from a temperature difference or form a change in the mechanical field. For the mechanical field, either thermal transfer or water movement can explicitly change the mechanical properties or the mechanical balance. This process could also occur in the opposite direction, that is, changes in the structure or the existence/changes of stress and strain can also influence the heat transfer and water movement. In summary, the couplings are strong in the abovementioned phenomena, and these strong couplings make the studies of thermohydromechanics (THM) a very popular topic in the multiphysics in porous materials.

The development of the modeling techniques for the THM benefited from a lot of studies for geomaterials. For example, there are a great number of papers dedicated to the coupled thermohydromechanical model or thermohydromechanochemical (THMC) model for rocks and soils (Thomas and King 1991; Thomas and He 1995, 1997; Thomas et al. 2009; Sahimi 2011; Noorishad et al. 1992; Stephanson et al. 1997; Rutqvist et al. 2001). Noteworthily, these models were developed either from the non-isothermal consolidation theory of deformable porous materials or the extension to Biot's phenomenological model. More recent models include the

Z. Liu, *Multiphysics in Porous Materials*,
https://doi.org/10.1007/978-3-319-93028-2_21

phase change of water (i.e., freezing/thawing phenomena) in porous media. These THM models are usually solved numerically, because an analytical solution is usually unavailable. High nonlinearity is common because of the complicated couplings between the involved monolithic physical fields.

Thermohydromechanics is very closely related to thermohydraulics and hydromechanics. In many studies, thermohydromechanics is extended from thermohydraulics or hydromechanics. For example, thermohydraulics models are widely adopted in soil sciences to predict water movement under varying ambient temperatures. In this case, moisture content is the major interest, while the temperature changes are the cause, though the influence of the water movement on the temperature variation could also be appreciable. "Thermal" is thus placed in front of "hydraulic." While in disciplines that more focus on stresses and strains such as geotechnical engineering or geophysics, the THM models are also obtained by adding the thermal field into existing hydromechanical models, i.e., poroelasticity theory.

In this chapter, the establishment of a typical THM model with the phase change of the pore liquid will be presented for unsaturated porous materials (Liu and Yu 2011). We will show the establishment of the model using a bottom-up approach in contrast to the top-down approach using in the previous chapter for poroelasticity. Accordingly, the governing equations and auxiliary relationships for the thermal field, hydraulic field, and mechanical field will be introduced sequentially. Boundary conditions will be briefly mentioned. Finally, a practice problem for the use of a THM model for predicting the behavior of a soil in a non-isothermal environment will be presented.

Thermal Field

The thermal field is usually the main cause of the whole thermohydromechanical process. For example, during a frost heave process, the subfreezing temperatures cause the advancement of the frost front, which in return induces fluid migration and soil deformation. This process generated by energy dissipation within a multiphase material can be alternatively explained as the result of the varied surface tension, soil moisture suction, kinetic energy changes associated with the hydrogen bond distribution, and thermally induced osmotic gradients (Cary 1965, 1966). Besides the temperature-induced moisture flux or deformation, the energy carried by the migratory fluid and the heat parameters influenced by the fluid transport and metamorphic solid skeleton are also assumed to be "reactions" to the thermal field. To precisely formulate the energy transport in porous materials, a modified Fourier's equation with both conduction and convection terms can be adopted:

$$C_{\mathrm{a}}\frac{\partial T}{\partial t}+C_{\mathrm{w}}\nabla(\mathbf{J}T)=\nabla\cdot(\lambda\nabla T), \tag{21.1}$$

where C_w is the heat capacity of unfrozen water, C_a is the apparent heat capacity, λ is the thermal conductivity, T is the temperature, t is time, and $\mathbf{J}$ is the water flux from the hydraulic field. Both C_a and λ are coupling variables. The moisture migration changes the soil composition and consequently C_a and λ, which in turn affects the heat transfer process.

The effects of the air phase and radiation are usually neglected as the phase transition of water dominates the process of energy conversion. The apparent volumetric heat capacity C_a in Eq. 21.1 takes into account the energy released/absorbed by the phase change of the pore water. Instead of being treated as an energy sink or source on the right-hand side of Fourier's equation, it is a common practice in partially frozen porous materials to incorporate the enthalpy change due to the phase change into the heat capacity to reduce the nonlinearity (Anderson et al. 1973)

$$C_a = C_s\theta_s + C_w\theta_w + C_i\theta_i + C_v(n - \theta_w - \theta_i) + L_f \frac{d\theta_i}{dt}, \tag{21.2}$$

where θ_s, θ_w, and θ_i denote the volumetric content of soil mass, unfrozen water, and ice, respectively (the volume change of soil skeleton is neglected here) and L_f is the latent heat. The same convention on subscripts applies to other parameters.

The thermal conductivity λ in Eq. 21.1 can be approximated by empirical relationships such as Eq. 21.3 (McInnes 1981; Hansson et al. 2004):

$$\lambda = \mathrm{C}_1 + \mathrm{C}_2(\theta_w + \theta_i F) - (\mathrm{C}_1\text{-}\mathrm{C}_4) \cdot \exp\left\{-[\mathrm{C}_3(\theta_w + \theta_i) + F\theta_i]^{\mathrm{C}_5}\right\} \tag{21.3}$$

$$F = 1 + \mathrm{F}_1\theta_i^{\mathrm{F}_2}, \tag{21.4}$$

where C_1, C_2, C_3, C_4, C_5, F_1, and F_2 are empirical curve fitting constants. There are also other similar relationships such as the equations proposed by Gardner (1958), Sawada (1977), and Campbell (1985).

Hydraulic Field

For variably unsaturated porous materials, the fluid movement is generally described by mixed-type Richards' equation, which has been proven to perform well in ensuring mass conservation (Celia and Binning 1992). To extend Richards' equation to allow for freezing/thawing, a term related to ice formation needs to be added to the left-hand side of Richards' equation to obtain

$$\frac{\partial\theta_w}{\partial t} + \frac{\rho_i}{\rho_w}\frac{\partial\theta_i}{\partial t} = \nabla \cdot (\kappa_{Lh}\nabla h + \kappa_{Lh}\mathbf{i} + \kappa_{LT}\nabla T), \tag{21.5}$$

where θ_w is the volumetric content of water, θ_i is the volumetric content of ice, ρ_w is the density of water, ρ_i is the density of ice, κ_{Lh} is the hydraulic conductivity, κ_{LT} is the hydraulic conductivity due to a thermal gradient, **i** is the unit vector along the direction of gravity, and h is the matric potential head (or pressure head). The matric potential head is the equivalent water head (unit: m) of the matric potential (negative water suction, unit: Pa). It is common to formulate the above equation in terms of the matric potential head, matric potential, total soil potential head, or total soil potential. But attention is needed to make sure the units of other terms are consistent with that of the selected dependent variable.

Based on the analogy between the freezing process and the drying process, it has been accepted that the concept of the soil-water characteristic curve (SWCC) can be extended to describe the relationship between the unfrozen water content and the matric potential (Koopmans and Miller 1966; Spans and Baker 1996). The hydraulic conductivity can be estimated by integrating the SWCC (Childs and Colis-George 1950; Mualem 1986; Fredlund et al. 1994). Most THM studies employed either simplified van Genuchten's equation (van Genuchten 1980, Hansson et al. 2004) (Eq. 21.6) or Fredlund and Xing's equation (Fredlund and Xing 1994) to mathematically describe the SWCC. Corresponding approaches are then adopted to predict the hydraulic conductivity with the SWCC. Taking van Genuchten's equation, for example, we have the following equation for the SWCC:

$$S_{\mathrm{e}} = \frac{\theta - \theta_{\mathrm{r}}}{\theta_{\mathrm{s}} - \theta_{\mathrm{r}}} = \left(1 + |\alpha h|^{n}\right)^{-m}, \tag{21.6}$$

where S_{e} is the effective saturation; θ_{s} and θ_{r} are the saturated and residual water content, respectively; and α, m, and n are empirical parameters.

κ_{Lh} and κ_{LT} are hydraulic conductivities related to the pore water head and temperature, respectively. One set of widely accepted relationships for these parameters is

$$\kappa_{\mathrm{Lh}} = \kappa_{\mathrm{s}} S_{\mathrm{e}} \left[1 - \left(1 - S_{\mathrm{e}}^{\frac{1}{m}}\right)^{m}\right]^{2}, \tag{21.7}$$

$$\kappa_{\mathrm{LT}} = \kappa_{\mathrm{Lh}} \left(h G_{\mathrm{wT}} \frac{1}{\gamma_0} \frac{d\gamma}{dT}\right), \tag{21.8}$$

where γ denotes the surface tension of soil-water, which is temperature-dependent and can be approximated using the following equation:

$$\gamma = 75.6 - 0.1425T - 2.38 \times 10^{-4} T^{2}, \tag{21.9}$$

where γ_0 is the value of γ at 25 °C, i.e., $\gamma_0 = 71.89 \times 10^{-3}\ \mathrm{kg} \cdot \mathrm{m}^{3}$.

As shown in Eq. 21.7, the hydraulic conductivity of a partially saturated or partially frozen soil, κ_{Lh}, is obtained by multiplying the saturated conductivity with a saturation-dependent "relative conductivity" term. The thermal-induced hydraulic conductivity κ_{LT} in Eq. 21.8 is developed from thermodynamics (Philip and de Vries 1957). G_{wT} is a gain factor, which has a value around 7 for coarse-grained soils (Noborio et al. 1996). The dependence of the viscosity on temperature can be neglected here to unify the equations for the intrinsic hydraulic conductivity and the hydraulic conductivity.

The role of ice as an impedance for fluid migration was first proposed by Harlan (1973) while studying the unsaturated hydraulic conductivity in partially frozen soils. This viewpoint was confirmed by subsequent researchers such as Jame and Norum (1980) and Hansson et al. (2004). An impedance factor was adopted to describe the effects of ice on the fluid migration. However, a few more recent publications reported that the impedance factor is unnecessary as long as the SWCC is precisely determined (Newman and Wilson 1997; Watanabe and Flury 2008).

When phase changes are involved, the generalized Clapeyron equation needs to be employed to describe the thermodynamic condition for the coexistence of water and ice. The local freezing point of the pore fluid can be obtained from the generalized Clapeyron equation:

$$\frac{dh}{dT} = \frac{L_{\mathrm{f}}}{\mathrm{g}T}, \tag{21.10}$$

where h is the water head, L_{f} is the latent heat of water, and g is gravitational acceleration.

Assuming thermodynamic equilibrium conditions are maintained at the ice-pore water interface within infinitesimal time intervals, the Clapeyron equation will determine the liquid water content via the phase composition curves (Liu and Yu 2013), while the frozen water content can be determined by the governing equation. Therefore, a change to the governing equation is needed at the highest freezing temperature of pore water. When such a numerical treatment is not feasible, the following approximation can be employed to determine the ice content via Eq. 21.11:

$$\frac{d\theta_{\mathrm{i}}}{dT} = \frac{L_{\mathrm{f}}}{\mathrm{g}T}\frac{d\theta}{dT}. \tag{21.11}$$

Mechanical Field

The governing equation for the mechanical field is Navier's equation, which, as introduced in the mechanical field, incorporates the equation of motion, the strain-displacement correlation, and the constitutive relationship. The equation equilibrium is introduced in a general tensor format as

$$\nabla \cdot \boldsymbol{\sigma} + \mathbf{F} = 0, \tag{21.12}$$

where $\mathbf{u}$ is the displacement vector and $\mathbf{F}$ is the body force vector.

The strain-displacement equation is

$$\boldsymbol{\varepsilon} = \frac{1}{2}\left[\nabla\mathbf{u} + (\nabla\mathbf{u})^T\right]. \tag{21.13}$$

The constitutive equation is

$$\boldsymbol{\sigma} = \mathbf{C}:\boldsymbol{\varepsilon}, \tag{21.14}$$

where $\boldsymbol{\sigma}$ is the Cauchy stress tensor, $\mathbf{C}$ is the fourth-order tensor of material stiffness, $\boldsymbol{\varepsilon}$ is the infinitesimal strain tensor, and the symbol "**:**" stands for double contraction.

In order to consider the influence of the thermal field and the hydraulic field on the mechanical field, the constitutive relationship for porous materials is formulated as

$$\boldsymbol{\sigma} = \mathbf{D}\boldsymbol{\varepsilon}_{\text{el}} + \boldsymbol{\sigma}_0, \tag{21.15}$$

where $\mathbf{D}$ is the stiffness matrix of soil skeleton, $\boldsymbol{\sigma}_0$ is the initial stress vector, and $\boldsymbol{\varepsilon}_{\text{el}}$ is the elastic strain which can be obtained from the following relationship:

$$\boldsymbol{\varepsilon} = \boldsymbol{\varepsilon}_{\text{el}} + \boldsymbol{\varepsilon}_{\text{th}} + \boldsymbol{\varepsilon}_{\text{tr}} + \boldsymbol{\varepsilon}_{\text{hp}} + \boldsymbol{\varepsilon}_0, \tag{21.16}$$

where $\boldsymbol{\varepsilon}_{\text{th}}$ is the strain caused by thermal expansion, e.g., $[\alpha(T - T_{\text{ref}}), \alpha(T - T_{\text{ref}}), 0]^{\text{T}}$ in 2D; $\boldsymbol{\varepsilon}_{\text{tr}}$ is the strain caused by the phase change of water, which was approximated as $[0.09Q, 0.09Q, 0]^{\text{T}}$ when a unit localization tensor in mixture theory is followed, where Q is the degree of water phase transition and 0.09 is the relative change of volume when water turns into ice; $\boldsymbol{\varepsilon}_0$ is the initial strain; and $\boldsymbol{\varepsilon}_{\text{hp}}$ is the strain resulting from the change in the matric potential, which is calculated using $[h/\mathbf{H}, h/\mathbf{H}, 0]^{\text{T}}$. $\mathbf{H}$ is the poroelastic expansion coefficient in unsaturated soils. The value of $\mathbf{H}$ can be obtained through experimental measurement. The use of $\mathbf{H}$ casts light on the independent role of matric potential in the constitutive relationship of unsaturated porous media as indicated in Biot's model for an unsaturated fluid with air bubble and in Fredlund's method to address volume change of unsaturated soil (Biot 1941; Fredlund and Rahardjo 1993). The exact formulation of the stress or strain terms to couple different phases in unsaturated porous materials or materials with even more fluid phases is still under debate. The formulations given above are just simple examples to illustrate the coupling of the hydraulic field with the phase change of the pore liquid to the mechanical field.

General Boundary Condition

The treatment of the boundary conditions in the thermohydromechanical field is not different from the ensemble of the treatments of the boundary conditions of the individual monolithic physical fields. However, in a multiphysical field involving several physical fields, it would be beneficial to treat all the boundary conditions in a general way. The multiphysics simulation package COMSOL provides a very convenient way for such a purpose.

The general boundary condition, which includes the special cases such as the Dirichlet (first-type), Neumann (second-type), and Robin (third-type) boundary conditions, is formulated by Eq. 21.17:

$$\mathbf{n} \cdot (c\nabla u + \varsigma u - \gamma) + qu = \delta - \mathbf{h}^{\mathrm{T}}\boldsymbol{\mu}, \tag{21.17}$$

where $\mathbf{n}$ is the outward normal unit vector of a boundary, u is the dependent variable of the field under consideration (temperature, matric potential, displacements, etc.), c is a conductivity term, ς is the conservative flux convection coefficient, γ is the source in the subdomain, q is the boundary absorption coefficient, δ is the boundary source, $\mathbf{h}^{\mathrm{T}}$ is a matrix designated for the flexibility of the constraint type, and $\boldsymbol{\mu}$ is the matrix of Lagrange multiplier.

Practice Problem

Problem Description

A cylindrical soil sample with a diameter of 0.06 m and a height of 0.26 m wrapped by thermal and hydraulic insulation materials on its side. The top and bottom are connected to solid objects at temperatures of 293.15 K and 343.15 K, respectively (Fig. 21.1).

The soil has a thermal diffusivity of $1 \times 10^{-7}\mathrm{m}^2/\mathrm{s}$; a saturated hydraulic conductivity of $3.2 \times 10^{-6}\mathrm{m/s}$; a residual and saturated water content of 0.05 and 0.535, respectively; a gain factor for thermally induced water flux of 5; Young's modulus of 0.1 MPa; Poisson's ratio of 0.3; a poroelastic expansion coefficient of 7653 Pa; and a thermal expansion coefficient of $0.8 \times 10^{-6}1/\mathrm{K}$. The SWCC and

Fig. 21.1 Experimental setup

unsaturated water content are formulated using van Genuchten's model, and the constants in the model are $n = 1.48$, $m = 0.2$, and $l = 0.5$. The initial temperature and volumetric water content are 293.15 K and 0.5, respectively. The constants for the computation of the thermal diffusivity are $C_1 = 0.55$, $C_2 = 0.8$, $C_3 = 3.07$, $C_4 = 0.13$, $C_s = 2\times10^6$, $C_w = 4.2\times10^6$e6, and $C_v = 1.2\times10^3$. The initial soil-water potential head is −7 m, and the initial temperature is 293.15 K. Please simulate the THM field to predict the temperature and water distributions at the end of 5 days, and compare them against the following experimental results. Also, simulate the variation of the axial deformation of the cylindrical sample within the 5 days.

Results

Figure 21.2 shows the measured distributions of volumetric water content and temperature at the time of 5 days.

Figure 21.3 illustrates the results from the numerical simulation.

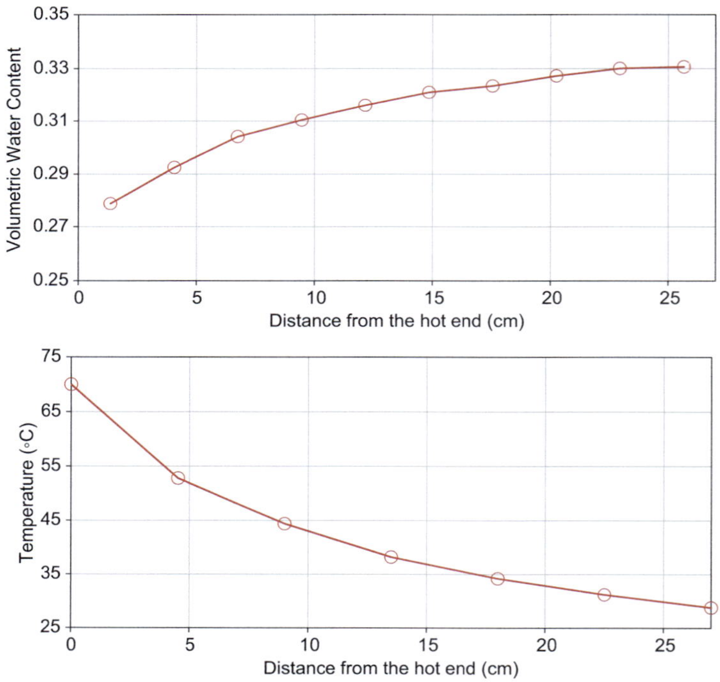

Fig. 21.2 Measured water content (upper) and temperature (lower) along the specimen at the end of the lab test (5 days)

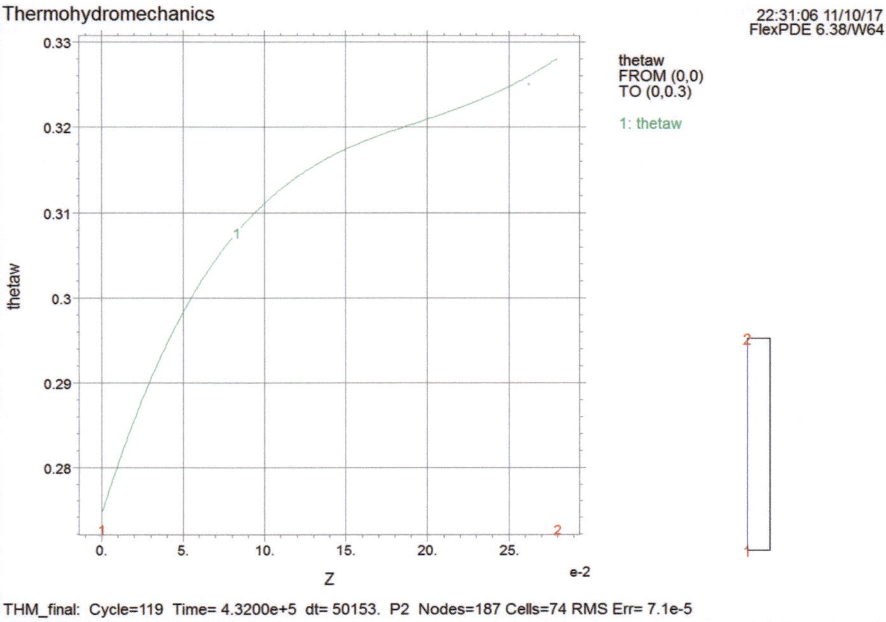

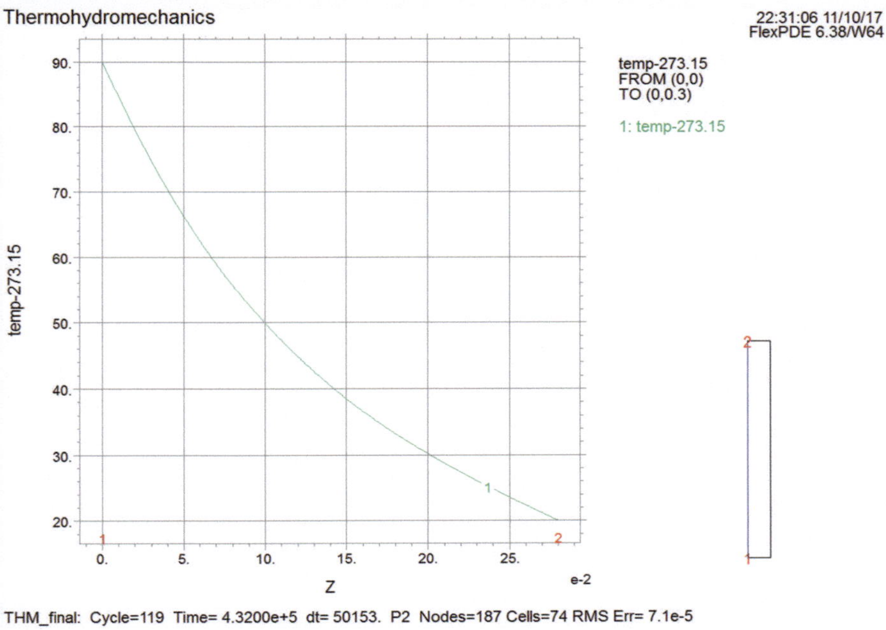

Fig. 21.3 Simulation results for water content (upper), temperature (middle), and axial deformation (lower)

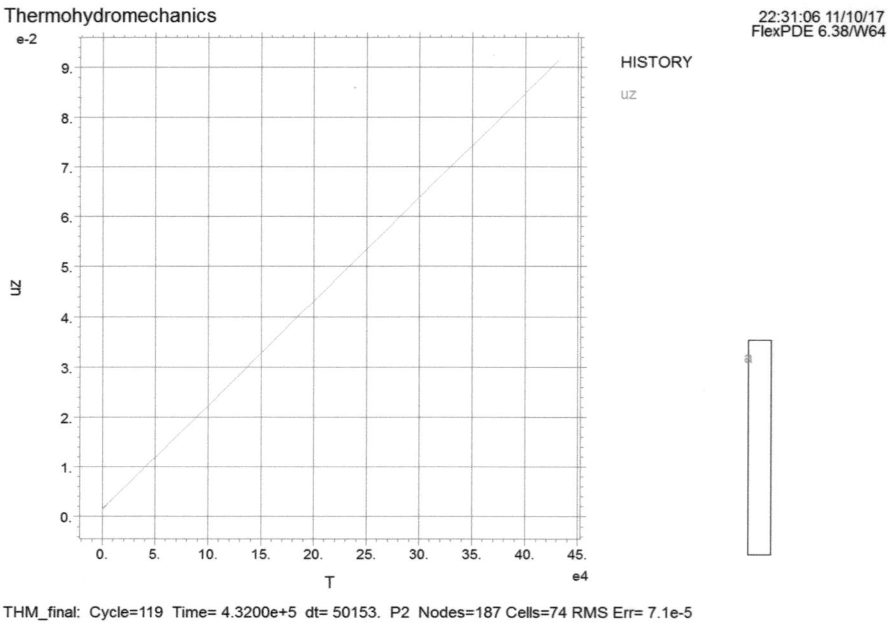

Fig. 21.3 (continued)

Appendix: FlexPDE Script File

```
{Thermohydromechanical process in a plane strain case. This program is
developed to simulate an THM test conducted by LiuRG (www.liurg.org)}
TITLE 'Thermohydromechanics'
COORDINATES YCYLINDER
VARIABLES
  u=VECTOR(ur,uz)
  Temp
  H
DEFINITIONS
  C1=0.55
  C2=0.8
  C3=3.07
  C4=0.13
  Cs=2e6
  Cw=4.2e6
  Cv=1.2e3
  thetar=0.05
  thetas=0.535
  Ks=3.2e-6
  alpha=1.11
  n=1.48
  m=0.2
  l=0.5
```

```
  gamma0=71.89e-3
  rhow=1000  !rhoi=931  ! rhos=2700
  E=1e5
  nu=0.3
  Hk=7653
  al=0.8e-6
  Se=(1+abs(alpha*H)^n)^(-m)
  thetaw=Se*(0.535-0.05)+0.05
  KH=Ks*Se*(1-(1-Se^(1/m))^m)^2
  KT=KH*(H*4*1/gamma0*(-0.1425-2.38e-4*2*Temp)*1e-3) mu=E/2/(1+nu)
  lambda=E*nu/(1+nu)/(1-2*nu)
  K=E/3/(1-2*nu)
  D_temp=(C1+C2*(thetaw)-(C1-C4)*exp(-C3*thetaw)) / (Cs*thetas
+Cw*thetaw) ! Thermal diffusivity
    delta=VECTOR(1,1)
  C_stiffness=TENSOR((2*mu+lambda,lambda,0),(lambda,2*mu
+lambda,0),(0,0,mu)) ! Stiffness matrix
  epsilon_total=VECTOR(dr(ur),dz(uz),0.5*(dz(ur)+dr(uz)))
  epsilon_thermal=VECTOR(al*(T-293.15),al*(T-293.15),0)
  epsilon_H=VECTOR((H-(-55))/HK,(H-(-55))/HK,0)
  epsilon_elastic=epsilon_total -epsilon_thermal-epsilon_H
sigma=dot( C_stiffness,epsilon_elastic)
  sigma_tensor=TENSOR((XCOMP(sigma),ZCOMP(sigma)),(ZCOMP
(sigma),YCOMP(sigma)))
INITIAL VALUES
  temp=293.15
  u=VECTOR(0,0)
  H=-7
EQUATIONS
  u: DIV(sigma_tensor)=VECTOR(0,0)
  temp: Dt(temp)+DIV(KH*GRAD(H+z)*temp)=DIV(D_temp*GRAD(temp))
  h: Dt(thetaw)=DIV(KH*GRAD(H+z))+DIV(KT*GRAD(Temp))
BOUNDARIES
  REGION 1
    START(0,0)
      NATURAL(ur)=0
      VALUE(uz)=0
      VALUE(temp)=293.15+70
      NATURAL(H)=0
    LINE TO (0.03,0)
      NATURAL(ur)=0
      NATURAL(uz)=0
      NATURAL(temp)=1*D_temp*(temp-293.15)
      NATURAL(H)=0
    LINE TO (0.03,0.28)
      NATURAL(ur)=0
      NATURAL(uz)=0
      VALUE(temp)=293.15
      NATURAL(H)=0
    LINE TO (0,0.28)
      VALUE(ur)=0
      NATURAL(uz)=0
      NATURAL(temp)=0
      NATURAL(H)=0
    LINE TO CLOSE
```

```
TIME 0 TO 3600*24*5 by 3600
MONITORS
PLOTS
  FOR CYCLE=1
    CONTOUR(thetaw)
    ELEVATION(thetaw) FROM (0,0) TO (0,0.3)
    ELEVATION(temp-273.15) FROM (0,0) TO (0,0.3)
HISTORY(uz) at (0,0.26)
END
```

Chapter 22
Electrokinetics

Introduction

Electrokinetics includes a family of phenomena resulting from the distributions of the charges and electric potential at interfaces between two different phases. In porous materials, these two phases primarily refer to the solid phase such as clay particles and the liquid phase such as water. Such two dissimilar phases usually have different chemical potentials. Upon contact, it is very likely that one of the phases, like water, is a polar liquid and will tend to orient its (dipolar) molecules in a particular direction at the interface and consequently generate a potential difference. In many cases, there are ions or excess electrons in one or both phases or ionogenic groups present; as a result, there will be a tendency for the electric charges to distribute in a nonuniform way in a small region close to the interface. All these distributions of the charges and electric potential will make the electric field in that small region close to the interface different from that inside the bulk materials. This small region can be divided into two layers according to the mobility of the liquid. A theory explaining such distributions of the charges and potential is called the double layer theory. The special redistribution of charges and potential is the origin of various electrokinetic phenomena.

The existence of this special electric field gives rise to various electrokinetic phenomena. This is because an external electric field can cause relative movements of phases via the electrostatic forces, whereas relative movement between the liquid and solid phases can lead to changes in the electric field in the double layer or even in the bulk of the liquid phase. The most common electrokinetic phenomenon is electroosmosis, in which the solid phase remains stationary while the liquid moves in response to an external electric field. This phenomenon usually occurs in fully or partially saturated porous materials, when the solid is still in the form of capillaries or porous plugs (Hunter 2013). As shown in Fig. 22.1 for soils, water moves as polarized water molecules are attached to cations, which are mobilized by an external electric field. Meanwhile, as the external electric field is also applied to

Z. Liu, *Multiphysics in Porous Materials*,
https://doi.org/10.1007/978-3-319-93028-2_22

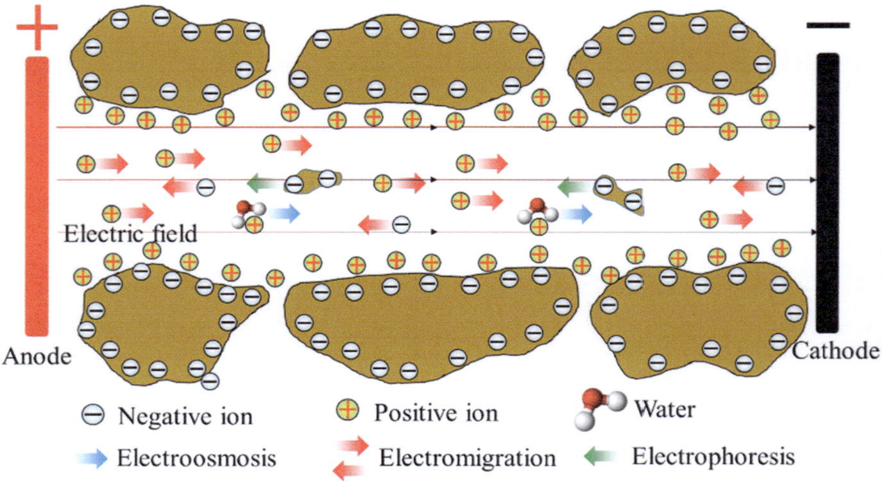

Fig. 22.1 Major electrokinetic mechanisms in porous materials

the particles with excess charges, the suspended particles can be moved, leading to another significant electrokinetic phenomenon: electrophoresis. In addition, charged ions and ion complexes also migrate due to the electric field, resulting in the phenomenon of electromigration.

There are several other types of electrokinetic processes. Similar to an external electric field, an external mechanical disturbance such as a pressure can also trigger coupled changes due to the existence of the electric field. If a pore liquid moves as a result of an external pressure gradient, the migratory water could carry the charges in it and thus result in an electric current. This phenomenon, termed streaming potential, usually happens in saturated or partially saturated porous materials, because the solid is relatively immobile compared to the liquid. By contrast, in a suspension, a solid phase can settle down due to gravity, and this relative movement of the solid to the liquid can cause an electric potential in a similar way, which is called the sedimentation potential. Introduced above are the five major types of electrokinetic phenomena.

In a broader sense, electrokinetic effects also include several other types of electrokinetic phenomena. Diffusiophoresis refers to the motion of particles under the influence of chemical potential gradients. Capillary osmosis describes the motion of a liquid in a porous body under the influence of chemical potential gradients. A colloid vibration current is an electric current generated by particles moving in a fluid under the influence of ultrasound. The electric sonic amplitude is used to describe the ultrasound generated by colloidal particles in an oscillating electric field.

Electrokinetics is a typical interdisciplinary subject which spans a wide breadth of disciplines. Related areas include colloid science, microfluidics, biochemistry, environment remediation techniques, and geo-engineering applications. The electrokinetic phenomena were first extensively studied in colloid chemistry in the early

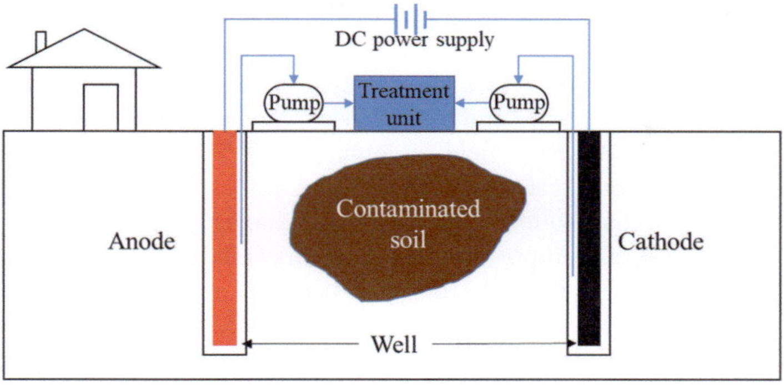

Fig. 22.2 Application of electrokinetics for soil remediation

twentieth century, since its discovery in the early part of the nineteenth century (Casagrande 1949). These days, these phenomena have attracted great attention in micro- and nano-fluidic lab-on-a-chip systems for manipulating both liquids and particles, in terms of electrokinetics over insulating surfaces and that over conducting surfaces (Zhao and Yang 2012). In biochemistry, interest has been placed on the theories and practice for separating a complex mixture of components with different motilities (Righetti et al. 1979). In environmental applications, low-level direct currents have been explored as a "cleaning agent" to remediate polluted soils and other contaminated porous materials using a setup similar to the one illustrated in Fig. 22.2. This technique demonstrated a great potential in saline soil restoration, nutrient recovery from wastes, and repair/maintenance of building structures (Huang et al. 2012). Attempts in geotechnical engineering were first pioneered by Casagrande (1949) in the 1930s, which has been developed into a common technique for strengthening soft clays and silts and stabilizing earth slopes and dams, e.g., Bjerrum et al. (1967), Casagrande (1983), and Lo et al. (1991).

In this chapter, we will first discuss the double layer theory, which sets the basis for understanding the electrokinetic phenomena. Then three sections will be devoted to electroosmosis, electromigration, and electrophoresis for the transport of water, free ions, and charged particles in soils, respectively, under the influence of an external electric field. Later, we will combine these theories into a complete framework for considering these three major types of electrokinetic phenomena. A practice problem developed based on a lab test will be introduced in the last section.

Double Layer Theory

The occurrence of electrokinetic phenomena is attributed to the potential and charge redistributions on the interface between the charged solid and its neighboring liquid containing opposite charges, which can be described using the double layer theory.

Therefore, it is necessary to understand the double layer theory to better comprehend electrokinetics. However, this could lead to a confusion between the double layer theory and electrokinetics. In fact, the physical mechanisms responsible for the double layer theory also represent a multiphysical process at the scale of the double layer: an equilibrium diffusion field is coupled with an electrostatic field, while adsorption and ionization reactions may also play a role. Multiphysics can also be used to analyze the distribution of particles and charges in the double layer area. Apparently, this type of multiphysics process occurs at the nano- or microscale. This multiphysical field is less common though it becomes an interest in microfluidics. Most effort that is devoted to the understanding of double layers was of an analytical nature, forming the major body of the double layer theory. By contrast, electrokinetic phenomena occur at a scale which is larger than the double layer: a cluster of particles or bulk porous materials instead of a single interface. This is the multiphysics to be discussed in this chapter. Despite the difference, electrokinetics borrows conclusions and understandings from the double layer theory. Considering the significance of the double layer theory, the classic theory of double layers will be briefly introduced in this subsection for the sake of completeness.

On solid-liquid interfaces, which are common in porous materials, the redistributions of electric potentials and charges on these interfaces are triggered by surface charges via a few common mechanisms: (1) preferential adsorption of ions in the solution, (2) adsorption-desorption of lattice ions, (3) direct dissociation or ionization of surface groups, and (4) charge-defective lattice: isomorphic substitution (Zhao and Yang 2012). These charges on the solid surface give rise to the redistribution of charges in the liquid phase, which are needed to achieve a charge balance in the vicinity of the solid-liquid interface.

As shown in Fig. 22.3, the liquid close to the interface can be divided into two layers according to the mobility of the charges within each. According to Stern (1924), the first layer, which is sometimes termed the Stern layer, includes one layer of ions adsorbed to the solid through a specific chemical adsorption potential. This layer is immobile and behaves more like a solid. The outer diffuse part of the double layer, or diffuse layer, starts at a distance of one layer of ions from the wall until a distance where the Coulombic force is negligible or where charge neutrality is reached in the solution. Within the diffuse layer, there is an imaginary surface which separates the diffuse layer into an inner stationary part and an outer part which could move in response to shear forces. The electric potential at this slipping plane relative to that at a point in the bulk fluid far from the interface is termed the electrokinetic potential or zeta potential, which is a characteristic potential of the whole double layer. The surface potential and the Stern potential are two other characteristic potentials on the surface and the outer surface of the Stern layer, respectively.

The electrostatic potential distribution is assumed to obey Poisson's equation in the system, which represents a typical electrostatic problem. Besides, the equilibrium diffusion field for the ion distribution is assumed to be determined by the Boltzmann equation. Combining these two mechanisms, we can obtain the so-called Poisson-Boltzmann equation. Detailed deductions of a typical double layer theory will be presented in the following paragraphs.

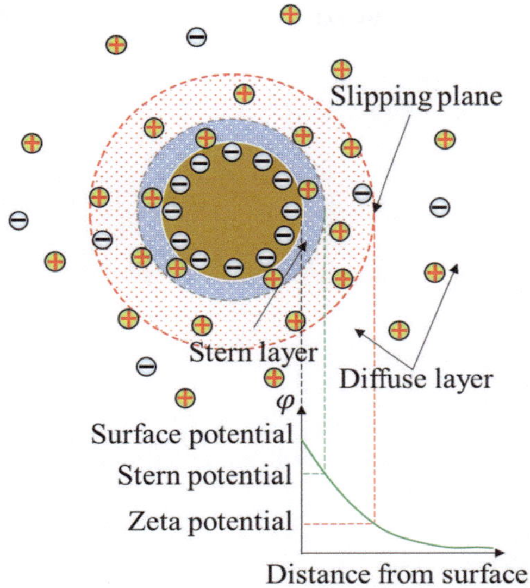

Fig. 22.3 Double layer and electric potentials

The separation of charges that occurs at the interface between two phases is called an electrical double layer because it consists of, ideally, two regions of opposite charges. The simplest model of such a system is usually attributed to Helmholtz in which both layers of charge are fixed in parallel planes to form a molecular condenser. According to Overbeek and Kruyt (1952), this model was first explicitly proposed by Perrin (1904). Although this model can be used to derive some important electrokinetic relationships, it has long been recognized as an inadequate representation of the situation. To obtain the Poisson-Boltzmann equation based on the model, Poisson's equation is first recalled:

$$\nabla^2 \psi = -\frac{\rho}{\varepsilon_0 \varepsilon_r}, \tag{22.1}$$

where ψ is the electric potential, ρ is the volumetric charge density, ε_0 is the dielectric constant of vacuum, and ε_r is the relative permittivity.

At equilibrium, the diffusional force and the electric force on an ion must balance out, which requires

$$\nabla \mu_i = -z_i e \nabla \psi, \tag{22.2}$$

where μ_i and z_i are the chemical potential and valence of ion type i, respectively, e is the elementary charge, and ψ is the electric potential.

Then we use the definition of the chemical potential per ion:

$$\mu_i = \mu_{i,0} + k_B T \ln(C_i), \tag{22.3}$$

where k_B is the Boltzmann constant, T is the absolute temperature, and C_i is the number concentration (see the chapter for the concentration field). Combining the above two equations and then integrating from a point in the bulk liquid lead to the Boltzmann equation:

$$C_i = C_{i,0}\exp\left(-\frac{z_i e\psi}{k_B T}\right). \tag{22.4}$$

The volumetric charge density in the neighborhood of the surface is given by

$$\rho = \sum_i C_i z_i e. \tag{22.5}$$

Substituting Eqs. 22.4 and 22.5 into Poisson's equation (Eq. 22.1), we obtain the complete Poisson-Boltzmann equation:

$$\nabla^2\psi = -\frac{1}{\varepsilon_0\varepsilon_\mathrm{r}}\sum_i\left[C_{i,0}z_i e\exp\left(-\frac{z_i e\psi}{k_B T}\right)\right]. \tag{22.6}$$

The major part of the double layer theory is to solve, improve, and validate this equation for predicting the distributions of the electric potential and charges within the double layer based on the above equations. These equations form the major part of the DLVO theory (Gouy 1910; Derjaguin and Landau 1941; Verwey 1947), which has been found to have many limitations, especially the limited applicability in relatively low concentration ranges. However, the theory describes the major mechanisms for the formation of the electric field close to the surface, which is responsible for the electrokinetic phenomena, and identifies critical concepts such as the slipping or shear plane.

Electroosmosis

The water flow in partially or fully saturated porous materials induced by an external electric field is referred as electroosmotic flow. Electroosmosis is the main mechanism underlying the water flow through fine-grained porous materials under the influence of an electric field.

The direction of the water movement depends on the nature of the double layer system around the charged particles. This water movement, or electroosmotic flow, is caused by the Coulombic force resulting from the net charges in the water and the external electric field. In soils, the clay particles are usually negatively charged. Therefore, there are more positive charges than negative charges in the diffuse layer to ensure an electrostatic balance in the double layer. The positive charges will move to the cathode due to the Coulombic forces between the charges and the external electric field. This movement will lead to the movement of the water, enabling water

Table 22.1 Values of electroosmotic hydraulic conductivity of soils

Number	Soil type	K_{EO} (m^2/ V. s)
1	London clay	5.8×10^{-9}
2	Boston Blue clay	5.1×10^{-9}
3	Kaolin	5.7×10^{-9}
4	Clayey silt	5.0×10^{-9}
5	Rock flour	4.5×10^{-9}
6	Na-montmorillonite	2.0×10^{-9}
7	Mica powder	6.9×10^{-9}
8	Fine sand	4.1×10^{-9}
9	Quartz powder	4.3×10^{-9}
10	As quick clay	$(20.0–2.5) \times 10^{-9}$
11	Bootlegger Cove clay	$(2.4–5.0) \times 10^{-9}$
12	Silty clay, West Branch Dam	$(3.0–6.0) \times 10^{-9}$
13	Clayey silt, Little Pic River, Ontario	1.5×10^{-9}

to move to the cathode in the above case. The water moves in the opposite direction if the solid particles are positively charged.

The electroosmotic flow can be modeled using the Helmholtz-Smoluchowski theory (Chung and Kang 1999). The flux of the electroosmotic flow is described by an equation analogous to Darcy's law (Casgrande 1949; Bjerrum et al. 1967; Alshawabkeh et al. 1999):

$$\mathbf{J}_{EO} = -K_{EO} \nabla \psi, \tag{22.7}$$

where $\mathbf{J}_{EO}$ is the electroosmotic flow flux (m/s), K_{EO} is the electroosmotic (hydraulic) conductivity $m^2/(V \cdot s)$, and ψ is the electric potential. The coefficient of electroosmotic conductivity is generally in the range of 10^{-9} to 10^{-10} $m^2/(V \cdot s)$ as shown in Table 22.1 (Mitchell and Soga 2005).

The electromagnetic flow occurs because water molecules move with cations, to which the water molecules are "attached" due to their polarity. It is usually accepted to assume that the water velocity, $\mathbf{v}$, is the same as the drift velocity of the ion, $\mathbf{v}_d$, under the influence of an external electric field, $\mathbf{E} = -\nabla\psi$. In this case, the electroosmotic flux $\mathbf{J}_{EO}$, or equivalently, a Darcy velocity due to electroosmosis, is equal to the drift velocity:

$$\mathbf{J}_{EO} = \mathbf{v} = \mathbf{v}_d = \mu_q \mathbf{E}. \tag{22.8}$$

This average drift velocity of the cations in the diffuse layer can be described using the equation proposed by Smoluchowski (1903):

$$\mu_q = -\frac{\varepsilon \zeta}{\eta}, \tag{22.9}$$

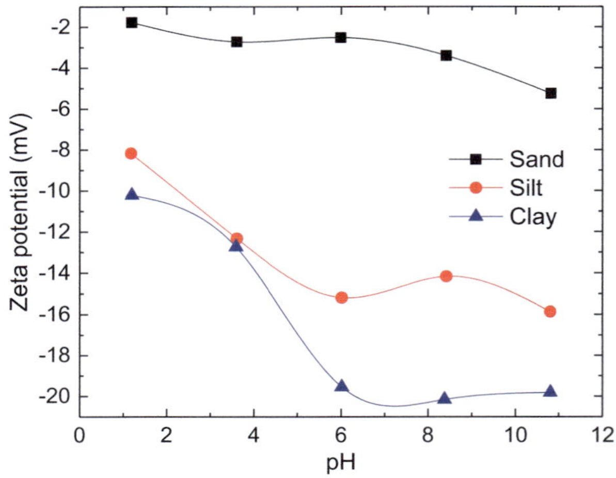

Fig. 22.4 Zeta potential values of typical soils (Qi and Zhang 2015)

where ζ is the zeta potential, ε is the (absolute) permittivity of the dispersion medium (i.e., pore water), and η is the dynamic viscosity. Please be aware of that in other chapters of this book, the dynamic viscosity is represented using the symbol μ, which is reserved for electrical ion mobility in this chapter. A negative sign is added to allow for the fact that ζ values for normal porous materials such as soils are usually negative. Typical values of the zeta potential values for different soils are shown in Fig. 22.4.

Combining the above three equations, we can obtain the formulation of K_{EO}:

$$K_{\mathrm{EO}} = -\frac{\varepsilon \zeta}{\eta}. \tag{22.10}$$

In porous materials, the formulation of the electroosmotic hydraulic conductivity needs to take into consideration of the porosity ϕ and tortuosity τ. But it is noted that tortuosity defined in the literature for electrokinetics is usually different from that is used in other areas, i.e., the definition in the previous chapters. Usually, τ is defined as the ratio between the length of the real flow streamline L and the corresponding straight line distance, d, as $\tau = L/d$; however, the tortuosity used in electrokinetics is defined as

$$\tau_{\mathrm{EK}} = (d/L)^2. \tag{22.11}$$

Therefore, the two definitions of the tortuosity can be related as

$$\tau_{\mathrm{EK}} = (1/\tau)^2. \tag{22.12}$$

The effective electrical ionic mobility in a porous material saturated with a pore liquid can be formulated as

$$\mu_q^{\text{eff}} = -\frac{\varepsilon\zeta}{\eta}\phi\tau_{\text{EK}}. \tag{22.13}$$

Accordingly, the electroosmotic hydraulic conductivity including the porosity and tortuosity is

$$K_{\text{EO}} = -\frac{\varepsilon\zeta}{\eta}\phi\tau_{\text{EK}}. \tag{22.14}$$

In unsaturated soils, it is more appropriate to replace the porosity with the volumetric water content as

$$K_{\text{EO}} = -\frac{\varepsilon\zeta}{\eta}\theta\tau_{\text{EK}}. \tag{22.15}$$

Despite the above analytical expression, for the numerical simulation purpose, it seems to be more practical to measure or even assume the value of the saturated electroosmotic hydraulic conductivity and extend this conductivity to unsaturated soils based on the degree of saturation as

$$K_{\text{EO}} = K_{\text{EO},0} \cdot a \cdot S^b, \tag{22.16}$$

where $K_{\text{EO},\,0}$ is the electroosmotic hydraulic conductivity of the saturated soil, S is the degree of saturation, and a and b are fitting constants. As can be seen, the above two equations predict a linear and a nonlinear relationship between the electroosmotic hydraulic conductivity and the water content, respectively. Despite these equations, the real nature of this relationship is not that clear and deserves further investigations.

Two characteristics differentiate the electroosmotic flow from the flow caused by the hydraulic gradients in porous materials. First, the electroosmotic flow is a plug flow, in which the velocity of the fluid is invariant across any cross section perpendicular to the flow direction. This leads to much less significant dispersive effects (see dispersion in Chap. 13). Second, the velocity of the electroosmotic flow is independent of the size of the water passway, as long as the thickness of the electrical double layer is much smaller than the characteristic length of the passway. This means that the electroosmotic flow could be significant or even dominant in small pores or fine-grained porous materials, in which the flow caused by hydraulic gradients diminished due to the small size of the water passway. Unlike a parabolic velocity profile generated by a pressure gradient in the flow, a plug flow's velocity profile is approximately planar, with slight variation near the electrical double layer. In a simple calculation to compare the two types of flows (Huweg 2013), we can assume they are equal:

$$-K_{\text{EO}}\nabla U = -K_{\text{H}}\nabla\psi. \tag{22.17}$$

If a K_H of 1.3×10^{-10} m/s and a K_{EO} of 5×10^{-9} m^2/(V·s) are used, plus an electric potential gradient of 50 V/m, which is common, the equivalence of the two flows will require a hydraulic gradient of 1923 m/m. This value is higher than the numbers we would encounter in the field by several orders of magnitude.

In recent years, in order to apply electroosmotic flow to electrokinetic remediation, the complex physical phenomena underlying the water flow in soils under an external electric field have been extensively investigated (Acar and Alshawabkeh 1993; Virkutyte et al. 2002). Besides altering the pore water pressure regime by inducing a time-dependent coupled flow and deformation (consolidation) process in the soil mass, the application of an electric field to a fine-grained soil promotes complex electrochemical phenomena, such as hydrolysis reactions at the electrodes, ion migration, changes in pore water pH, aqueous phase reactions, precipitation and dissolution, and adsorption and desorption. These phenomena can, in turn, induce permanent modifications to the mechanical properties of the soil (Esrig and Gemeinhardt 1900; Gray 1970). Therefore, it is worthwhile to mention that investigations into the electroosmotic phenomenon may not just be about electroosmosis but, instead, be associated with a complex multiphysical process.

Electromigration

An electric field applied to porous materials can cause the pore liquid containing ions to move, i.e., electroosmotic flow. Due to the same reason, the external electric field can also lead to the movement of the ions themselves. This movement of dissolved salt ions due to the external electric field is called electromigration. Ions can still move due to diffusion, dispersion, and advection as we introduced in the concentration field. However, at field moisture contents we mostly encounter, the dominant ionic transport mechanism during the electrokinetic process is usually electromigration (Alshawabkeh et al. 1999). Experiments have led to a conclusion that the rate of electromigration is proportional to the ion concentration (Kim et al. 2005) and salt type (Cho et al. 2009). Together with the more easily observable relationship between the magnitude of the electric field and the rate the electromigration, we can reach the mathematical formulation of the electromigration in a bulk solution:

$$\mathbf{J}_{\mathrm{EM}} = \mathbf{v}C = -\mu_q \nabla \psi c, \tag{22.18}$$

where $\mathbf{J}_{\mathrm{EM}}$ is the electromigration (ion) flux (mol/(m^2 · s)), μ_q is the electric mobility of the ion (m^2/(V · s)), and c is the (molar) concentration (mol/m^3). As introduced in the previous section, the ionic mobility can be defined as the average velocity of an ion species moving under a unit electric field (1 V/m).

The above equation is valid for any ion species. Accordingly, μ_q is the electric mobility of the ion species under consideration, which is different from the average mobility discussed in the electroosmosis section. In order to use the above equation, the electric mobility of the dispersed ion needs to be related to the material properties that we are more familiar with, such as diffusivity. This goal can be achieved via the Einstein relationship or Einstein-Smoluchowski relation:

$$D_{\text{diff}} = \frac{\mu_q k_B T}{q}, \tag{22.19}$$

where D_{diff} is the coefficient of diffusivity, k_B is the Boltzmann constant, T is the absolute temperature (in Kelvin), and q is the charge of each ion. The above equation can be reformulated into

$$\mu_q = \frac{D_{\text{diff}} q}{k_B T} = \frac{D_{\text{diff}} \cdot z \cdot F}{RT},$$

where F is Faraday's constant and $F = e \cdot N_a$ (96485.3329 C/mol), in which e is the elementary charge and N_a is Avogadro's constant; R is the universal gas constant and $R = k_B N_a$ (J /K mol); and z is the valence of the ion species under consideration and $q = z \cdot e$.

Similar to the extension of the bulk solution theory to allow for porous materials in the previous section, the effective electrical ionic mobility considering the porosity and tortuosity of (unsaturated) porous materials is:

$$\mu_q^{\text{eff}} = \frac{D_{\text{diff}} \cdot z \cdot F}{RT} \theta \tau_{\text{EK}}. \tag{22.20}$$

Then the flux of the ion due to electromigration can be calculated as

$$\mathbf{J}_{\text{EM}} = \mathbf{v} c = -\frac{D_{\text{diff}} \cdot z \cdot F}{RT} \theta \tau_{\text{EK}} \nabla \psi c. \tag{22.21}$$

In the above two equations, if the diffusivity coefficient has already considered the porosity and tortuosity, i.e., in terms of effective diffusivity $D_{\text{diff}}^{\text{eff}}$, then $\theta \tau_{\text{EK}}$ can be left out:

$$\mu_q^{\text{eff}} = \frac{D_{\text{diff}}^{\text{eff}} \cdot z \cdot F}{RT}. \tag{22.22}$$

The ionic mobility of salt ions is at least one order of magnitude higher than the diffusion coefficient of the same ion. In addition, the value of the ionic mobility is higher than that of the electroosmotic hydraulic conductivity, typically by two orders of magnitude (Sullivan 2008; Huweg 2013). In other words, the velocity of electromigration of ions is typically higher than the electroosmotic flow rate. More significantly, the rate of electromigration is not directly affected by the characteristics of the double layer, such as the zeta potential.

The direction of the electromigration is determined by the charge of the ion: an ion moves to the electrode with the opposite charge. This direction is reflected in the mobility term. Positively charged ions, i.e., cations, have a positive charge number, enabling them to move in the direction of the electric field. On the contrary, negatively charged ions, i.e., anions, move against the electric field. As the electroosmotic flow can also advect particles, it is possible that the direction of this advection is opposite to the direction of the electromigration for some ions. But as mentioned, the velocity of electromigration usually far surpasses that of the electroosmotic flow; therefore, the electromigration should dominate. The coexistence of multiple ions will further complicate the formulation of the mobility due to the interaction between different ions. Some further information will be offered in later sections.

Electrophoresis

Electrophoresis is the movement of dispersed particles relative to the fluid in contact with them, due to the electric field generated by the double layer between the solid and the liquid. This phenomenon usually happens in suspensions because suspended particles are more mobile compared to the liquid. This is because the particles are the dispersed phase, while the liquid as the dispersion medium is predominant in volume and forms the skeleton of the composite. By contrast, in soils with a fixed solid skeleton, especially unsaturated soils, the influence of this mechanism on the ion transport is not appreciable.

The motion of the dispersed particles is caused by an applied electric field. The electrostatic forces caused by this external electric field act on both the solid and bulk liquid. From this viewpoint, electrophoresis is the "opposite" process of electroosmosis: the solid particles move when they are dispersed in a relatively stagnant water, e.g., colloid, whereas in electroosmosis the liquid moves if the solid particles are fixed.

The suspended particles suffer from this force due to the existence of electric charges on their surfaces. Also, considering that the Stern layer and a very small part of the diffuse layer will also move along with the particles, the electrostatic forces will also be affected by surface-adsorbed species. In the meanwhile, the electric field also exerts a force on the ions in the diffuse layer which has a direction opposite to that acting on the surface charge. Though this force is not actually applied to the particle, it is transferred indirectly to the particle surface through viscous stress. This force is termed the electrophoretic retardation force, which also contributes to the movement of suspended particles. Both the direct and indirect parts contributing to the motion of the particles are of the Coulombic force nature.

The motion caused by the electrostatic Coulombic forces will contribute to the development of an average velocity field. Typically, the average velocity is directly proportional to the applied force. In a colloid, the water velocity can be described using the Smoluchowski equation or drift-diffusion equation introduced in the section for electroosmosis:

$$\mathbf{v} = -\mu_q \mathbf{E} = -\nabla\psi \frac{\varepsilon\varsigma}{\eta}, \tag{22.23}$$

where ε is absolute permittivity of the dispersion medium (fluid), ς is the zeta potential, and η is the dynamic viscosity. The negative sign in front of **E** indicates that, if ς is negative, the dispersed particles move against the electric field: from the cathode to the anode, and vice versa.

It is noted that the above formulation of the velocity overlooks many effects such as the Debye length and surface conductivity. When the Debye length is much larger than the particle radius, i.e., a thick double layer, the following equation can be used instead:

$$\mathbf{v} = -\nabla\psi\frac{2\varepsilon\varsigma}{3\eta}. \tag{22.24}$$

This velocity field will be an independent mechanism contributing to the overall movement of the solid particles. Usually, the movement is of interest for particle transport in terms of a concentration field. Accordingly, the above velocity attributed to electromigration will function in a way similar to the velocity in the ordinary advection mechanism. Due to the same reason, the electrophoresis flux as a flux for ion transport is

$$\mathbf{J}_{\mathrm{EP}} = \mathbf{v}c = -\frac{\varepsilon\varsigma}{\eta}c\nabla\psi, \tag{22.25}$$

or

$$\mathbf{J}_{\mathrm{EP}} = -\frac{2\varepsilon\varsigma}{3\eta}c\nabla\psi \tag{22.26}$$

for thick double layers. It is noted that c is the concentration of the dispersed particles that are mobile in response to the external electric field.

As mentioned, the mechanism is not significant in porous materials as movable solid particles with double layers may only account for a very small portion of the solid phase with double layers. Therefore, the mathematical formulation including the porosity (or water content) and tortuosity is usually not of interest. Here, we still give out a general equation for the electrophoretic flux in porous materials to show the idea:

$$\mathbf{J}_{\mathrm{EP}} = -\frac{\varepsilon\varsigma}{\eta}c\nabla\psi\theta\tau_{\mathrm{EK}}. \tag{22.27}$$

Mathematical Formulation

This section presents a general mathematical formulation of the multiphysics associated with typical electrokinetic processes including the conservation of pore fluid mass (h: water head), electrostatics (ψ: electric potential), and ion transport (c: molar

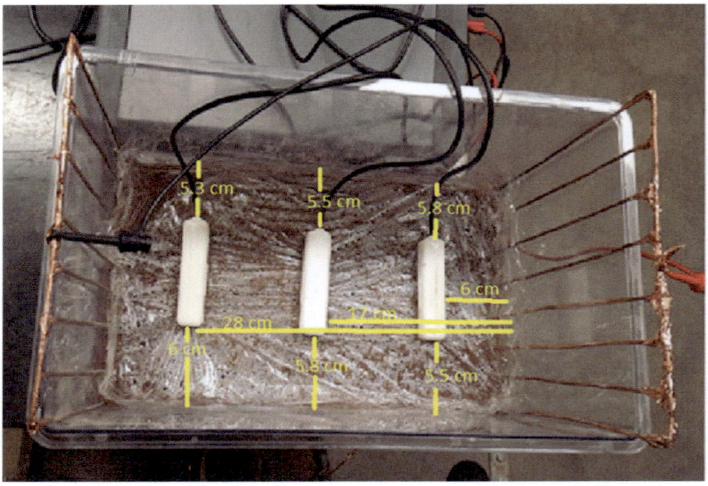

Fig. 22.5 Setup of an electrokinetics experiment with soils

concentration). Figure 22.5 shows a typical experimental setup to generate such processes. The external electric field applied via the two copper electrodes on two sides of a soil can trigger a multiphysical process including but not limited to the above physical fields. For electrokinetics, mechanisms to be considered include electroosmosis, electromigration, and electrophoresis in addition to other major particle transport mechanisms, i.e., diffusion, dispersion, and advection.

Governing Equation

Unlike electrophoresis, electroosmosis contributes to the mass transfer of the continuous phase instead of the dispersed phase. Accordingly, the water movement equation should be used. For a more general variably saturated porous material, an extension to Richards equation (Richards 1931) for unsaturated water flow could be expressed as

$$\phi \frac{\partial S_w}{\partial h} \frac{\partial h}{\partial t} + \nabla \cdot \mathbf{J}_\mathrm{m} = 0, \tag{22.28}$$

where S_w is the degree of saturation, t is the time, ϕ is the porosity, h is the hydraulic pressure head, and $\mathbf{J}_\mathrm{m}$ is flux or the extended Darcy velocity. In this case, let us consider the water flow caused by both the water pressure/potential gradient and the electric potential gradient:

$$\mathbf{J}_\mathrm{m} = \mathbf{J}_\mathrm{H} + \mathbf{J}_\mathrm{EO} = -K_{\mathrm{H},0} K_{\mathrm{H}r} \nabla (h + z) - K_{\mathrm{EM},0} K_{\mathrm{EM},r} \nabla \psi \tag{22.29}$$

where $K_{\text{H, 0}}$ is the intrinsic (saturated) hydraulic conductivity (due to water head), $K_{\text{H}r}$ is the coefficient of relative hydraulic conductivity, z is the coordinate along the gravity direction, $K_{\text{EM, 0}}$ is the intrinsic (saturated) electroosmotic conductivity, $K_{\text{EM, }r}$ is the coefficient of relative electroosmotic conductivity, and ψ is the electric potential. Substituting Eq. 22.29 into Eq. 22.28, the governing equation for the coupled unsaturated water flow including electroosmosis turns into

$$\phi \frac{\partial S_w}{\partial h} \frac{\partial h}{\partial t} = \nabla \cdot [K_{\text{H},0} K_{\text{H}r} \nabla (h + z) + K_{\text{EM},0} K_{\text{EM},r} \nabla \psi]. \tag{22.30}$$

According to Ohm's law, the electric current density $\mathbf{J}_{\text{E}}$ can be expressed as (Ohm 1827)

$$\mathbf{J}_{\text{E}} = -\sigma \nabla \psi, \tag{22.31}$$

where σ is the conductivity of the soil. Due to the conservation of electric charges, the governing equation for the electric potential could be obtained by placing a divergence operator on the two sides of the Ampère equation, leading to

$$-\nabla \cdot \mathbf{J}_{\text{E}} = \varepsilon \frac{\partial \psi}{\partial t}, \tag{22.32}$$

where ε is the absolute permittivity of the soil, which is very small. Substituting Eq. 22.31 into Eq. 22.32, the governing equation for the electrostatics in soils could be obtained as follows:

$$\nabla \cdot (\sigma \nabla \psi) = \varepsilon \frac{\partial \psi}{\partial t}. \tag{22.33}$$

In addition to the water movement and electrostatics, the conservation of ions is also usually of major interest in electrokinetics applications. The governing equation for ion transport can be obtained by extending the equation introduced in the chapter for the concentration field:

$$\frac{\partial c}{\partial t} = \underbrace{-\nabla \cdot (-D_{\text{diff}} \nabla c)}_{\text{diffusion+dispersion}} - \underbrace{\nabla \cdot (\mathbf{J}_{\text{EM}} + \mathbf{J}_{\text{EP}})}_{\text{dispersion}} - \underbrace{\nabla \cdot (\mathbf{v}c)}_{\text{advection}}. \tag{22.34}$$

All the flux and velocity terms in the above equation have been introduced in the previous sections.

Typical Auxiliary Relationships

One big challenge in electrokinetics simulation is the nonlinearity in the material properties, that is, the complicated mathematical descriptions of the undetermined parameters in the governing equations: the relationships between these parameters and their dependence on the dependent and independent variables. Most of these physical relationships are associated with the consideration of the unsaturated regime in the computational domain. The (relative) hydraulic conductivity, (relative) electroosmosis conductivity, and (relative) electric conductivity could be expressed as a function of the degree of saturation. There is no consensus on the adoption of the auxiliary relationships. However, to offer an example, in this subsection, some relatively new equations for the auxiliary relationships in electrokinetics, which were adopted in Mattson et al. (2002), Tamagnini et al. (2010), and Yuan and Hicks (2013), together with some other physically based equations, which have been tested by the author with numerical simulations, are presented. Typical values for the fitting constants in the relationships are also given, which will be used in the later section for a practice problem.

The degree of saturation could be expressed as a function of the hydraulic pressure head:

$$S = [1 + (-\alpha h)^n]^{-m}, \tag{22.35}$$

where typical values for the fitting constants are $\alpha = 8.51\ \mathrm{m}^{-1}$, $n = 1.31$, and $m = 1 - \frac{1}{n}$.

It is also common to express the relative hydraulic conductivity as a function of the degree of saturation. Here we use a simple power rule:

$$K_{\mathrm{H}r} = a_{\mathrm{H}}(S)^{b_{\mathrm{H}}}, \tag{22.36}$$

where typical values for the fitting constants are $a_{\mathrm{H}} = 1.0$ and $b_{\mathrm{H}} = 5$.

Similarly, the electroosmotic conductivity could be expressed as a function of the degree of saturation:

$$K_{\mathrm{EO},r} = a_{\mathrm{EO}}(S)^{b_{\mathrm{EO}}}, \tag{22.37}$$

where typical values for the fitting constants are $a_{\mathrm{EO}} = 1.0$ and $b_{\mathrm{EO}} = 3.2$.

The electric conductivity deserves special attention in soil-related applications especially electrokinetics. In some literature, the electric conductivity of a soil is assumed to be the product of the electric conductivity of the saturated soil σ_0 and the relative electric conductivity σ_r, which is similar to the treatment of the hydraulic conductivity. Accordingly, a relationship between the electric conductivity and the degree of saturation has also been proposed:

$$\sigma_r = a_\sigma(S_w)^{b_\sigma}, \tag{22.38}$$

where a_σ and b_σ are fitting constants. Typical values are $a_{\mathrm{E}} = 0.109$ and $b_{\mathrm{E}} = 2$. However, this relationship may not work well in electrokinetic applications, because the variation of ion concentrations under an external electric field can also significantly alter the electric conductivity. Therefore, a more accurate formulation of the electric conductivity should also include the ion concentration(s). In a simple way, we can obtain such a formulation by considering the origin of the electric conductivity. The electric conductivity of a soil has contributions from both the liquid and the solid phases

$$\sigma = \sigma_\mathrm{l} + \sigma_\mathrm{s}, \tag{22.39}$$

where σ_l and σ_s are the electric conductivity components contributed by the liquid and the solid, respectively. σ_l is termed the bulk electric conductivity of the pore liquid in some literature. This quantity changes with both the degree of saturation, tortuosity, and the types and concentrations of the ions in the liquid. Theoretically, σ_l can be formulated based on the theory of electromigration:

$$\sigma_\mathrm{l} = \sigma_w \theta \tau_\mathrm{EK} = \mu_q^\mathrm{eff} cF = \frac{D_\mathrm{diff} \cdot z \cdot F}{RT} \theta \tau_\mathrm{EK} cF, \tag{22.40}$$

where σ_w is the true electric conductivity of the pore water and $\mu_q^\mathrm{eff} c$ gives out the number of ions (in mol) passing through a unit area per unit time under a unit electric field. The use of Faraday's constant turns the mass into a charge amount (C), leading to the electric conductivity, i.e., amount of charge passing through a unit area per unit time under a unit electric field. In strong electrolytes with high concentrations, Kohlrausch's law instead of a linear function is more appropriate for describing the relationship between σ_l and the concentration:

$$\sigma_\mathrm{l} = \mathrm{a}_\sigma c - \mathrm{b}_\sigma c^{1.5}, \tag{22.41}$$

where a_σ and b_σ are fitting constants.

Besides, the electric conductivity in the above context needs to be differentiated from some other electric conductivity concepts in soil-related applications. In fact, the σ discussed above is the apparent electric conductivity of the whole soil, which is represented using the symbol EC_a in many studies (Corwin and Plant 2005). It is the value we would obtain using an electric conductivity sensor. Another electric conductivity that is frequently used in soil chemistry and agricultural applications is the extract conductivity σ_e. σ_e is the saturation extract electric conductivity. This electric conductivity is obtained by first saturating a soil with deionized water and then measuring the electric conductivity of the extracted water from this saturated soil. This is the traditional way to measure the electric conductivity of a pore liquid. Therefore, a majority of the electric conductivity values reported in the literature are, in fact, σ_e. σ_e is different from the true electric conductivity of the pore water σ_w. If the salts stay dissolved in water as the water content changes, then the following relationship is valid:

$$\sigma_\mathrm{e} = \sigma_\mathrm{w} \cdot S. \tag{22.42}$$

One more auxiliary relationship is still needed to enable numerical simulation of electrokinetics in porous materials for tortuosity. Mattson et al. (2002) employed a simple relationship between the tortuosity and the water content as follows:

$$\tau_{\text{EK}} = \frac{F_{\lambda}(\theta - \theta_{\text{r}})^{n+1}}{\theta_{\text{sat}}}. \tag{22.43}$$

where θ, θ_{r}, and θ_{sat} are the water content, immobile water content, and saturated water content, respectively, and F_{λ} is a constant, i.e., 1.58.

A more complicated way is available to consider the mobility of various ion species in porous materials. In this way, the effective ionic mobility is related to that in a bulk solution to consider the influence of other ions on the mobility of the ion under consideration, i.e., species i:

$$\mu_{q,i}^{\text{eff}} = \mu_{q,i} \cdot \theta \cdot \tau_{\text{EK}} \cdot 10^{-0.5058 \cdot z^2 \cdot \left(\frac{\sqrt{I}}{1+\sqrt{I}} - 0.3I\right)}, \tag{22.44}$$

where I is the ionic strength, which can be related to the electric conductivity of the free solution σ_{w} via the following function, and z is the valence of the ion. This function is obtained by fitting the test results of 124 river water samples (Ponnamperuna et al. 1966):

$$I = 0.127 \cdot \sigma_{\text{w}} - 0.003. \tag{22.45}$$

Practice Problem

Problem Description

An experiment was conducted using a mixture of a play sand (poorly graded fine sand) and a clay (high plasticity) with a dry mass ratio of 1:1. The mixture was compacted gently using a miniature compactor, and the final specific gravity is 1.9. The compacted soil sample sat for 24 h to allow for the water content to become uniform, and the sample was also covered with a plastic sheet to reduce the effects of evaporation (see Fig. 22.5).

Electrodes made of copper robs were placed on two sides of the plastic tank to trigger the electrokinetic process. A voltage of 20 V was placed on the two electrodes: 10 V and − 10 V on the two electrodes. The tank had (inner) dimensions of 37 cm (length) by 21 cm (width) by 22.5 cm (height). The electrodes extend to the bottom of the soil layer (also tank), which filled slightly over one half of the tank height to generate an approximately 1D problem.

Three water content sensors (Decagon GS3), connected to a data logger (Decagon EM50), were placed along the x-axis of the tank, whose positions are displayed in Fig. 22.6. The sensors, from the left to the right, are named Sensor 1, Sensor 2, and Sensor 3. A program called ECH20 Utility was used to record the volumetric moisture content and the electric conductivity automatically. The GS3 water content sensors measure the variation of the electric conductivity between two of its probes.

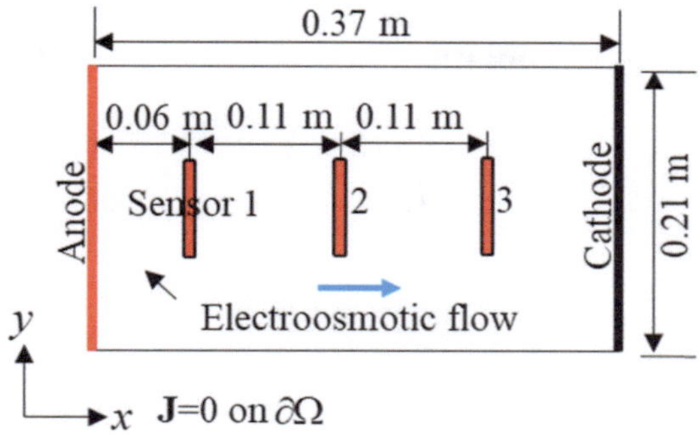

Fig. 22.6 2D simplification of an electrokinetics experiment using a 1:1 sand and clay mixture

The program took measurements every minute. Data were recorded for about 417 h. Temperatures of the soil sample were recorded using Type-K thermocouples and the built-in temperature sensors in the water content sensor.

The saturated hydraulic conductivity and saturated electroosmotic hydraulic conductivity are 5×10^{-8} m/s and 1.5×10^{-9} m/s, respectively. The soil water characteristic curve, relative hydraulic conductivity, and relative electroosmotic conductivity are formulated using Eqs. 22.35, 22.36, and 22.37 with the given default values. The tortuosity is calculated using Eq. 22.43. The initial soil water pressure head is −0.08 m. The saturated and residual water contents are assumed to be 0.40 and 0, respectively. The concentrations of cations and anions in the pore water are both 8 mol/m^3. The average diffusivities of cations and anions are 0.5×10^{-9} m^2/s and 1.5×10^{-9} m^2/s. The average valence is assumed to be 1 for both the cations and anions. The electric conductivity can be calculated using Eqs. 22.39 and 22.40 with $\sigma_s = 0.0032$ S/m.

Results

The results of the sand-clay mixture electrokinetics experiment were shown in terms of variations of the volumetric water content and electric conductivity at Sensors 1, 2, and 3 in Figs. 22.7 and 22.8. Shown in the figures are the raw data. As can be seen, the initial water contents measured by the sensors are much different from each other. The differences were most likely caused by the lack of the calibration. However, the trends in the curves can still reflect some major mechanisms. The electric conductivity measurements are more consistent.

If you try a few simulations for electrokinetics, it will not be difficult for you to find out that electrokinetics can easily generate the most highly nonlinear phenomena. The involved physical fields are strongly dependent on each other in a variety of

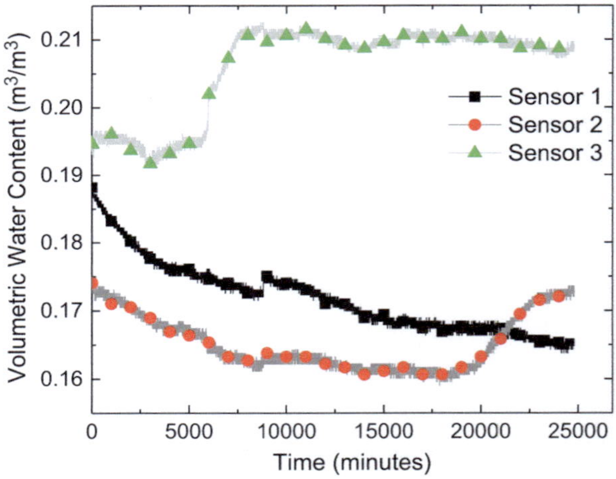

Fig. 22.7 Measured water content measured at the three sensors

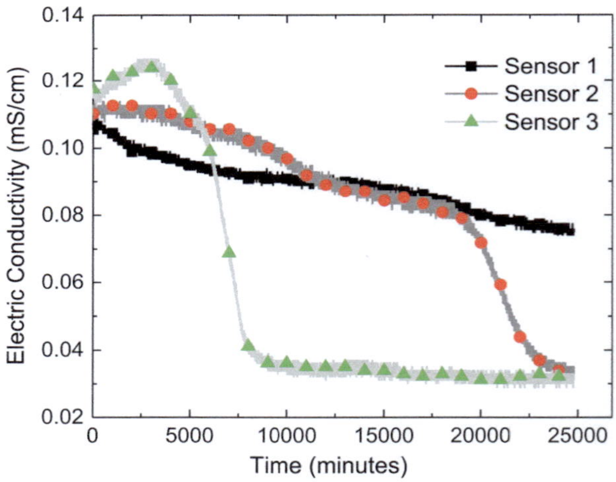

Fig. 22.8 Measured electric conductivity at the three sensors

ways. Numerical results for the water content and electric conductivity are shown in Figs. 22.9 and 22.10. Points a, b, and c are the locations of Sensor 1, Sensor 2, and Sensor 3, respectively. We did not conduct a quantitative analysis here due to the missing information about the soil properties. Despite the lack of information, we can still see the simulation with the theory presented in this chapter successfully captured the major trends in the experiments.

Besides, a detailed explanation of the above experiment can be obtained by analyzing the numerical results as shown in Figs. 22.11 and 22.12. The water content variations were primarily attributed to the pore water movement toward and water

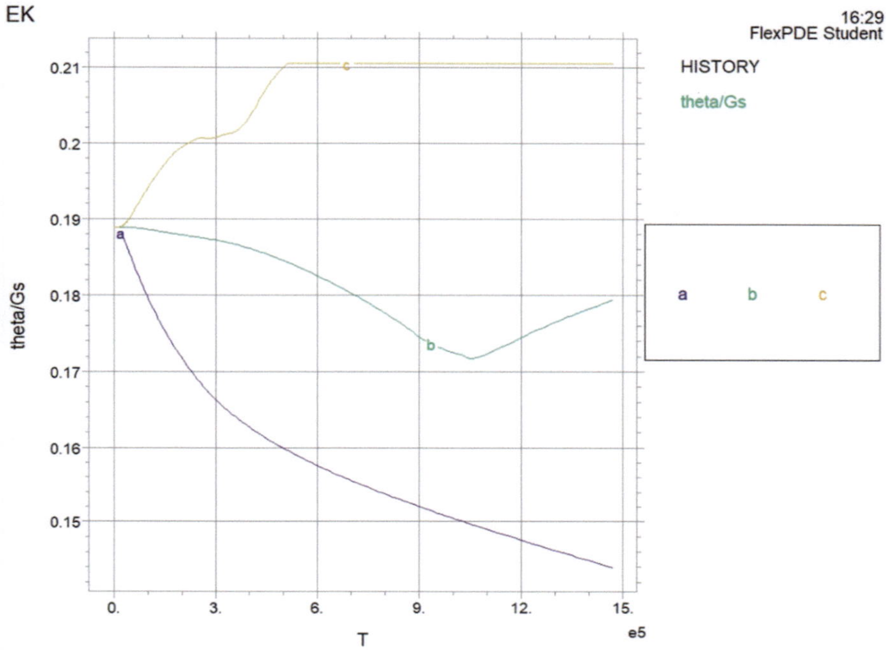

Fig. 22.9 Simulated water content at the three sensors

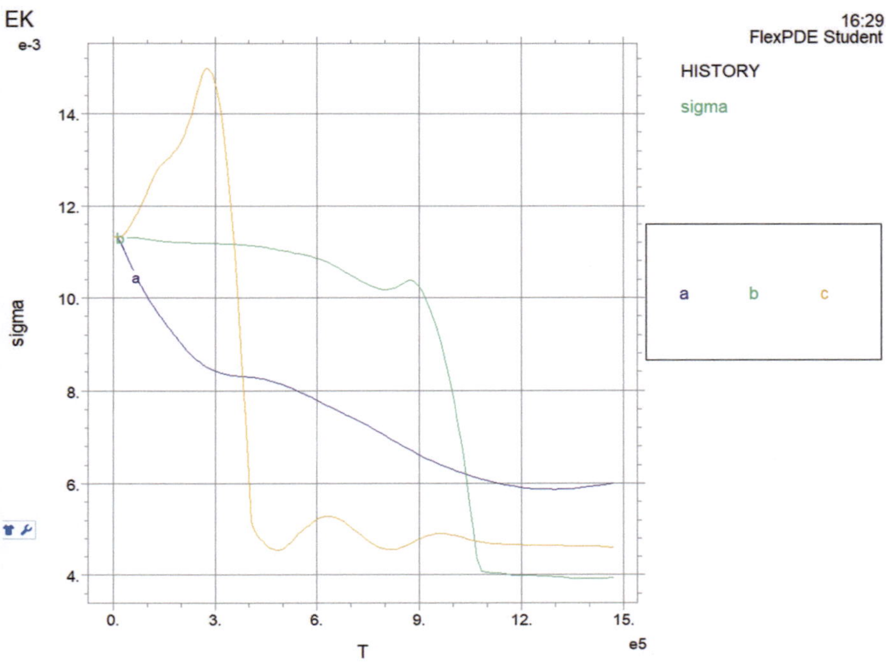

Fig. 22.10 Simulated electric conductivity at the three sensors

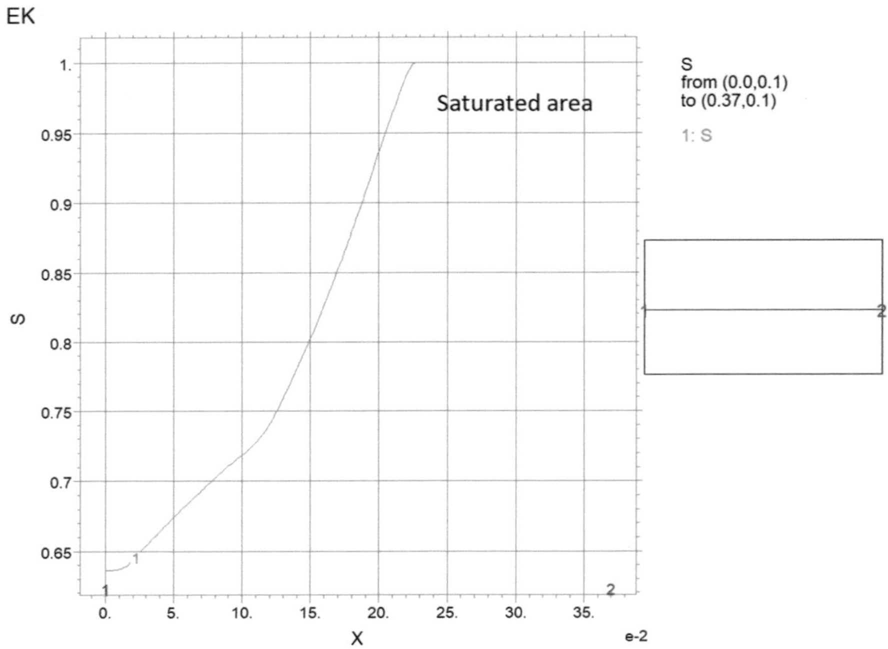

Fig. 22.11 Final distribution of the degree of saturation

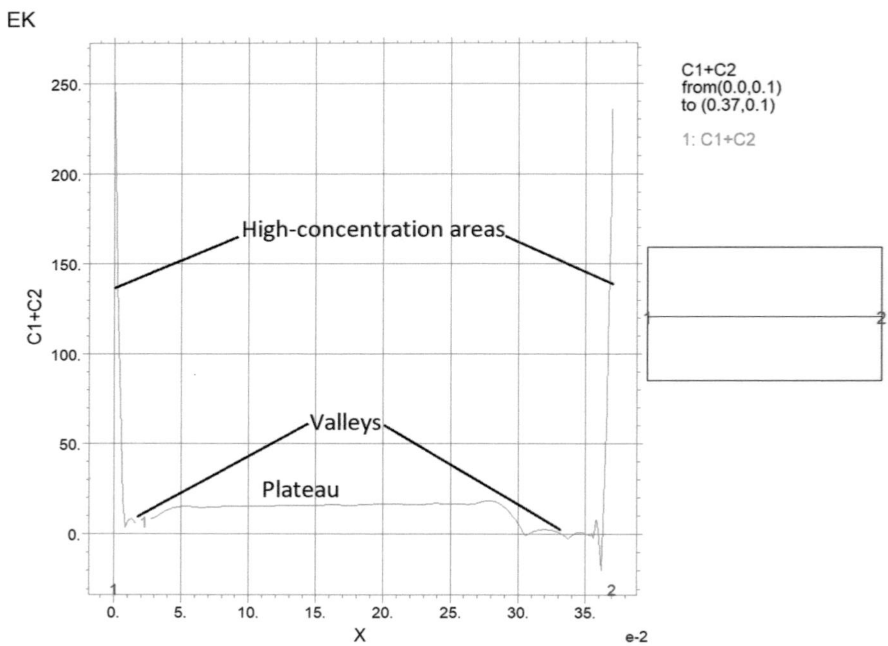

Fig. 22.12 Typical distribution of the total concentration in the process

accumulation at the cathode. The water movement leads to a decrease in the water content at the three sensors. As more and more water accumulates at the cathode, as shown in Fig. 22.11, the saturated region in the vicinity of the cathode gets bigger and bigger. Then the left boundary of this saturated region reaches Sensor 3 first and causes a significant increase in the water content there. The redistribution of the electric conductivity is mainly caused by the migrations of cations and anions toward the cathode and anode, respectively. As illustrated in Fig. 22.12, the ion migrations will produce two areas with very high ion concentrations in the neighborhoods of the two electrodes. In addition, there will be two valleys between these two high-concentration zones and the plateau occupying the middle area between the two electrodes. The plateau area shrinks, and its height decreases as more and more ions are attracted to the electrodes.

Appendix: FlexPDE Script File

```
{ This program is developed to simulate an electrokinetics test
conducted by LiuRG (www.liurg.org) }
TITLE 'EK'
VARIABLES
  H ! Pressure head
  V ! Potential, "phi" in the book
  C1 ! Anion concentration (negatively charged)
  C2 ! Cation concentration (positively charged)
SELECT
  !ERRLIM 0.001
  !NGRID 30
DEFINITIONS  { parameter definitions }
  zeta=-10e-3 ! Zeta potential
  eta=1e-3 ! Dynamic viscosity
  R=8.314 ! Universal gas constant
  F=96487 ! Faraday's constant
  Zn=1 ! Charge number
  epsilon=8.854e-12*81 ! Permittivity of dispersion fluid (water)
  Gs=1.9 ! Specific gravity of soil

  Sw=(1+(-8.51*H)^(1.31))^(-0.23) { Original definition of degree of
saturation }
  S= IF H>=0 THEN 1.0 ELSE (1+(-8.51*H)^(1.31))^(-0.23) { Extended Sw
to allow for full saturation }
  Sr= IF H>=0 THEN 1.0 ELSE -0.23*1.31*(-8.51)*Sw*0.4 { 0.4 is
porosity }
  Kw=50e-9*S^5
  Ke=1.5e-9*S^3.2 { Ke0*Kre }
  z=0 { Gravity is not considered }
  theta=S*0.40 ! Volumetric water content
  Flambda=1.58
  tau=Flambda*(theta-0.0)^(1+1.31)/0.4 ! Tortuosity
  Di1=1.5e-9 ! Molecular diffusion coefficient
```

```
  Di_eff1=theta*tau*Di1
  Di2=0.5e-9 ! Molecular diffusion coefficient
  Di_eff2=theta*tau*Di2
  u=VECTOR(-Kw*dx(H+z)-Ke*dx(V),-Kw*dy(H+z)-Ke*dy(V)) {Water flux }
  uh=VECTOR(-Kw*dx(H+z),-Kw*dy(H+z))
  Cb1=IF C1<=0 THEN 0 ELSE C1 { Use Cb to set a lower bound for C: C cannot
be negative }
  Cb2=IF C2<=0 THEN 0 ELSE C2
  sigma= Di_eff1*Zn*F/(R*293.15)*Cb1*F + Di_eff2*Zn*F/(R*293.15)
*Cb2*F +0.0032 { 0.0032 is the electric conductivity of solid }
INITIAL VALUES
  H=-0.08
  V=10-20*x/0.37
  C1=8
  C2=8
EQUATIONS
  H: DIV(Kw*GRAD(H+z)+Ke*GRAD(V))=Sr*dt(H+z)
  V: DIV(sigma*GRAD(V))=epsilon*dt(V)
  C1: -DIV( -Di_eff1*GRAD(Cb1)) - DIV( Di_eff1*Zn*F/(R*293)*Cb1
*GRAD(V) )=dt(C1)+DIV(u*Cb1)  { JEp=- epsilon*zeta/eta*GRAD(V)*Cb
is excluded }
  C2: -DIV( -Di_eff2*GRAD(Cb1)) - DIV( Di_eff2*Zn*F/(R*293)*Cb2
*GRAD(V) )=dt(C2)+DIV(u*Cb2)  { JEp=- epsilon*zeta/eta*GRAD(V)*Cb
is excluded }
BOUNDARIES
  REGION 1
     START (0.0,0)
        NATURAL(V)=0
        NATURAL(H)=0
        NATURAL(C1)=0.0
        NATURAL(C2)=0.0
     LINE TO (0.37,0)
        VALUE(V)=-10
        NATURAL(H)=0.0
        NATURAL(C1)=0.0
        NATURAL(C2)=0.0
    LINE TO (0.37,0.21)
        NATURAL(V)=0.0
        NATURAL(H)=0.0
        NATURAL(C1)=0.0
        NATURAL(C2)=0.0
    LINE TO (0.0,0.21)
        NATURAL(H)=0
        VALUE(V)=10
        NATURAL(C1)=0.0
        NATURAL(C2)=0.0
    LINE TO CLOSE
TIME 0 TO 3600*24*17 BY 10
MONITORS
FOR Cycle=1
  CONTOUR(S)
  CONTOUR(V)
  CONTOUR(C1+C2)
```

```
  VECTOR(u)
  VECTOR(uh)
  CONTOUR(-Di_eff1*Zn*F/(R*293.15)*dx(V)/(epsilon*zeta/eta*dx(V)))
  CONTOUR(-Di_eff1*Zn*F/(R*293.15)*dx(V)/ (-Kw*dx(H+z)-Ke*dx(V)))
  Elevation(S) from (0.0,0.1) to (0.37,0.1)
  Elevation(C1+C2) from(0.0,0.1)to (0.37,0.1)
  HISTORY(theta/Gs) AT (0.06,0.105) AT (0.17,0.105)  AT
(0.28,0.105) { theta/Gs is the gravimetric water content }
  HISTORY(sigma) AT (0.06,0.105) AT (0.17,0.105)  AT (0.28,0.105)
END
```

Chapter 23
Electromagnetics

Introduction

Electromagnetics, or electromagnetism, deals with physical processes in which both the electric field and magnetic field are involved. Unlike an individual electrical field and an individual magnetic field, which exists independently as electrostatics and magnetostatics, respectively, the electrical field and magnetic field in electromagnetics are varying (dynamic), and each can cause the other, leading to a fully coupled multiphysical process. From a viewpoint of physics, electromagnetics is the study of the electromagnetic force between electrically charged particles as one of the four fundamental physical interactions. The electromagnetic force usually appears in the form of electromagnetic fields, such as electric fields, magnetic fields, and light. Aside from the electromagnetic force, the other three fundamental interactions are the strong interaction, weak interaction, and gravitation.

Electromagnetic phenomena originate from the electromagnetic force, which is also called the Lorentz force. The electromagnetic force includes both electricity and magnetism as two elements of one mechanism. The electromagnetic force primarily determines the internal properties of most objects encountered in daily life. Ordinary materials take their form due to the intermolecular forces between molecules in the materials which originate from electromagnetics. Even inside an atom, electrons are bounded by electromagnetic wave mechanics within orbitals around the atomic nuclei. However, the electromagnetics discussed in this book will be focused on the macroscopic field theories in the context of continuum mechanics instead of pure forces between microscopic particles.

Maxwell's equations, together with the Lorentz force law, form the foundation of classical electrodynamics, classical optics, and electric circuits. For the electromagnetic field of interest here, Maxwell's equations are the primary mathematical descriptions. Maxwell's equations in the tensor form include four partial differential equations and a few auxiliary constitutive relationships between several terms in these partial differential equations. Maxwell's equations describe how electric and

Z. Liu, *Multiphysics in Porous Materials*,
https://doi.org/10.1007/978-3-319-93028-2_23

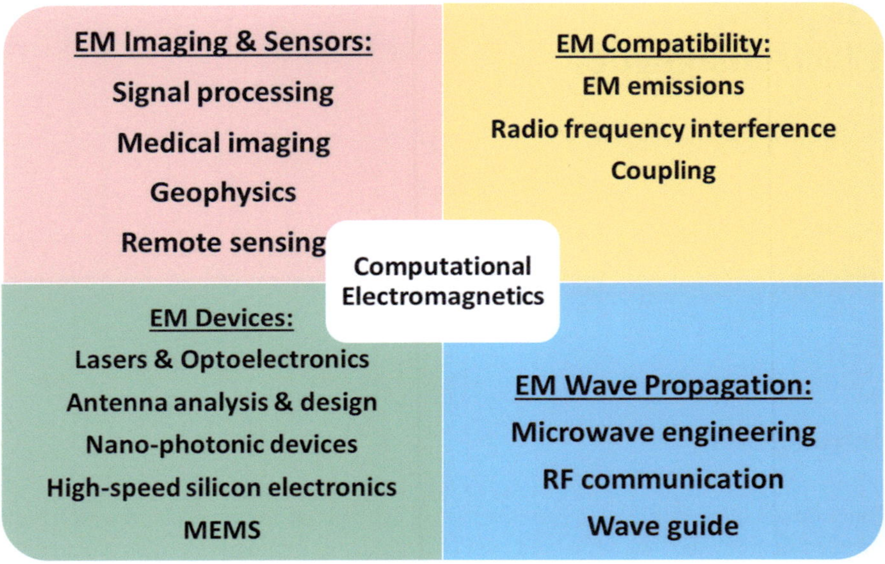

Fig. 23.1 Common application areas of computational electromagnetics

magnetic fields are generated and altered by each other and by charges and currents. These equations are named after the physicist and mathematician James Clerk Maxwell, who published an early form of the equations between 1861 and 1862. The numerical solution of Maxwell's equations forms an important study area called computational electromagnetics, which also involves the development and application several numerical methods that are not that well known in the general scope of numerical simulation. As shown in Fig. 23.1, computational electromagnetics has applications spanning over most traditional areas involving the electromagnetic field.

Electromagnetics is not a major topic in the traditional scope of porous materials and related research. However, it is still of interest as electromagnetic wave-based sensors are widely used in geophysics, petroleum engineering, and geotechnical engineering. For example, electromagnetic induction techniques have been used since the late 1970s to characterize the spatial variability of soil properties including the soil salinity, water content, flow patterns, soil texture, compaction, organic matter content, pH, and depth to subsurface horizons and stratigraphic layers or bedrock (Doolittle and Brevik 2014). Also, ground penetration radar, time-domain reflectometry sensor, and frequency domain sensors have been widely used in transportation engineering, geotechnical engineering, and soil sciences. It is interesting to notice that the applications of these electromagnetic techniques have been gaining rapidly increasing acceptance. Moreover, the electromagnetics-related explorations in microfluidics and bio-environments will also prompt the application of electromagnetics in porous materials. Electromagnetics is turning into an essential part of the multiphysics in porous materials.

In the following sections, Maxwell's equations as the governing equation, together with auxiliary relationships needed to close the mathematical description, will be introduced first. Later, the governing equations will be further explained with physics regarding polarization and magnification, several variants of Maxwell's equations, and an analogy between the electromagnetic fields and the mechanical field. More discussions will be made to facilitate numerical simulation with the equation system for both vacuum and special conditions like conductive materials and to explain the treatments of boundary conditions. Electromagnetic waves will be introduced by deriving the wave equation from Maxwell's equations. To facilitate numerical implementations, formulations of Maxwell's equations using potentials and phasors (for waveguides) will be detailed in two separate sections. The implementation of Maxwell's equations using both the field vector and potentials will be achieved via a simple practice problem in the last section.

Conventional Maxwell's Equations

Maxwell's equations have been well developed and validated for the description of the electromagnetic field. Considering the fact that they are so well known, a conventional form of the equations will be given first. Then the four PDEs will be explained one by one regarding their physical meanings and mathematical constructions. Strictly speaking, Maxwell's equations are not exact laws of the universe such as in phenomena related to photons or virtual photons. Instead, they are a classical approximation to the more accurate and fundamental theory of quantum electrodynamics. However, deviations from Maxwell's equations are immeasurably small. These deviations are generally not considered in the context of multiphysics.

The equations in SI units can be formulated as follows:

Gauss's law

$$\nabla \cdot \mathbf{D} = \rho_{\mathrm{e}}^{\mathrm{f}} \tag{23.1}$$

Gauss's law of magnetism

$$\nabla \cdot \mathbf{B} = 0 \tag{23.2}$$

Faraday's law of induction

$$\nabla \times \mathbf{E} = -\frac{\partial}{\partial t}\mathbf{B} \tag{23.3}$$

Ampère's circuital law

$$\nabla \times \mathbf{H} = \mathbf{J}_{\mathrm{e}}^{\mathrm{f}} + \frac{\partial}{\partial t}\mathbf{D} \tag{23.4}$$

where **D** is the electric displacement (field), ρ_e^f is the free charge density, **E** is the electric field, **B** is the magnetic field, **H** is the magnetizing field or the auxiliary magnetic field, and $\mathbf{J}_e^f$ is the free current density.

The first two equations are the governing equations for electrostatics and magnetostatics, both of which can be called Gauss's law. The first equation indicates that the electrical field is a field with a source, whereas the second equation implies that the magnetic field is a field without a source. That is, there are positive or negative charges but no magnetic charge (or monopole). The third and fourth equations describe how a varying magnetic field generates an electric field and how a varying electric field generates a magnetic field, respectively. The above four equations contain seven unknowns. Therefore, three more constitutive relationships are needed to close the mathematical description:

$$\mathbf{D} = \varepsilon \mathbf{E} \tag{23.5}$$

$$\mathbf{B} = \mu \mathbf{H} \tag{23.6}$$

$$\mathbf{J}_e^f = \sigma_e \mathbf{E}, \tag{23.7}$$

where ε is the absolute permittivity, μ is the permeability, and σ_e is the electrical conductivity. The last constitutive equation (Eq. 23.7) is Ohm's law.

The above Maxwell's equations were obtained based on the assumption of the absence of magnetic charges or monopoles. However, even though the existence of magnetic charges has not been proven, it is often convenient to introduce magnetic charges and currents into the equations. This treatment will lead to symmetric equations for the electric and magnetic fields. If magnetic charges do exist, Maxwell's equations would become

$$\nabla \cdot \mathbf{D} = \rho_e^f \tag{23.8}$$

$$\nabla \cdot \mathbf{B} = \rho_m \tag{23.9}$$

$$\nabla \times \mathbf{E} = -\mathbf{J}_m - \frac{\partial}{\partial t}\mathbf{B} \tag{23.10}$$

$$\nabla \times \mathbf{H} = \mathbf{J}_e^f + \frac{\partial}{\partial t}\mathbf{D}, \tag{23.11}$$

where ρ_m is the magnetic charge density and $\mathbf{J}_m$ is the magnetic current density, which can be related to other quantities in a way similar to their electric counterparts:

$$\nabla \cdot \mathbf{J}_m + \frac{\partial \rho_m}{\partial t} = 0, \tag{23.12}$$

$$\mathbf{J}_m = \sigma_m \mathbf{H}. \tag{23.13}$$

Physical Understanding of Maxwell's Equations

The Maxwell's equations involve complicated relationships between **D**, **E**, **B**, **H**, ρ_e^f, ρ_m, $\mathbf{J}_e^f$, and $\mathbf{J}_m$. These relationships involved both the constitutive relationships within each field and the interaction between the two fields. To further understand the relationships, we first need to recall more detailed constitutive relationships in the electric and magnetic fields:

$$\mathbf{D} = \varepsilon_0 \mathbf{E} + \mathbf{P} \tag{23.14}$$

$$\mathbf{B} = \mu_0 \mathbf{H} + \mu_0 \mathbf{M}, \tag{23.15}$$

where **P** and **M** are the electric polarization and magnification (or magnetic polarization), respectively.

The above two equations can be understood in a similar way. That is, the electric/magnetic displacement currents or electric/magnetic charges consist of two parts: the part independent of material type, i.e., the first term on the right-hand side of each equation, and the part depending on material properties, i.e., polarization and magnification for electric and magnetic fields, respectively. For the former part, only one parameter related to the vacuum, ε_0 or μ_0, is involved for each field. We can substitute the second constitutive equation (Eq. 23.15) into the fourth Maxwell's equations to obtain

$$\begin{aligned} \nabla \times \left(\frac{\mathbf{B}}{\mu_0}\right) &= \underbrace{\mathbf{J}_e^f}_{\text{free current}} + \underbrace{\nabla \times \mathbf{M}}_{\text{magnefication}} + \underbrace{\frac{\partial}{\partial t}\mathbf{P}}_{\text{polarization}} + \underbrace{\frac{\partial}{\partial t}(\varepsilon_0 \mathbf{E})}_{\text{vacuum}}. \\ &= \mathbf{J}_t + \frac{\partial}{\partial t}(\varepsilon_0 \mathbf{E}) \end{aligned} \tag{23.16}$$

The total electric current pertaining to the material is thus composed of the following components:

$$\mathbf{J}_t = \mathbf{J}_e^f + \nabla \times \mathbf{M} + \frac{\partial \mathbf{P}}{\partial t} = \mathbf{J}_e^f + \mathbf{J}_e^m + \mathbf{J}_e^p, \tag{23.17}$$

where $\mathbf{J}_t$ is the total electrical current, $\mathbf{J}_e^m$ is the magnification current, and $\mathbf{J}_e^p$ is the polarization current. The sum of $\mathbf{J}_e^m$ and $\mathbf{J}_e^p$ is the bound current. To obtain the relationship between the current and the other terms, we can try to carry out the curl operation on both sides of the fourth Maxwell's equation, leading to

$$\nabla \cdot \mathbf{J}_e^f + \frac{\partial \rho_e^f}{\partial t} = 0, \tag{23.18}$$

which is the variation of the first Maxwell's equation (Gauss's equation)

$$\nabla \cdot \mathbf{J}_{\mathrm{e}}^{\mathrm{f}} + \nabla \cdot \frac{\partial \mathbf{D}}{\partial t} = 0. \tag{23.19}$$

However, the above equation does not imply that

$$\mathbf{J}_{\mathrm{e}}^{\mathrm{f}} + \frac{\partial \mathbf{D}}{\partial t} = 0. \tag{23.20}$$

This is because we cannot remove the divergence operator ($\nabla \cdot$) from the two sides of the equation without impairing the equality. Alternatively, we have the following relationship:

$$\mathbf{J}_{\mathrm{e}}^{\mathrm{D}} = \frac{\partial \mathbf{D}}{\partial t} = \frac{\partial (\varepsilon_0 \mathbf{E})}{\partial t} + \frac{\partial \mathbf{P}}{\partial t}, \tag{23.21}$$

where $\mathbf{J}_{\mathrm{e}}^{\mathrm{D}}$ is defined as the displacement current.

Similar relationships and concepts can be derived for the magnetic field as well. However, it is not common to do so because the magnetic charge is an "imaginary" object and thus not widely used. Notwithstanding, the following equation is still helpful:

$$\nabla \cdot \mathbf{J}_{\mathrm{m}} + \frac{\partial \rho_{\mathrm{m}}}{\partial t} = 0. \tag{23.22}$$

On the other hand, an analogy can be drawn between both the electric and magnetic fields and the stress-strain field as listed in the following table. As can be seen in Table 23.1, **D** and **B** are quantities that can be applied without specifying the material type, and their units imply the meaning of "quantity per unit area"; they are thus analogous to the "stress," i.e., force per area. **E** and **H** can be caused by **D** and **B**, respectively, and their definitions are related to the material type. Therefore, they are analogous to the concept of strain. The relationships between the "strains," e.g., **E** and **H**, to the "stresses," e.g., **D** and **B**, are defined as material properties, e.g., permittivity and permeability, respectively. These two properties are analogous to the compressibility of the material in the mechanical field. $\rho_{\mathrm{e}}^{\mathrm{f}}$ and $\rho_{\mathrm{e}}^{\mathrm{f}}$ serve the roles of sources or body forces. $\mathbf{J}_{\mathrm{e}}^{\mathrm{f}}$ and $\mathbf{J}_{\mathrm{m}}$ are the "stresses" arising due to the resistance to the external loading; therefore, they are analogous to the effective stress in porous materials. U and Q directly correspond to **E** and **D**, respectively, and have the meaning of "total" instead of "per unit area"; therefore, they are analogous to strain and stress, respectively.

Table 23.1 Analogy between the electromagnetic field and mechanical field

U	Q	$\mathbf{D}$	$\mathbf{P}$	$\mathbf{B}$	$\mathbf{E}$	$\mathbf{H}$	$\mathbf{M}$	$\rho_{\mathrm{e}}^{\mathrm{f}}$	ρ_{m}	$\mathbf{J}_{\mathrm{e}}^{\mathrm{f}}$	$\mathbf{J}_{\mathrm{m}}$
V	C	$\frac{\mathrm{C}}{\mathrm{m}^2}$	$\frac{\mathrm{C}}{\mathrm{m}^2}$	$\frac{\mathrm{N}}{\mathrm{Am}}$	$\frac{\mathrm{V}}{\mathrm{m}}$	$\frac{\mathrm{A}}{\mathrm{m}}$	$\frac{\mathrm{A}}{\mathrm{m}}$	$\frac{\mathrm{C}}{\mathrm{m}^3}$	$\frac{\mathrm{N}}{\mathrm{Am}^2}$	$\frac{\mathrm{C}}{\mathrm{m}^2\mathrm{s}}$	$\frac{\mathrm{N}}{\mathrm{Ams}}$
L	**F**	$\boldsymbol{\sigma}$	$\boldsymbol{\sigma}$	$\boldsymbol{\sigma}$	$\boldsymbol{\varepsilon}$	$\boldsymbol{\varepsilon}$	$\boldsymbol{\varepsilon}$	$\mathbf{f}$	$\mathbf{f}$	$\boldsymbol{\sigma}'$	$\boldsymbol{\sigma}'$
m	N	$\frac{\mathrm{N}}{\mathrm{m}^2}$	$\frac{\mathrm{N}}{\mathrm{m}^2}$	$\frac{\mathrm{N}}{\mathrm{m}^2}$	$\frac{\mathrm{m}}{\mathrm{m}}$	$\frac{\mathrm{m}}{\mathrm{m}}$	$\frac{\mathrm{m}}{\mathrm{m}}$	$\frac{\mathrm{N}}{\mathrm{m}^3}$	$\frac{\mathrm{N}}{\mathrm{m}^3}$	$\frac{\mathrm{N}}{\mathrm{m}^3\mathrm{s}}$	$\frac{\mathrm{N}}{\mathrm{m}^3\mathrm{s}}$

These analogies could help readers who are more familiar with the concepts in the stress and strain field to understand the concepts in the electromagnetic field. However, it is worth pointing out that the concepts in the electromagnetic field can be understood in many other different ways. In the table, macroscopic quantities such as the potential and charges are mentioned for the electric field while no corresponding quantities are listed for the magnetic field.

Equation System for Numerical Simulations

Independent Governing Equations

In the four Maxwell's equations without magnetic charges, there are only two independent equations after including two constitutive relationships, i.e., Eqs. 23.14 and 23.15. Generally, Faraday's law and Ampère's law are chosen as the independent governing equations:

$$\nabla \times \mathbf{E} = -\frac{\partial}{\partial t}\mathbf{B}, \tag{23.23}$$

$$\nabla \times \mathbf{H} = \mathbf{J}_{\mathrm{e}}^{\mathrm{f}} + \frac{\partial}{\partial t}\mathbf{D}. \tag{23.24}$$

The above two equations are in some cases called Maxwell's equations. Gauss's equations for both the electric and magnetic fields can be deduced from the above two equations by taking the divergence of Eqs. 23.23 and 23.24. An auxiliary equation is required to close the equation system for solution. This equation should be able to relate the electric charge flux and the charge density. In fact, this is the equation describing the conservation of charges at any point, which is frequently referred to as the equation of continuity or the equation of the electric charge density conservation:

$$\nabla \cdot \mathbf{J}_{\mathrm{e}}^{\mathrm{f}} + \frac{\partial \rho_{\mathrm{e}}^{\mathrm{f}}}{\partial t} = 0. \tag{23.25}$$

The two governing equations can also be combined to obtain a second-order governing equation in terms of **E** as

$$\varepsilon\mu\frac{\partial^2 \mathbf{E}}{\partial t^2} + \nabla \times \nabla \times \mathbf{E} = -\mu\frac{\partial \mathbf{J}_{\mathrm{e}}}{\partial t}, \tag{23.26}$$

or in terms of **H** as

$$\varepsilon\mu\frac{\partial^2 \mathbf{H}}{\partial t^2} - \nabla \times \nabla \times \mathbf{E} = -\nabla \times \mathbf{J}_{\mathrm{e}}.$$

Conductive Materials

Special attention is needed to apply Maxwell's equations to conductive materials like metals. It is well known that the charges will vanish from an interior point of a metal body and appear at the surface almost immediately. To get a better understanding, let us first combine the equation for the conservation of electric charge, Ohm's law, and Gauss's equation into the following equation:

$$\frac{\partial \rho_{\mathrm{e}}^{\mathrm{f}}}{\partial t} + \sigma_{\mathrm{e}} \frac{\rho_{\mathrm{e}}^{\mathrm{f}}}{\varepsilon} = 0. \tag{23.27}$$

The electric charge density at any time is thus equal to

$$\rho_{\mathrm{e}}^{\mathrm{f}} = \rho_{\mathrm{e}}^{\mathrm{f0}} e^{-\frac{\sigma_{\mathrm{e}}}{\varepsilon} t}, \tag{23.28}$$

where $\sigma_{\mathrm{e}}/\varepsilon$ characterizes the relaxation time of the material and $\rho_{\mathrm{e}}^{\mathrm{f0}}$ is the free electric charge density at the beginning. According to the above solution, charges within a conductive material decay exponentially with time. For metal, with a typical conductivity of the order 10^6 Siemens/m and a permittivity ε of 10^{-11} As/Vm, the decaying time will be of the order 10^{-17} s. Even for a relatively poor conductor such as distilled water, the relaxation time is less than 10^{-6} s. Therefore, we can believe that no charge will be accumulated inside the conductor even at high frequencies. Consequently, the equation of electric charge density conservation can be approximated by

$$\nabla \cdot \mathbf{J}_{\mathrm{e}}^{\mathrm{f}} = 0. \tag{23.29}$$

Similarly, for Ampère's equation, we can assume a harmonic field with an angular frequency of ω. Then the free electric current flux and displacement current flux are $\mathbf{J}_{\mathrm{e}}^{\mathrm{f}}$ and $\omega\varepsilon/\sigma_{\mathrm{e}}$, respectively. The comparison of these two fluxes (the latter one is very small even at high frequencies) indicates that it is reasonable to neglect the displacement current in a conductive material even at high frequencies in the way as

$$\nabla \times \mathbf{H} = \mathbf{J}_{\mathrm{e}}^{\mathrm{f}}. \tag{23.30}$$

The approximate equation, i.e., Eq. 23.29, for the electric charge density can be obtained by taking the divergence of Eq. 23.30 because the divergence of a curl is zero.

Boundary Conditions

The boundary conditions can be derived from the governing equations by ensuring the validity of the equations on the boundary. The derivation is more complicated than other physical fields due to (1) the number of equations, (2) the vector form of the dependent variables, and (3) the involvement of the cross product operators.

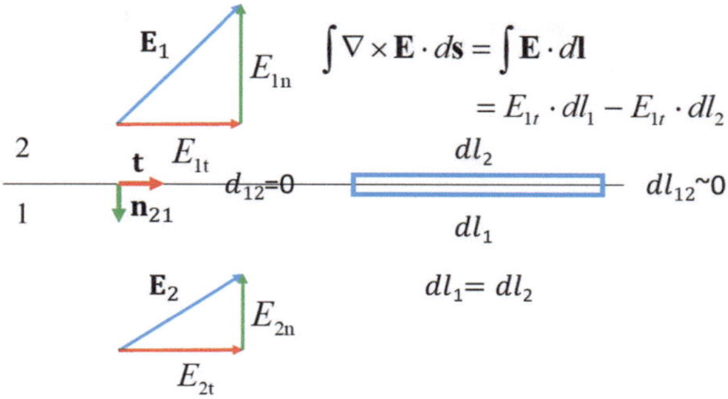

Fig. 23.2 Boundary conditions of electromagnetic field

Figure 23.2 illustrates a simple derivation for Faraday's equation. Let us consider a small slender rectangular area bounded by the blue lines along the boundary between the two regions. Application of the Stokes theorem to Faraday's equation will lead to $E_{1t} \cdot dl_1 - E_{1t} \cdot dl_2 = 0$ if the area is infinitely slender, i.e., $dl_{12} \approx 0$. Because $dl_1 = dl_2$, we obtain $E_{1t} = E_{1t}$ (or $\mathbf{E}_2^{||} - \mathbf{E}_1^{||} = 0$), which is equivalent to $\mathbf{n}_{12} \times (\mathbf{E}_2 - \mathbf{E}_1) = 0$. The other boundary conditions can be derived in similar ways.

Depending on the electric and magnetic properties, the electromagnetic fields can be discontinuous or continuous on each side of the common boundary (interface) between two different materials. In summary, on an interface between two materials, the field must satisfy the following conditions:

$$\mathbf{n}_{12} \times (\mathbf{E}_2 - \mathbf{E}_1) = 0 \Leftrightarrow \mathbf{E}_2^{||} - \mathbf{E}_1^{||} = 0 \tag{23.31}$$

$$\mathbf{n}_{12} \cdot (\mathbf{D}_2 - \mathbf{D}_1) = \rho_e \Leftrightarrow \mathbf{D}_2^{\perp} - \mathbf{D}_1^{\perp} = 0 \tag{23.32}$$

$$\mathbf{n}_{12} \times (\mathbf{H}_2 - \mathbf{H}_1) = \mathbf{J} \Leftrightarrow \mathbf{H}_2^{||} - \mathbf{H}_1^{||} = \mathbf{J} \tag{23.33}$$

$$\mathbf{n}_{12} \cdot (\mathbf{B}_2 - \mathbf{B}_1) = 0 \Leftrightarrow \mathbf{B}_2^{\perp} - \mathbf{B}_1^{\perp} = 0, \tag{23.34}$$

where $\mathbf{n}_{12}$ is the unit normal vector pointing from Region 1 to Region 2 and $\mathbf{u}^{||}$ and $\mathbf{u}^{\perp}$ are the tangential and normal components, respectively.

The above four equations indicate that (1) the electric field's tangential components are continuous across the interface, (2) the electric flux's normal component is discontinuous across the interface between materials and the difference is the charge density, (3) the tangential components of the magnetic field strength are discontinuous across the interface and the "jump" equals an electric current flux, and (4) the magnetic flux density's normal component is continuous across the interface (Fisk 2008). ρ_e and $\mathbf{J}_e^f$ vanish under the conditions of no electric charge and no electric current, respectively.

For external boundaries, we only consider the region inside the boundary. Then Eqs. 23.31, 23.32, 23.33 and 23.34 reduce into

$$\mathbf{n} \times \mathbf{E} = 0 \tag{23.35}$$

$$\mathbf{n} \cdot \mathbf{D} = -\rho_e \tag{23.36}$$

$$\mathbf{n} \times \mathbf{H} = -\mathbf{J}_e^f \tag{23.37}$$

$$\mathbf{n} \cdot \mathbf{H} = 0. \tag{23.38}$$

Electromagnetic Waves

The four Maxwell's equations can be consolidated into two wave equations in special cases. Such a consolidation can be used to unify the theories of electromagnetism and optics. The 3D diagram in Fig. 23.3 shows a plane linearly polarized wave propagating from left to right.

The special conditions that are required for the reformulation of the original Maxwell's equations into wave equations are (1) no charges ($\rho = 0$) and (2) no currents ($\mathbf{J}_e^f = 0$), such as the condition in vacuum. In such cases, Maxwell's equations reduce to

$$\nabla \cdot \mathbf{D} = 0 \tag{23.39}$$

$$\nabla \cdot \mathbf{B} = 0 \tag{23.40}$$

$$\nabla \times \mathbf{E} = -\frac{\partial}{\partial t}\mathbf{B} \tag{23.41}$$

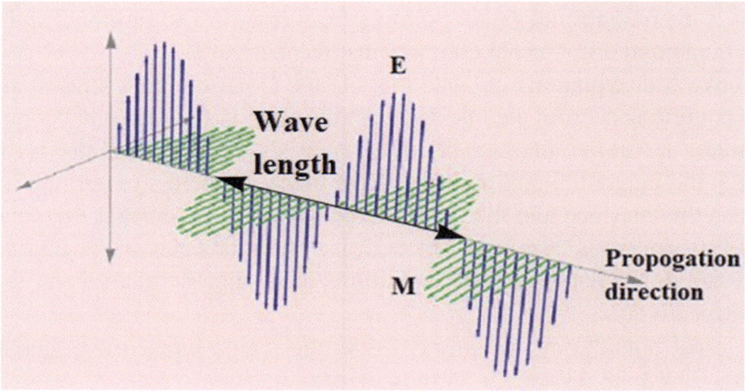

Fig. 23.3 Visualization of magnetic wave

$$\nabla \times \mathbf{H} = \frac{\partial}{\partial t}\mathbf{D}. \tag{23.42}$$

Taking the curl ($\nabla\times$) of the curl operation on the last two equations and using the curl of the curl identity, i.e., $\nabla \times (\nabla \times u) = \nabla(\nabla \cdot u) - \nabla^2 u$, we obtain the wave equations:

$$\frac{\partial^2 \mathbf{D}}{\partial t^2} = \frac{1}{\varepsilon_0 \mu_0} \nabla^2 \mathbf{D} \tag{23.43}$$

$$\frac{\partial^2 \mathbf{B}}{\partial t^2} = \frac{1}{\varepsilon_0 \mu_0} \nabla^2 \mathbf{B}, \tag{23.44}$$

where

$$c = \sqrt{\frac{1}{\varepsilon_0 \mu_0}} = 2.9979 \times 10^8 \mathrm{m/s} \tag{23.45}$$

is the speed of light in free space. In materials with relative permittivity ε_r and relative permeability μ_r, the phase velocity of light becomes

$$c = \sqrt{\frac{1}{\varepsilon_0 \varepsilon_r \mu_0 \mu_r}}, \tag{23.46}$$

which is usually less than c in vacuum.

In addition, **D** and **B** are mutually perpendicular and both are normal to the direction of wave propagation. **D** and **B** are also in phase with each other. A sinusoidal plane wave is one special solution to these equations. Maxwell's equations explain how these waves can physically propagate through space. A changing magnetic field creates a changing electric field through Faraday's law. In turn, that electric field creates a changing magnetic field by following Ampère's law with Maxwell's addition. This perpetual cycle allows these waves, which is now known as electromagnetic radiation, to move through space at the velocity of c.

Formulation Using Potentials

Governing Equation

The formulation of the equation system using the field vectors was introduced in a previous section. In fact, the electromagnetic field can be formulated using the electric and magnetic vector potentials, which are more commonly used due to the advantages of this type of formulation in boundary condition treatments and in the convergence of the numerical solution. As introduced in the chapters for electrostatics and magnetostatics, the electric and magnetic field satisfy a zero curl and zero divergence condition, respectively; we thus can introduce a vector potential first for the magnetic field:

$$\mathbf{B} = \nabla \times \mathbf{A}, \tag{23.47}$$

where **A** is the magnetic vector potential. From Maxwell's equations, it is known that

$$\nabla \times \mathbf{E} = -\frac{\partial \mathbf{B}}{\partial t} = -\frac{\partial}{\partial t}(\nabla \times \mathbf{A}) = -\nabla \times \frac{\partial \mathbf{A}}{\partial t} \tag{23.48}$$

or

$$\nabla \times \left(\mathbf{E} + \frac{\partial \mathbf{A}}{\partial t}\right) = 0. \tag{23.49}$$

Because $\nabla \times \nabla\psi = 0$, one solution to Eq. 23.48 is

$$\mathbf{E} = -\frac{\partial \mathbf{A}}{\partial t} - \nabla\psi. \tag{23.50}$$

The negative sign in front of $\nabla\psi$ implies that the electric potential decreases along electric lines. Multiplying the electric field with σ_e, we will get

$$\mathbf{J}_e^f = \sigma_e \mathbf{E} = -\sigma_e \frac{\partial \mathbf{A}}{\partial t} - \sigma_e \nabla\psi = -\sigma_e \frac{\partial \mathbf{A}}{\partial t} + \mathbf{J}_e^s, \tag{23.51}$$

where $\mathbf{J}_e^s$ is the source current density in the induction coil. It follows from the Ampère's equation with a free charge current that

$$\nabla \times \frac{1}{\mu}\mathbf{B} = \mathbf{J}_e^f + \frac{\partial D}{\partial t}. \tag{23.52}$$

Substituting Eqs. 23.47 and 23.51 into Eq. 23.52, we obtain

$$\nabla \times \nabla \times \mathbf{A} = -\mu\sigma_e \frac{\partial \mathbf{A}}{\partial t} + \mu \mathbf{J}_e^s - \varepsilon\mu \frac{\partial^2 \mathbf{A}}{\partial t^2} - \varepsilon\mu \frac{\partial}{\partial t}\nabla\psi \tag{23.53}$$

for a homogeneous and isotropic medium independent of field intensity. Then let us expand and impose the Coulomb gauge condition, i.e., $\nabla \cdot \mathbf{A} = 0$, the above equation is simplified into

$$\varepsilon\mu \frac{\partial^2 \mathbf{A}}{\partial t^2} - \nabla^2 \mathbf{A} = \mu \mathbf{J}_e^s - \sigma_e \mu \frac{\partial \mathbf{A}}{\partial t} - \varepsilon\mu \frac{\partial}{\partial t}\nabla\psi, \tag{23.54}$$

which is called the diffusion equation. The equation can also be expressed in the same form but for different vectors: instead of **A**, we write the equations in terms of **J**, **E**, **H**, and **B**.

The relation of the vector potential to the current is

$$\mathbf{A} = \frac{\mu_0}{4\pi}\int \frac{\mathbf{J}_e^s}{r}\,dV, \tag{23.55}$$

where **J** is the current density.

The following Poisson's equation in terms of the electric scalar potential will complement the above the equation:

$$\nabla^2\psi = \rho_e^f. \tag{23.56}$$

Boundary Conditions

The treatment of the boundary conditions in the potential formulation is generally simpler. The following equations show the boundary conditions in the potential formulation and their counterparts in the field vector formulation:

$$\mathbf{n}_{12} \times (\mathbf{E}_2 - \mathbf{E}_1) = 0 \Leftrightarrow \psi_1 = \psi_2 \tag{23.57}$$

$$\mathbf{n}_{12} \cdot (\mathbf{D}_2 - \mathbf{D}_1) = \rho_e \Leftrightarrow -\mathbf{n}_{12} \cdot (\nabla\psi_2 - \nabla\psi_1) = \rho_e \tag{23.58}$$

$$\mathbf{n}_{12} \times (\mathbf{H}_2 - \mathbf{H}_1) = \mathbf{J}_e^f \Leftrightarrow -\mathbf{n}_{12} \cdot \left(\frac{1}{\mu_1}\nabla\mathbf{A}_2 - \frac{1}{\mu_1}\nabla\mathbf{A}_1\right) = \mathbf{J}_e^f \tag{23.59}$$

$$\mathbf{n}_{12} \cdot (\mathbf{B}_2 - \mathbf{B}_1) = 0 \Leftrightarrow \mathbf{A}_1 = \mathbf{A}_2. \tag{23.60}$$

The equivalence can be proven easily using the definitions of the potentials. For external boundaries, Eqs. 23.57, 23.58, 23.59, and 23.60 reduce to

$$-\mathbf{n} \times \mathbf{E} = 0 \Leftrightarrow \psi = 0 \tag{23.61}$$

$$-\mathbf{n} \cdot \mathbf{D} = \rho_e^f \Leftrightarrow \mathbf{n} \cdot \nabla\psi = \rho_e^f \tag{23.62}$$

$$-\mathbf{n} \times \mathbf{H} = \mathbf{J}_e^f \Leftrightarrow \mathbf{n} \cdot \frac{1}{\mu}\nabla\mathbf{A} = \mathbf{J}_e^f \tag{23.63}$$

$$-\mathbf{n} \cdot \mathbf{B} = 0 \Leftrightarrow \mathbf{A} = \mathbf{0}. \tag{23.64}$$

Phasor Formulation (Frequency Domain) and Waveguide

Phasor Formulation

The phasor formulation of Maxwell's equations is also called the time-harmonic form. This formulation is not only convenient but also necessary in many cases. For example, when considering heating due to electromagnetic waves, it is very likely to be impractical to analyze the problem in the time domain. This is because the time-domain electromagnetic wave simulation requires a very small time step, which is shorter than the period of the wave, e.g., a few nanoseconds to capture the temporal variations, while heating processes usually involve a time duration from seconds to hours. Therefore, the consideration of the coupled process requires a very small time step and a large number of time steps, which poses a great challenge on numerical simulation. Despite the necessity, the phasor formulation can also bring great advantages. For example, the formulation enables the analysis of the specific frequencies and the search for the primary frequencies (i.e., eigenvalue problems). Essentially, this formulation converts a time-dependent problem into an equilibrium or an eigenvalue problem, which can remarkably reduce the level of difficulty from a numerical simulation perspective. In addition, the phasor formulation can also significantly simplify the consideration of the energy.

The idea of this formulation is related to the Fourier transform, in which a signal in time is rewritten as the sum of sinusoids. When the fields and sources in Maxwell's equations are all monochromatic functions of time, the equations can be reformulated using the phasor, which has a general form as

$$\mathbf{u} = \tilde{\mathbf{u}} e^{i\omega t}, \tag{23.65}$$

where $\mathbf{u}$ represents the field vectors, e.g., $\mathbf{E}$, $\mathbf{D}$, $\mathbf{H}$, and $\mathbf{B}$. $\tilde{\mathbf{u}}$ thus expresses the spatial distribution, while $e^{i\omega t}$ stores the temporal variation information. One major advantage of using this phasor form is that the temporal derivative has the following form:

$$\frac{\partial}{\partial t} = i\omega. \tag{23.66}$$

Substituting the phasors into Maxwell's equations, we obtain

$$\nabla \cdot \tilde{\mathbf{D}} = \tilde{\rho}_{\mathrm{e}}^{\mathrm{f}} \tag{23.67}$$

$$\nabla \cdot \tilde{\mathbf{B}} = 0 \tag{23.68}$$

$$\nabla \times \tilde{\mathbf{E}} = -i\omega\tilde{\mathbf{B}} \tag{23.69}$$

$$\nabla \times \tilde{\mathbf{H}} = \tilde{\mathbf{J}}_{\mathrm{e}}^{\mathrm{f}} + i\omega\tilde{\mathbf{D}}. \tag{23.70}$$

The electromagnetic wave equations can be developed using phasors and the general field vector as

$$\nabla^2\tilde{\mathbf{u}} + \varepsilon\mu\omega^2\tilde{\mathbf{u}} = 0 \tag{23.71}$$

or

$$\nabla^2\tilde{\mathbf{u}} + k^2\tilde{\mathbf{u}} = 0, \tag{23.72}$$

where $k = \omega\sqrt{\varepsilon\mu} = \omega/c$ is the wave number, in which c is the wave velocity in the medium.

Complex numbers are needed to numerically solve some of the above equations or in the later calculations of other variables using the solutions to the equations. The treatment of complex numbers in numerical simulation is similar to that of vectors, in which multiple components are needed for one quantity (one complex number or one vector). Many numerical simulation programs now come with the ability to deal with complex-dependent variables. However, it is worthwhile to point out that the imaginary part i is not always necessarily included for numerical simulation. This is because the imaginary part is used to express the phase relationship between different fields (the relative position in time). However, in some cases, we are interested in the spatial distribution, the amplitude, or the energy dissipation, in which the phase information or temporal variation is not needed. In these cases, we can conduct simulations using the above equation(s) without the imaginary part. This concept can be seen in the practice example in Chap. 26 for microwave heating.

Waveguide

Waveguides are devices for transporting waves, which can be used for both mechanical and electromagnetic waves. The special dimensions and boundary conditions associated with such devices ensure that most energy will propagate along the longitudinal direction of the waveguides and thus achieve the goal of transmitting either power or communication signals with minimum loss. Waveguides function as a high-pass filter, in which most of the energy above a cutoff frequency will pass while that below the frequency will be attenuated.

A unique feature of waveguides is that they only transmit transverse waves, such as TEM (transverse electric and magnetic), TE (transverse electric), and TM (transverse magnetic) wave. TEM, TE, and TM mean the waves have no E&M, E, and M components along the longitudinal direction, respectively. A common waveguide is a metal tube with a rectangular cross section. The metal sidewall of the tube represents a perfect electric conductor boundary condition, $\mathbf{n} \times \mathbf{E} = 0$, which will lead to a TE or a TM wave. To facilitate the formulation, we split the Laplacian operator into two components for the transverse (t: x and y) and the longitudinal directions (z):

$$\nabla^2 = \nabla_t^2 + \nabla_z^2. \tag{23.73}$$

Taking TE, for example, we still have H_z. The solution to the EM wave equation in TE should have the form of

$$\mathbf{H} = \mathbf{H}(x, y)e^{-ikz}. \tag{23.74}$$

This is because, in the frequency domain, the **H** distribution on the transverse plane does not change with time, while it changes with z as a propagating wave e^{-ikz}. The negative sign means the wave is moving forward. Substituting the solution and the split Laplacian operator into the governing equation, we can obtain the governing equation for the TE wave:

$$\nabla^2 H_z + k^2 H_z = 0, \tag{23.75}$$

or

$$\nabla^2 H_z + \left(k^2 - k_z^2\right) H_z = 0. \tag{23.76}$$

Likewise, the TM wave has the following governing equation:

$$\nabla^2 E_z + k^2 E_z = 0, \tag{23.77}$$

or

$$\nabla^2 E_z + \left(k^2 - k_z^2\right) E_z = 0. \tag{23.78}$$

Once H_z or E_z is obtained, the transverse components can be obtained as follows:

$$\mathbf{E}_t = \frac{-ik_z}{k^2 - k_z^2} \nabla_t E_z + \frac{i\omega\mu}{k^2 - k_z^2} \mathbf{e}_z \times \nabla_t H_z \tag{23.79}$$

$$\mathbf{H}_t = \frac{-i\omega\varepsilon}{k^2 - k_z^2} \mathbf{e}_z \times \nabla_t E_z + \frac{-ik_z}{k^2 - k_z^2} \nabla_t H_z, \tag{23.80}$$

where $\mathbf{e}_z$ is the unit vector along the z (longitudinal) direction and k_z is the wave number along the z direction.

The phasor formulation using an auxiliary potential vector is also available (Balanis 1999). The practice of the above theory will be shown in Chap. 26 with microwave heating.

Practice Problem

This section is prepared to offer practice for computational electromagnetics using a very simple problem: the magnetic field generated by an electric current in a metal wire. Due to the simple PDEs involved, we here require to solve the problem using both the field vector formulation (original Maxwell's equations) and the potential formulation. A more complicated simulation of the electromagnetic field using phasors can be found in Chap. 26.

Problem Description

As shown in Fig. 23.4, in this benchmark problem, we consider a wire, which can be viewed as a perfect conductor, with an infinite length and a constant current of 0.01 A/m^2 in it. The material surrounding this wire is air. The wire is aligned in the z-direction and has a diameter of 1 m. Please simulate the **H** field using 2D numerical simulation. Please simplify the governing equation(s) from 3D to 2D first. This is a good practice to get familiar with Maxwell's equations. Please compare the distributions of the potential along the radial direction obtained by the potential method and field vector method.

Governing Equations Using Vector and Potential Formulations

The governing equations written using FlexPDE script are as follows:

Field formulation

$$\begin{aligned} \text{Hx} &: \text{dx(Hy)} - \text{dy(Hx)} = \text{Jz} \\ \text{Hy} &: \text{dx(Hx)} + \text{dy(Hy)} = 0 \end{aligned}$$

Potential formulation

$$\text{DEL2(Az)} = 0.$$

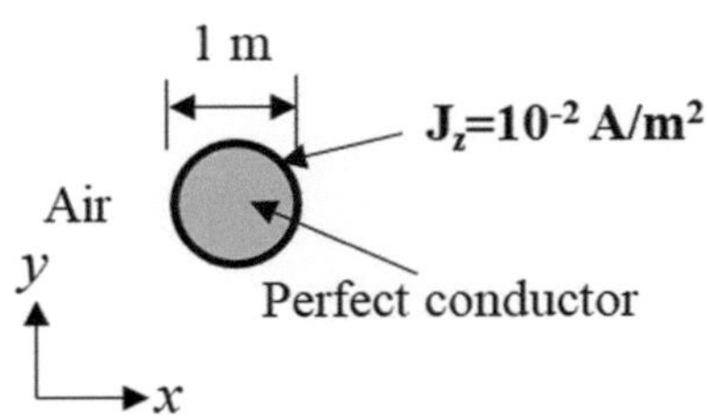

Fig. 23.4 Computation domain and boundary conditions

Results

The major results are shown in the following figures. The utilization of the two formulations of Maxwell's equations can lead to the same simulation results. However, from a numerical perspective, it takes much more effort for the field vector formulation method to reach convergence, at least with the finite element method. The treatments of the boundary conditions for both methods require special attention. The field vector method is more convenient when the magnitudes of the field vector components are known and easy to formulate, whereas the potential method provides the easier boundary condition treatment when the potential is known or can be easily calculated.

Figures 23.5 and 23.6 show the **H** field and potential obtained using the potential method, which are very similar to the results obtained with the field vector method. To quantitatively compare the two methods, the distributions of the potential along the radial direction obtained by the potential method and field vector method are depicted in Figs. 23.7 and 23.8, respectively. As can be seen, the results obtained by the two methods agree very well.

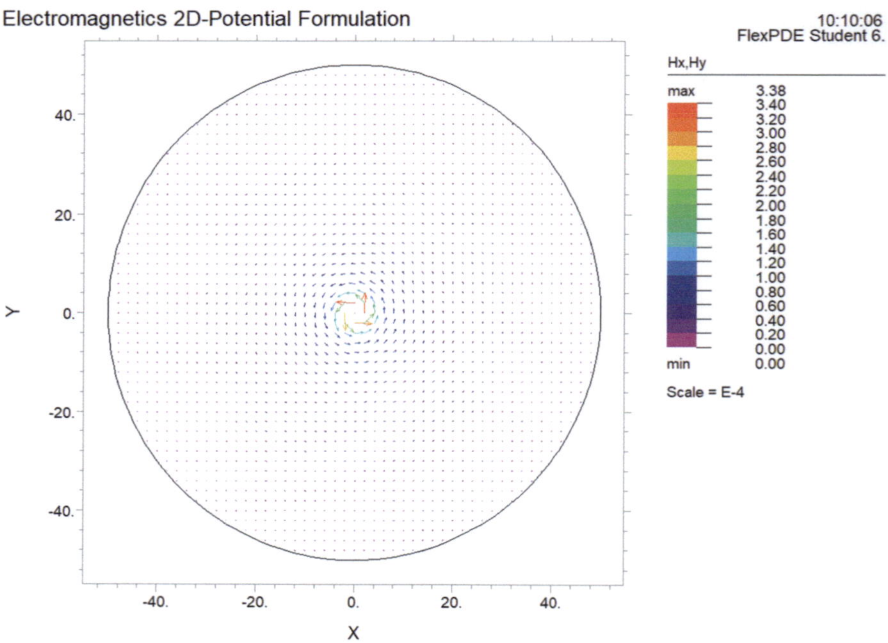

Fig. 23.5 H field

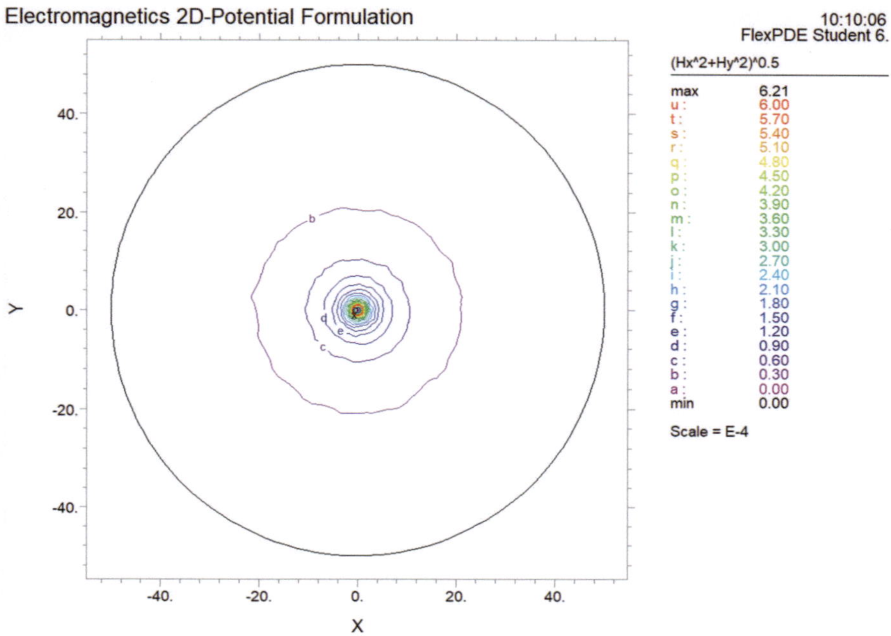

Fig. 23.6 Distribution of potential

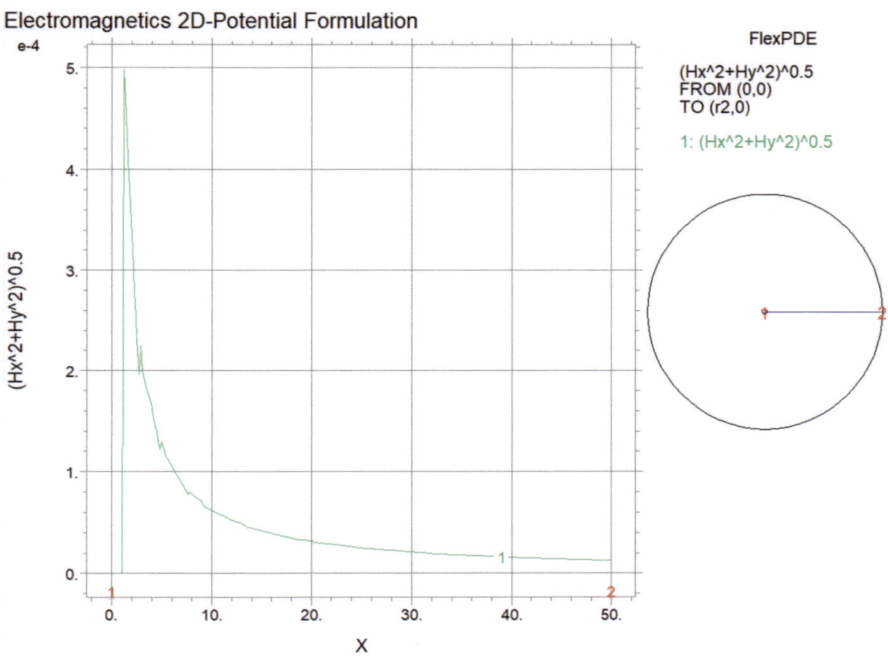

Fig. 23.7 Radial distribution of potential obtained by the potential method

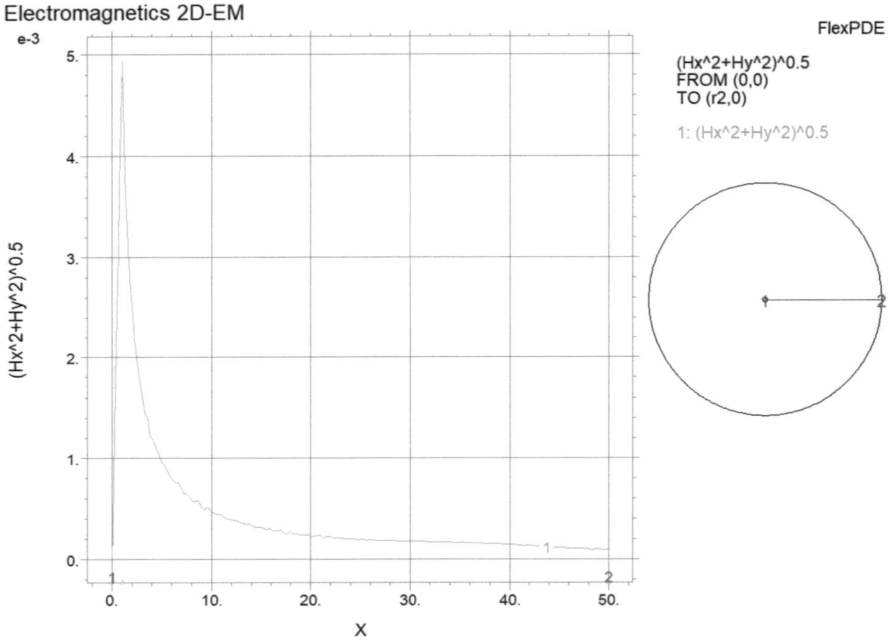

Fig. 23.8 Radial distribution of potential obtained by the field vector method

Appendix: FlexPDE Script File

Field Vector Formulation

```
{ Electromagnetics using E H vector formulation }
TITLE 'Electromagnetics 2D-EM' { the problem identification }
SELECT { The following error control is needed }
  ERRLIM 0.00005
  NGRID=35
  !REGRID OFF
VARIABLES
  Hx (1e-8)
  Hy (1e-8)
DEFINITIONS
  r1=1
  r2=50
  mu=4*pi*1e-7
  epsilon=8.654e-12
  c=1/sqrt(mu*epsilon)
  B_x=Hx*mu
  B_y=Hx*mu
  omega=10*pi
  Jz=0
```

```
EQUATIONS
  Hx: dx(Hy)-dy(Hx)=Jz
  Hy: dx(Hx)+dy(Hy)=0
BOUNDARIES
  REGION 1
    Jz=0
    START(r2,0)
    Value(Hx)=-1e-4*y/(x^2+y^2)^0.5
    Value(Hy)=1e-4*x/(x^2+y^2)^0.5
    ARC(CENTER=0,0) ANGLE=360
  REGION 2
    !Inactive(Hx,Hy)
    Jz=1e-2
    epsilon=1e12
    START(r1,0)
    NATURAL(Hx)=0
    NATURAL(Hy)=0
    ARC(CENTER=0,0) ANGLE=360
MONITORS
  FOR CYCLE=1
    VECTOR(Hx,Hy)
    CONTOUR((Hx^2+Hy^2)^0.5)
    ELEVATION((Hx^2+Hy^2)^0.5) FROM (0,0) TO (r2,0)
PLOTS
  FOR CYCLE=1
    VECTOR(Hx,Hy)
    CONTOUR((Hx^2+Hy^2)^0.5)
END
```

Potential Formulation

```
{ Electromagnetics 2D using potential formulation }
TITLE 'Electromagnetics 2D-Potential Formulation'
SELECT
  ERRLIM 0.00005
VARIABLES
  Az
DEFINITIONS
  r1=1
  r2=50
  mu=4*pi*1e-7
  epsilon=8.654e-12
  c=1/sqrt(mu*epsilon)
  Hx=dy(Az)/mu
  Hy=-dx(Az)/mu
  B_x=Hx*mu
  B_y=Hx*mu
EQUATIONS
  Az: DEL2(Az)=0
BOUNDARIES
  REGION 1
    START(r2,0)
    VALUE(Az)=mu/(4*PI)*1e-2*PI*r1^2/r2
    ARC(CENTER=0,0) ANGLE=360
```

```
  REGION 2
    START(r1,0)
    VALUE(Az)=mu/(4*PI)*1e-2*PI*r1^2/r1
    ARC(CENTER=0,0) ANGLE=360
MONITORS
  FOR CYCLE=1
    VECTOR(Hx,Hy)
    CONTOUR((Hx^2+Hy^2)^0.5)
    ELEVATION((Hx^2+Hy^2)^0.5) FROM (0,0) TO (r2,0)
PLOTS
  FOR t=2*pi
    VECTOR(Hx,Hy)
    CONTOUR((Hx^2+Hy^2)^0.5)
    ELEVATION((Hx^2+Hy^2)^0.5) FROM (0,0) TO (r2,0)
END
```

Chapter 24
Fluid Dynamics

Introduction

As a branch of fluid mechanics, fluid dynamics deals with fluid flow, that is, liquids and gasses in motion. Fluid dynamics encompasses two major sub-areas: aerodynamics for air and other gasses and hydrodynamics for liquids. Due to historical reason, hydrodynamics was used interchangeably with fluid dynamics in most studies before the twentieth century. This use can still be seen from the names of fluid dynamics topics such as magnetohydrodynamics and hydrodynamic stability, which in fact can also be applied to gasses. In this chapter, aerodynamics and hydrodynamics are differentiated. However, effort will also be directed to the introduction to aerodynamics when necessary, though this chapter is referred to as "hydrodynamics" in many places. In fact, there is a legitimate reason for putting hydrodynamics and aerodynamics together, that is, all the fluid dynamics problems are associated with the same governing equations, i.e., Navier-Stokes (N-S) equations. The differences originate from the different natures of the liquid flow and gas flow, such as flow velocity, compressibility, and viscosity. These differences lead to different forms of the Navier-Stokes equations and, consequently, different modeling and solution techniques. Hydrodynamics is more associated with low-velocity and incompressible flows, either viscous or inviscid, while aerodynamics is usually deemed to have a relatively high velocity, high compressibility, and thus negligible viscosity. However, these factors are not independent of each other, for example, the consideration of compressibility and viscosity needs to be done with reference to the velocity.

Fluid dynamics, as a study area, has a much broader scope than the fluid dynamics discussed in the context of multiphysics. The solution to a fluid dynamics problem typically involves the calculation of the spatial (and temporal) variations of various properties of a fluid, such as flow velocity, pressure, density, and temperature. These calculations are built on the conservation of mass, momentum, and energy, especially the first two. These axioms of conservation lead to the

Z. Liu, *Multiphysics in Porous Materials*,
https://doi.org/10.1007/978-3-319-93028-2_24

mathematical formulation of the fluid flow: Navier-Stokes equations. As the foundation of fluid dynamics, the Navier-Stokes equations can be used to model problems involving fluid flows, such as weather, ocean currents, aircraft design, scour, design of power station, the study of blood pressure, nebulae in interstellar space, and modeling fission weapon detonation. Some of the principles underlying the Navier-Stokes equations could even be used in modeling traffic and crowd dynamics. The equations can also be used to model and study magnetohydrodynamics when used with Maxwell's equations. Despite the theoretical and numerical efforts, traditional fluid dynamics also embraces empirical and semiempirical laws derived from flow measurements and thus enables theoretical and numerical tools for solving practical problems. However, the fluid dynamics in the following introduction will be confined within multiphysics.

The possible dramatic motion of liquid particles makes fluid dynamic problems deviated from the direct use of the classic conservation equations in solid mechanics. As we know, fluids are composed of molecules that collide with each other and with other objects in the fluids. However, the classic conservation equations treat fluids as a continuum rather than a discrete substance. As a result, quantities for the fluids' properties and states such as density, pressure, velocity, and temperature are defined at infinitesimally small points and thus are continuous from point to point. Therefore, there is a discrepancy between the nature of the material and the way we analyze it. In solids, this issue also exists but not as obvious as in fluids. This is because the fast mass and momentum transfer in liquids significantly magnifies the couplings between scales. The relative movement of fluids occurs at a wide range of scales, from that at the molecular scale, such as the viscous dissipation discussed in thermomechanics, to the Kolmogorov microscale which defines the smallest scales in turbulent flow, to the scale of interest. i.e., scale comparable to that of the computational domain (or real objects). A solution to the governing equations merely at macroscopic scale of the object/domain (defined by the grid/mesh) will overlook the details at smaller scales. The exclusion of the lower-scale processes will be a big issue when the flow is turbulent. This explains why, in computational fluid dynamics, a significant amount of effort has been made to the establishment of various turbulent simulation techniques, such as various turbulent models, which can be majorly classified into Reynolds average methods and large eddy methods.

In the following sections, different types of flows will be introduced first without heavily referring to the mathematical equations. The purpose is to offer a clear picture regarding the basic concepts in fluid dynamics and computational fluid dynamics (CFD) simulations. Presented next will be the introduction to the Navier-Stokes (N-S) equations. To make the introduction clear and complete, the general equations will be explained first, followed by the derivations of various common forms of the N-S equations, which correspond to different types of flows. Both the mass and momentum equations will be derived from the basic conservation law. Different constitutive relationships will be added to show different forms of the N-S equations for different fluids. Then, a section will be used to describe many concepts more in-depth with more mathematical details. The turbulent models will also be briefly introduced to complete the theoretical framework. Finally, a benchmark problem will be provided for practicing the numerical simulation of fluid dynamics.

Flow Types

Compressible and Incompressible Flow

All fluids are compressible to some extent, that is, the fluids' density will change in response to the changes in the pressure or temperature. However, in many situations, the changes in the pressure and temperature are so small that the subsequent changes in density are negligible. In these conditions, the flow can be simplified into the incompressible flow. Otherwise, the compressibility needs to be considered in the N-S equations and with an additional thermodynamic equation of state.

A thermodynamic equation of state is usually needed for the compressible flow to relate the pressure to the other thermodynamic variables, i.e., temperature and density, in addition to N-S equations for the mass, momentum, and energy conservation. A simple example is the perfect gas equation of state:

$$p = \frac{\rho RT}{M}, \tag{24.1}$$

where p is the pressure, ρ is the density, R is the gas constant, M is the molar mass, and T is the temperature.

By contrast, in the incompressible flow, the density of the fluid does not change with time:

$$\frac{D\rho}{Dt} = 0, \tag{24.2}$$

where D/Dt is the substantial (or material) derivative, which includes both the local and convective derivatives. Further simplifications can be obtained by assuming the fluid has a uniform and constant density.

Different criteria are used for liquids and gasses for the choice of compressible and incompressible fluid models. For liquids, this choice depends on the fluid properties, e.g., the critical pressure and temperature of the fluid and the flow conditions. For example, a criterion could be constructed based on the comparison between the critical pressure and the actual flow pressure. For gasses, the Mach number of the flow is usually adopted. The Mach number is the ratio between the flow velocity relative and the local speed of sound:

$$M_{\mathrm{a}} = \frac{v}{c}, \tag{24.3}$$

where M_a is the Mach number, v is the local flow velocity with respect to the boundaries, and c is the speed of sound in the material. As a rough guide, a gas can be treated as incompressible when the Mach numbers are below 0.3. Depending on the Mach number, there are subsonic, transonic, supersonic, and hypersonic flows. This classification will not be discussed in this book due to its relatively weak connections to porous materials.

It is worthwhile to point out that compressibility needs to be considered in acoustic problems. This is because sound waves are compressive waves involving changes in the pressure and density of a material through which the waves propagate. Therefore, the consideration of compression and compressibility is a prerequisite.

Inviscid, Newtonian, and Non-Newtonian Fluids

In a broad scope, fluids can be categorized as in Fig. 24.1. The time-independent fluid is of more interest in traditional computational fluid dynamics. Therefore, we will focus on this category in the following introduction.

In these fluids, fluid flows also can be classified according to the effect of the viscosity. Viscous fluids are those in which viscosity, which quantifies the fluid's friction, has a considerable effect on the fluid motion. Viscous fluids can be further categorized into Newtonian and non-Newtonian fluids according to the relationship between the viscosity and velocity gradient. On the contrary, the effect of the viscosity is negligible in inviscid fluids. The above classification of the fluids is relative and dependent on the flow conditions. Therefore, it is also common to

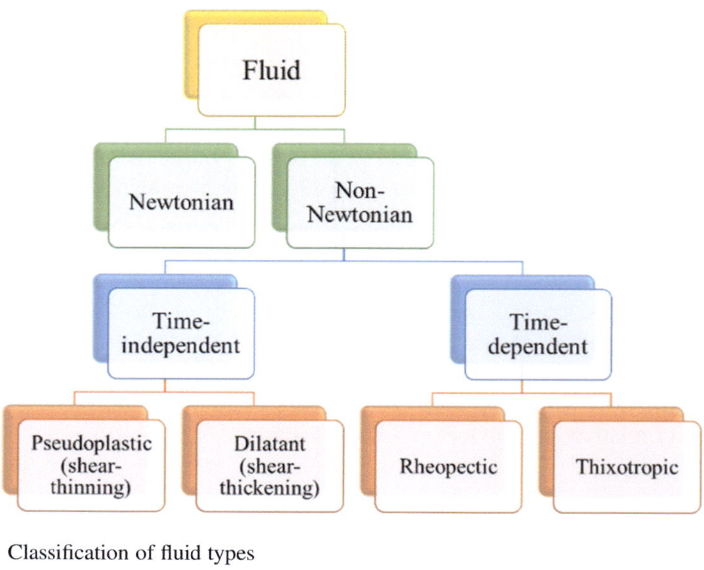

Fig. 24.1 Classification of fluid types

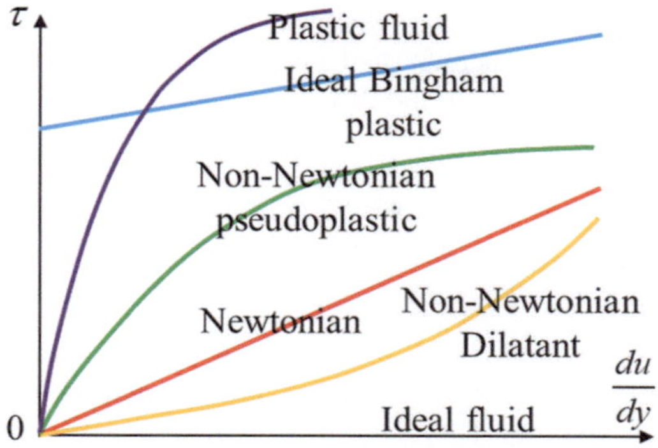

Fig. 24.2 Typical fluid types and their constitutive relationships

classify fluid flow into Stokes, viscous, and inviscid flows by considering both material properties and flow conditions. These three flow types use different governing equations which are simplified versions of the general N-S equations.

The Reynolds number is usually adopted to determine whether a flow is viscous or inviscid. The Reynolds number, *Re*, is the ratio between the inertial force and viscous force (the equation will be given in a later subsection). Stokes flow is the flow with very low Reynolds numbers, i.e., $Re \ll 1$. In Stokes flow, inertial forces can be neglected compared to viscous forces due to the low flow velocity. As the velocity increases, the Reynolds number increases, which leads to non-negligible inertial forces compared to viscous forces. Stokes flow is predominant in porous materials due to the low velocity of the flow through the pore spaces, whereas viscous flow is the most common one that we would encounter in surface water.

In viscous flow, the stress is related to the rate of strain. As illustrated in Fig. 24.2, in many fluids, such as water and air, the relationship is approximately linear. The proportionality between the stress and strain rate is the viscosity. Thus, those liquids maintaining a linear relationship or a constant viscosity are classified as Newtonian liquids, which are named after Isaac Newton who reported the relationship. However, there are also many other liquids who maintain a much more complicated nonlinear relationship between the stress and strain rate, such as honey, blood, and some polymers. More complicated non-Newtonian stress-strain behaviors are studied in the area of rheology.

If the velocity keeps increasing, the inertial forces can increase to values way above that of the viscous forces. Under such a condition, the viscous forces may be negligible, leading to inviscid flow. However, even for inviscid flow, viscosity still needs to be included for certain problems such as those involving solid boundaries: viscosity often cannot be neglected near solid boundaries because the no-slip condition can generate a thin region with large strain rates, i.e., boundary layer. The standard equations of inviscid flow are the Euler equations. It is also common to use the Euler equations in regions far from the boundary together with special boundary layer equations to allow for viscosity in regions close to the solid boundary. The Euler equations are a simplification of the N-S equations. The Euler equations can be integrated along a streamline to get Bernoulli's equation. When the flow is irrotational and inviscid, i.e., potential flows, Bernoulli's equation can be used throughout the flow field.

Laminar and Turbulent Flow

As mentioned above, the discrete nature of the fluid flow can significantly deviate the flow condition from what would be predicted using the N-S equations discretized at large scales. This is because the randomness and chaos caused by the discrete nature can change the flow patterns via couplings between different scales. The resultant recirculation, eddies, and apparent randomness are characterized as turbulence. Turbulence will become more obvious as the Reynolds number increases, which accounts for both geometry and flow velocity. Flow without noticeable turbulence is called laminar. When turbulence is noticeable to even dominant, the flow becomes turbulent. In Fig. 24.3, a laminar flow comes from the left. When the flow encounters an obstacle, i.e., porous fence, turbulent flow develops.

The above statement does not imply that the N-S equations are no longer valid in turbulent flow but, instead, suggests that N-S equations need to be discretized to very small spatial and temporal scales to capture the turbulence. This is essentially the idea of the direct numerical simulation (DNS) (Moin and Mahesh 1998; Moser et al. 1999). Therefore, no turbulent models are needed in the DNS. For the purpose, all the spatial scales of the turbulence must be resolved based on the computational mesh, from the Kolmogorov microscales (smallest dissipative scales) (Kolmogorov 1962) to the integral scale L. For the time scale, the time step needs to be small enough to ensure that a fluid particle moves only a fraction of the mesh spacing within each step. Existing practice has enabled DNS to simulate turbulent flows at moderate Reynolds numbers. The application is restricted by the power of the computer and the efficiency of the solution algorithm. The DNS has been validated against experimental data for some flows.

Other than the direct use of the N-S equations via the DNS, modeling flow with significant turbulence with normal meshes/grids is also possible. This goal can be achieved by introducing spatial or time averaging to decompose the flow into an

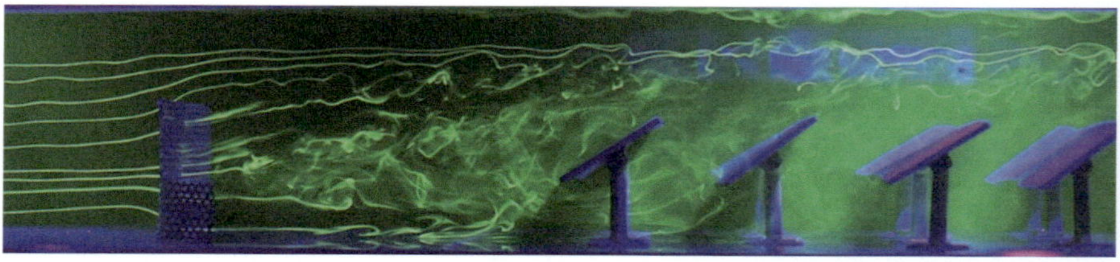

Fig. 24.3 Visualization of turbulence (behind a porous fence with 50% opening) [youtube.com/watch?v=bMO2IXmJa_Y]

average component and a perturbation component. The temporal decomposition, i.e., Reynolds decomposition, is more common. Reynolds-averaged Navier-Stokes (RANS) equations together with turbulent models provide an effective way to model the turbulent flow. The additional momentum transfer and the consequent extra mass and heat transfer are considered via the Reynolds stresses from turbulence. For space averaging, the large eddy simulation (LES) (Sagaut 2006) ignores the smallest length scales, which are the most computationally expensive to resolve, via low-pass filtering of the N-S equations. This treatment can be viewed as a time and spatial averaging. RANS and DES can be combined via the detached eddy simulation (DES) (Spalart 2009).

Most flows of interest including those industrial applications have Reynolds numbers far exceed which is affordable with the DNS. Therefore, the DNS is still a high-end auxiliary tool, though it will possibly become more and more popular and predominant as the available computational power multiplies in the incoming decades. For example, any vehicle large enough to carry a human and moving faster than 20 m/s will lead to a Reynolds number larger than four million, which is far beyond the limit of DNS simulation. In the following section, the N-S equations as the foundation of all the above simulation techniques will be introduced. Then focus will be placed on the RANS and the turbulent models.

Navier-Stokes Equations

The Navier-Stokes equations, named after Claude-Louis Navier and George Gabriel Stokes, describe the mass and momentum balance in moving viscous fluids. The mass balance equation, also called continuity equation, shared the same mass balance concept as those in groundwater and concentrations, but its formulation is still much different. The momentum balance equation, similar to Navier's equation in the mechanical field, traces back to Newton's second law. The major differences are as follows:

1. The momentum N-S equation additionally considers advection.
2. The momentum N-S equation usually contains an accumulation (inertial) term (time-dependent), while the Navier's equation only carries this term in dynamic problems.
3. The stress formulation is different: the stress in Navier's equation (solid) is related to the strain, while that in N-S equations is related to the strain rate. This leads to a distinction in terminology – we talk about elastic/plastic in solids but viscous/ Newtonian/non-Newtonian in fluids.

The N-S equations can be simplified into the Stokes equations by removing the inertial term in the momentum equation for Stokes flow with extremely low flow velocity. On the other extreme, the N-S equations can be simplified into the Euler equations by eliminating the viscous term for inviscid flow with a very high flow velocity.

A basic solution to the Navier-Stokes equations includes a field for the flow velocity and a field for the pressure. These two fields are highly coupled in the governing equations and thus are considered as one field in many places. A heat or energy balance equation and a concentration equation are also commonly added to obtain a comprehensive understanding of the fluids and its constituents. Trajectories of the fluid particles or solutes can be obtained once the flow velocity and its temperature variation are obtained.

The N-S equations are established based on the following assumptions:

1. Mass and momentum are conserved. In fact, all balance laws can be expressed in the following general form:

$$\frac{\partial u}{\partial t} = \nabla \cdot [u(\mathbf{v}_{\partial\Omega} - \mathbf{v})] - \nabla \cdot \mathbf{J} + g, \tag{24.4}$$

   where u is the conserved quantity which can be tensors of any order, $\mathbf{v}_{\partial\Omega}$ is the velocity of the moving boundary, $\mathbf{J}$ is the flux, and g is the source.

2. The fluid is continuous. Therefore, the dependent variables, i.e., flow velocity and pressure, are continuous and differentiable, at least weakly.
3. The control volume remains fixed in space (Euler description) or can move with the fluid (Lagrangian description). But the N-S equations that we commonly see adopt a control volume fixed in space (Euler description).

Conservation of Mass

The mass conservation equation can be easily obtained from the above general conservation equation by removing the flux and source terms in a computational domain and the non-moving boundaries. Since mass is the conserved quantity, the dependent variable will be the mass per unit volume, i.e., density. The general conservation equation turns into

$$\frac{\partial \rho}{\partial t} + \nabla \cdot (\rho \mathbf{v}) = 0, \tag{24.5}$$

where ρ is the mass density (mass per unit volume) and $\mathbf{v}$ is the flow velocity. In incompressible flow, ρ is a constant and the above equation reduces to

$$\nabla \cdot \mathbf{v} = 0. \tag{24.6}$$

According to the introduction in previous chapters, the above equation indicates that the $\mathbf{v}$ field is an equivoluminal field.

Conservation of Momentum

The quantity to be conserved for the momentum balance is the momentum per unit volume $\rho\mathbf{v}$. The $\mathbf{J}$ now is the negative of the Cauchy stress and the source is the body force. Substituting $\rho\mathbf{v}$ into the general balance equation, we obtain

$$\frac{\partial(\rho\mathbf{v})}{\partial t} + \nabla\cdot(\rho\mathbf{v}\mathbf{v}) = \nabla\cdot\boldsymbol{\sigma} + \mathbf{f}, \tag{24.7}$$

where $\mathbf{v} = (u, v, w) = \left(\frac{dx}{dt}, \frac{dy}{dt}, \frac{dz}{dt}\right)$, in which (x, y, z) are the coordinates of a fluid point and $\mathbf{f}$ is a vector source: body force. Thus, when applying Newton's second law to this point or a REV, we will have the material derivative. However, the computation domain is fixed as we use the Euler description. As a result, the material derivative splits into the two terms on the left-hand side of the momentum balance equation. For further derivations, let us first reformulate the left-hand side of the equation as

$$\mathbf{v}\left(\frac{\partial\rho}{\partial t} + \nabla\cdot(\rho\mathbf{v})\right) + \rho\left(\frac{\partial\mathbf{v}}{\partial t} + \mathbf{v}\cdot\nabla\mathbf{v}\right) = \nabla\cdot\boldsymbol{\sigma} + \mathbf{f}. \tag{24.8}$$

Recalling the mass balance equation, we have

$$\rho\left(\frac{\partial\mathbf{v}}{\partial t} + \mathbf{v}\cdot\nabla\mathbf{v}\right) = \nabla\cdot\boldsymbol{\sigma} + \mathbf{f}. \tag{24.9}$$

We can also write the above equation using the material derivative as

$$\rho\frac{D\mathbf{v}}{Dt} = \nabla\cdot\boldsymbol{\sigma} + \mathbf{f}. \tag{24.10}$$

As can be seen, the equation appears similar to the expression of Newton's second law ($m\mathbf{a} = \mathbf{F}$). The sum of the two terms on the right-hand side of the equation is the resultant force on the REV. In fact, this equation is the formulation of Newton's second law for continua, which thus can be applied to describe the nonrelativistic momentum conservation of any continuum that conserves mass. This so-called Cauchy momentum equation can be further derived into the general N-S equation by decomposing the stress term. In fluids, it is common to differentiate the pressure from the deviatoric stress:

$$\begin{aligned}\boldsymbol{\sigma} &= \sigma_{\mathrm{m}}\mathbf{I} + \boldsymbol{\tau}\\ &= \begin{pmatrix}\sigma_{\mathrm{m}} & 0 & 0\\ 0 & \sigma_{\mathrm{m}} & 0\\ 0 & 0 & \sigma_{\mathrm{m}}\end{pmatrix} + \begin{pmatrix}\sigma_{xx}+\sigma_{\mathrm{m}} & \tau_{xy} & \tau_{xz}\\ \tau_{yx} & \sigma_{yy}+\sigma_{\mathrm{m}} & \tau_{yz}\\ \tau_{zx} & \tau_{zy} & \sigma_{zz}+\sigma_{\mathrm{m}}\end{pmatrix},\end{aligned} \tag{24.11}$$

where $\mathbf{I}$ is the 3×3 identity matrix, $\boldsymbol{\tau}$ is the deviatoric stress tensor, and $-\sigma_{\mathrm{m}}$ is the mean normal stress, which is equal to the negative of the mechanical pressure p (or hydrostatic pressure in the context of hydrodynamics):

$$\sigma_{\mathrm{m}} = -p = -\frac{1}{3}\left(\sigma_{xx} + \sigma_{yy} + \sigma_{zz}\right). \tag{24.12}$$

The motivation for separating the two stress terms is that the pressure is typically a variable of interest, and also, this treatment simplifies the later application to specific fluid families because the rightmost tensor $\boldsymbol{\tau}$ in the above equation must be zero for a fluid at rest. The Navier-Stokes equation may now be written in the most general form as

$$\rho\left(\frac{\partial \mathbf{v}}{\partial t} + \mathbf{v} \cdot \nabla \mathbf{v}\right) = -\nabla p + \nabla \cdot \boldsymbol{\tau} + \mathbf{f}. \tag{24.13}$$

In the above equation, $\mathbf{v}$, p, and $\boldsymbol{\tau}$ are still unknown. Therefore, this equation is still not the N-S equation that we can use to analyze practical problems. In fact, $\mathbf{v}$ and p are dependent variables of the N-S equation. To derive solvable N-S equations, we just need to find constitutive relationships to relate the stress $\boldsymbol{\tau}$ to the velocity (or pressure). This will be discussed in the following section for the N-S equations for different fluids.

Navier-Stokes Equations for Different Fluids

The constitutive relationship to complete the general N-S equations is normally established based on the viscous behavior of the fluid. Specific forms of the Navier-Stokes equations will be obtained for different types of fluid flow. In the following, the Newtonian fluids and their flow will be discussed as this is the most common type. Non-Newtonian fluids will be briefly introduced as a more complicated and rarer case in viscous fluids. Then Stokes equations for creeping flow and Euler equations for the inviscid flow will be conveniently obtained by simplifying the above N-S equations for viscous flow.

Viscous Flow of Newtonian Fluid

The constitutive relationship for Newtonian fluids is established with the following three assumptions:

1. The fluid is isotropic.
2. The stress tensor is a linear function of the strain rate.

3. For a fluid at rest, $\nabla \cdot \boldsymbol{\tau}$ must be zero to ensure the existence of the hydrostatic pressure.

These assumptions will lead to

$$\tau_{ij} = \mu\left(\frac{\partial u_i}{\partial x_j} + \frac{\partial u_j}{\partial x_i} - \frac{2}{3}\delta_{ij}\frac{\partial u_k}{\partial x_k}\right) + \delta_{ij}\lambda\frac{\partial u_k}{\partial x_k}, \tag{24.14}$$

where μ is called the first coefficient of viscosity (usually just called "viscosity") and λ is the second coefficient of viscosity (related to the bulk viscosity). The value of λ quantifies the viscous effect associated with the volume change and is very difficult to determine, including its sign. This volume change effect could be negligible in compressible flows but be significant in nearly incompressible flows, leading to a matter of controversy. The most commonly used values of λ are 0 and $-2\mu/3$.

Compressible Newtonian Fluid

Directly substituting the above constitutive relationship into the general N-S equations yields the N-S equations for the flow of compressible Newtonian fluids:

$$\begin{aligned}\rho\left(\frac{\partial \mathbf{v}}{\partial t} + \mathbf{v}\cdot\nabla\mathbf{v}\right) &= -\nabla p + \nabla\cdot\left\{\mu\left[\nabla\mathbf{v} + (\nabla\mathbf{v})^T\right]\right\} \\ &\quad + \nabla\left(-\frac{2\mu}{3}\nabla\cdot\mathbf{v}\right) + \rho\mathbf{g},\end{aligned} \tag{24.15}$$

where p is used to represent the pressure and the gravity is considered as the only body force, i.e., $\mathbf{f} = \rho\mathbf{g}$. The associated mass continuity equation is

$$\frac{\partial \rho}{\partial t} + \nabla\cdot(\rho\mathbf{v}) = 0. \tag{24.16}$$

The above form of the N-S equations can cover the dynamics of most gasses and liquids with a good equation of state and good functions for the dependence of parameters such as viscosity on the variables.

Incompressible Newtonian Fluid

The N-S equations for incompressible Newtonian fluids can be derived from Eqs. 24.15 and 24.16 for compressible fluids with two further simplifications:

1. The viscosity μ is a constant.
2. The second viscosity coefficient diminishes: $\lambda = 0$.

As the fluid is incompressible, the mass balance equation is simply transformed into

$$\nabla \cdot \mathbf{v} = 0. \tag{24.17}$$

Attention is needed for the deduction of the viscous term on the right-hand side of the momentum balance equation. Taking the u component of the velocity $\mathbf{v}(u, v, w)$, for example, we have

$$\begin{aligned}
&\frac{\partial}{\partial x}\left(2\mu \frac{\partial u}{\partial x} - \frac{2\mu}{3}\nabla \cdot \mathbf{v}\right) + \frac{\partial}{\partial y}\left[\mu\left(\frac{\partial u}{\partial y} + \frac{\partial v}{\partial x}\right)\right] + \frac{\partial}{\partial z}\left[\mu\left(\frac{\partial u}{\partial z} + \frac{\partial w}{\partial x}\right)\right] \\
&\quad = 2\mu \frac{\partial^2 u}{\partial x^2} + \mu \frac{\partial^2 u}{\partial y^2} + \mu \frac{\partial^2 v}{\partial y \partial x} + \mu \frac{\partial^2 u}{\partial z^2} + \mu \frac{\partial^2 w}{\partial z \partial x} \\
&\quad = \mu \frac{\partial^2 u}{\partial x^2} + \mu \frac{\partial^2 u}{\partial y^2} + \mu \frac{\partial^2 u}{\partial z^2} + \mu \frac{\partial^2 u}{\partial x^2} + \mu \frac{\partial^2 v}{\partial y \partial x} + \mu \frac{\partial^2 w}{\partial z \partial x} \\
&\quad = \mu \nabla^2 u + \mu \frac{\partial}{\partial x}(\nabla \cdot \mathbf{v}) \\
&\quad = \mu \nabla^2 u.
\end{aligned} \tag{24.18}$$

Similarly, we can obtain $\mu\nabla^2 v$ and $\mu\nabla^2 w$ for the y and z directions. Then the N-S equations for incompressible Newtonian fluids are

$$\nabla \cdot \mathbf{u} = 0 \tag{24.19}$$

$$\rho\left(\frac{\partial \mathbf{u}}{\partial t} + \mathbf{u} \cdot \nabla \mathbf{u}\right) = -\nabla p + \mu \nabla^2 \mathbf{u} + \rho \mathbf{g}. \tag{24.20}$$

It is also common to write the momentum equation in the following form

$$\frac{\partial \mathbf{u}}{\partial t} + \mathbf{u} \cdot \nabla \mathbf{u} = -\nabla e + \nu \nabla^2 \mathbf{u} + \mathbf{g}, \tag{24.21}$$

where e is the specific (per unit mass) thermodynamic work and $\nu = \mu/\rho$ is the kinematic viscosity. e can also be used in the incompressible flow, in which the following identity holds if a uniform density can be assured:

$$\nabla e = \nabla\left(\frac{p}{\rho_0}\right) = \frac{1}{\rho_0}\nabla p. \tag{24.22}$$

Non-Newtonian Fluids

In non-Newtonian fluids, e.g., ketchup, custard, toothpaste, starch suspensions, paint, blood, and shampoo, the viscosity is not necessarily dependent on the shear rate or shear rate history. It is also possible that the viscosity is related to the high-order temporal derivative of displacement. For those non-Newtonian fluids with a shear rate-dependent viscosity, the relation between the shear stress and the shear rate is not linear anymore. The constitutive relationships for a few common and simple non-Newtonian fluids are given here to provide basic information.

In Bingham fluids, the following relationship holds:

$$\frac{\partial u}{\partial y} = \begin{cases} 0 & , \tau < \tau_0 \\ (\tau - \tau_0)/\mu, & \tau \geq \tau_0 \end{cases} \quad (24.23)$$

Therefore, Bingham fluids, e.g., toothpaste and some saturated clays, are capable of bearing some shear before they start flowing.

In addition, power laws are useful for approximating the behavior of many common non-Newtonian fluids including shear thinning fluids, e.g., latex paint, and shear thickening fluids, e.g., cornstarch-water mixture. This constitutive relationship has the following form:

$$\tau = K\left(\frac{\partial u}{\partial y}\right)^n. \quad (24.24)$$

Stokes Flow

Stokes flow, also termed creeping flow, is a type of fluid flow where advective inertial forces are negligible compared with viscous forces. It is the flow condition in pore water. Removing the inertial term from the momentum balance equation, the N-S equations then reduce to the Stokes equations:

$$\frac{\partial \rho}{\partial t} + \nabla \cdot (\rho \mathbf{v}) = 0, \quad (24.25)$$

$$0 = -\nabla \sigma_{\mathrm{m}} + \nabla \cdot \boldsymbol{\tau} + \mathbf{f} = -\nabla p + \mu \nabla^2 \mathbf{u} + \rho \mathbf{g}. \quad (24.26)$$

Figure 24.4 shows the Stokes flow passes a solid particle. This is the normal flow condition in granular materials without other interphase phenomena such as double layers.

Inviscid Flow

In fluid dynamics, the Euler equations are a set of quasilinear hyperbolic equations governing adiabatic and inviscid flow. The viscous effect is negligible in the inviscid flow. Accordingly, the incompressible Euler equations are as follows:

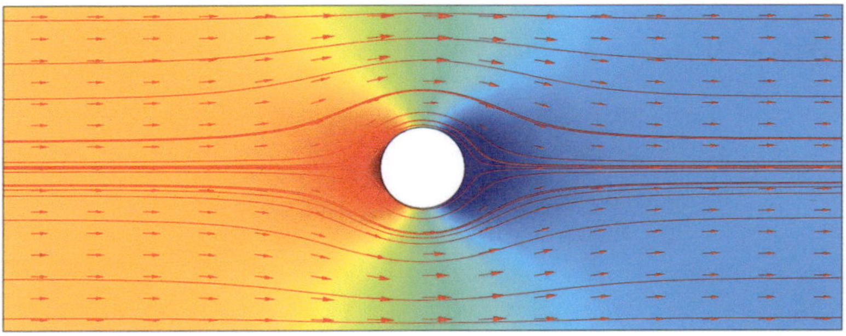

Fig. 24.4 Flow field of the Stokes flow against of solid obstacle

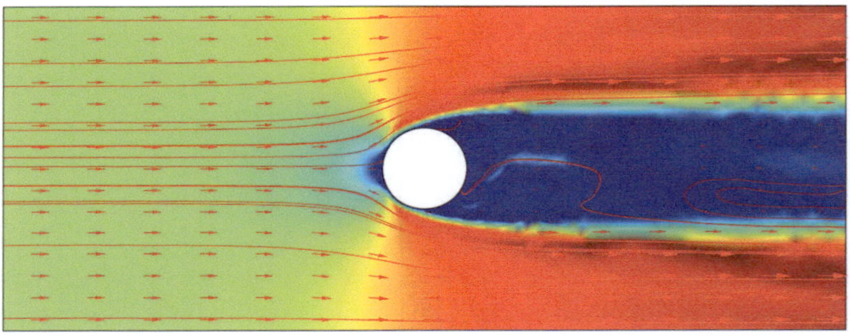

Fig. 24.5 Flow field of the inviscid flow against of solid obstacle

$$\frac{\partial \rho}{\partial t} + \nabla \cdot (\rho \mathbf{v}) = 0, \tag{24.27}$$

$$\rho\left(\frac{\partial \mathbf{v}}{\partial t} + \mathbf{v} \cdot \nabla \mathbf{v}\right) = -\nabla p + \rho \mathbf{g}. \tag{24.28}$$

Figure 24.5 demonstrates the inviscid flow passing a solid particle. The exclusion of the viscous term, which functions as a diffusion mechanism in momentum transfer, will raise the level of difficulty in numerically solving the equations. The solution in Fig. 24.5 exhibits the instability and error in the area behind the solid particle.

Turbulent Models

Reynolds Number: From Laminar to Turbulent

The Reynolds number can be defined in different ways by incorporating fluid properties including the density, viscosity, flow velocity, and a characteristic length (or characteristic dimension). However, the choice of this length is a matter of convention. For example, the radius or diameter is usually chosen to describe spheres or circles, while the length or width is commonly used for aircrafts or ships. An equivalent diameter is usually defined for non-spherical objects. More complicated rules need to be applied in fluids with a variable density such as non-Newtonian fluids. Mathematically, the Reynolds number is usually defined as below:

$$R_e = \frac{\text{inertial forces}}{\text{viscous forces}} = \frac{\rho \mathbf{v} L}{\mu} = \frac{\mathbf{v} L}{\nu}, \tag{24.29}$$

where L is a characteristic linear dimension.

For "fully developed" flow in a pipe with a diameter, D, experimental observations show that laminar flow occurs when $R_e = \frac{\rho \mathrm{v} D}{\mu} < 2100$ and turbulent flow occurs when $R_e > 4000$. "Transition" flows exist in the R_e interval between 2300 and 4000, in which both laminar and turbulent flows are possible, depending on other factors such as pipe roughness and flow uniformity. Figure 24.6 shows the transition from a laminar flow to a turbulent flow as Reynolds number increases. However, it is worthwhile to point out that the Reynolds number in the figure is different from what is defined in a pipe due to the existence of the obstacles and the channel. Therefore, the threshold values are also different.

The primary use of the Reynolds number is to evaluate the relative significance of inertial and viscous forces in fluid flow and to differentiate laminar flow from turbulent flow based on that. Additionally, Reynolds numbers are very useful for scaling of fluid dynamics problems to assess the dynamic similitude between two different fluid flows.

Simulation of Turbulent Flows

Turbulent flow occurs in the high Reynolds number range. Consequently, this type of flow is dominated by inertial forces and tends to produce chaotic eddies, vortices, and other flow instabilities. In fact, most fluid flows occurring in nature are turbulent to some extent. Turbulence can be described as a state of continuous instability in the flow and is characterized by the irregularity in the flow, increased diffusivity, and energy dissipation. Turbulent flow always has 3D and time-dependent characteristics, even if the boundary conditions of the flow do not change with time.

As introduced in the previous section, the key idea to separate turbulence from the mean flow is averaging, for which the following three different ways exist:

1. Time averaging in a fixed point of space, for stationary turbulence
2. Space averaging for a fixed moment in time in the case of homogeneous turbulence
3. Ensemble averaging for a series of identical experiments (the most general form of averaging)

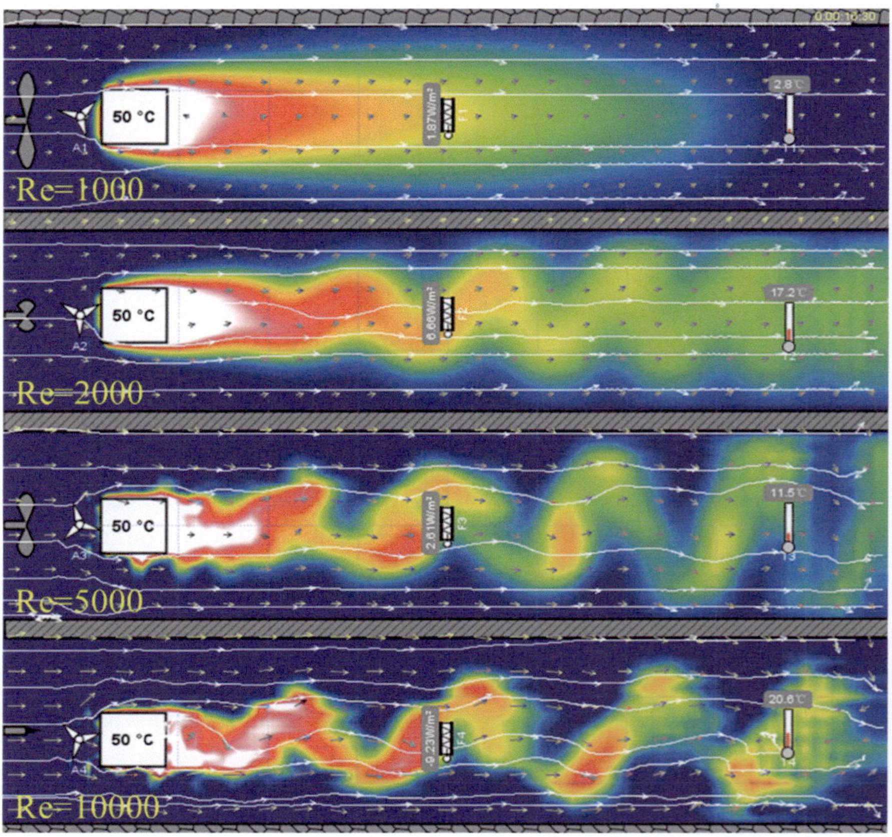

Fig. 24.6 Transition from laminar to turbulent flows: the characteristic length is the width of the flow channel (obtained using Energy2D web tool [http://energy.concord.org/energy2d/reynolds.html])

All these methods can appear in two forms, unweighted (Reynolds) and weighted. The Reynolds averaging technique operates as follows:

$$\phi(\mathbf{x}, t) = \overline{\phi}(\mathbf{x}, t) + \phi^{'}(\mathbf{x}, t), \tag{24.30}$$

where $\phi(\mathbf{x}, t)$ is decomposed into an average term $\overline{\phi}(\mathbf{x}, t)$ and an oscillating term $\phi^{'}(\mathbf{x}, t)$. $\phi^{'}(\mathbf{x}, t)$ denotes the fluctuation around the mean value, which is defined as

$$\overline{\phi}(\mathbf{x}, t) = \lim_{N \to \infty} \frac{1}{N} \sum_{i=1}^{N} \phi_i(\mathbf{x}, t), \tag{24.31}$$

where N is the number of identically performed experiments.

Applying the above averaging procedure to the incompressible N-S equations, the following form of the averaged equations can be obtained:

$$\nabla \cdot \bar{\mathbf{v}} = 0, \tag{24.32}$$

$$\frac{\partial \bar{\mathbf{v}}}{\partial t} + \nabla \cdot (\bar{\mathbf{v}}\bar{\mathbf{v}}) = \mathbf{g} - \nabla \bar{p} + \nabla \cdot \left(\nu \nabla \bar{\mathbf{v}} + \overline{\mathbf{v}'\mathbf{v}'}\right), \tag{24.33}$$

where the term $\overline{\mathbf{v}'\mathbf{v}'}$ is called the Reynolds stress tensor. In order to close the system, further modeling is necessary to obtain the Reynolds stress.

The goal of Reynolds-averaged turbulence modeling is to express the Reynolds stress tensor in terms of the known quantities. There are two widely accepted approaches (Wilcox 1998). The first approach formulates and solves the transport equation for the Reynolds stress tensor (and higher moments, depending on the order of the closure). It is, however, still necessary to model some of the terms, since the number of unknowns increases faster than the number of equations. The second and a more popular approach prescribes a relationship between the Reynolds stress and the mean velocity gradient. Although models prescribing nonlinear relations have been proposed recently, the most popular approach is to use the Boussinesq approximation (Boussinesq eddy viscosity hypothesis) (Schmitt 2007), which adopts a linear relation of the form

$$\overline{\mathbf{v}'\mathbf{v}'} = -\nu_t \left[\nabla \bar{\mathbf{v}} + (\nabla \bar{\mathbf{v}})^T\right] + \frac{2}{3} k \mathbf{I}, \tag{24.34}$$

where $k = \frac{1}{2} \overline{\mathbf{v}'} \cdot \overline{\mathbf{v}'}$.

The kinematic eddy viscosity ν_t can be evaluated in many different ways, ranging from algebraic relations and local equilibrium assumptions to the solution of transport equations. The most popular approach is to express ν_t as a function of the turbulent kinetic energy k and its dissipation rate ε, leading to a "two-equation" turbulence model:

$$\nu_t = C_\mu \frac{k^2}{\varepsilon}, \tag{24.35}$$

$$\varepsilon = \nu \overline{\mathbf{v}'\mathbf{v}' : \nabla \mathbf{v}'}. \tag{24.36}$$

In this type of turbulent model, k and ε are obtained by the solution of their respective transport equations. The first model of this kind was proposed by Harlow and Nakayama (1968). Derivation of the transport equations for k and ε can be found in Launder and Spalding (1974). In fact, a wide variety of k-ε models exist, while the most noteworthy one is the standard k-ε model proposed by Launder and Spalding (1974) and the RNG k-ε variant proposed by Yakhot and Orszag (1986) and Yakhot et al. (1992).

Standard k-ε *Turbulence Model*

The exact k-ε equations contain many unknown and unmeasurable terms. For a much more practical approach, the standard k-ε turbulence model (Launder and Spalding 1974) is used. This model was established based on our best understanding of the relevant processes and thus minimizes the number of unknowns and leads to a set of equations which can be applied to a large number of turbulent applications.

For turbulent kinetic energy k, we have

$$\frac{\partial(\rho k)}{\partial t}+\frac{\partial(\rho k v_i)}{\partial x_i}=\frac{\partial}{\partial x_j}\left(\frac{\mu_t}{\sigma_k}\frac{\partial k}{\partial x_j}\right)+2\mu_t E_{ij}E_{ij}-\rho\varepsilon. \tag{24.37}$$

For dissipation ε, a similar transport equation can be used:

$$\frac{\partial(\rho\varepsilon)}{\partial t}+\frac{\partial(\rho\varepsilon v_i)}{\partial x_i}=\frac{\partial}{\partial x_j}\left(\frac{\mu_t}{\sigma_\varepsilon}\frac{\partial \varepsilon}{\partial x_j}\right)+C_{1\varepsilon}\frac{\varepsilon}{k}2\mu_t E_{ij}E_{ij}-C_{2\varepsilon}\rho\frac{\varepsilon^2}{k}, \tag{24.38}$$

where v_i is the velocity component in the corresponding direction; E_{ij} is the component of the rate of deformation; and μ_t is the eddy viscosity, which can be calculated as $\mu_t=\rho C_\mu\frac{k^2}{\varepsilon}$.

The above equation states the following balance:

Rate of change of k or ε+ transport of k or ε by convection = transport of k (or ε) by diffusion + rate of production of k (or ε) – rate of the destruction of k (or ε).

Equations 24.37 and 24.38 also consist of some adjustable constants: σ_k, σ_ε, $C_{1\varepsilon}$, and $C_{2\varepsilon}$. The values of these constants have been obtained by numerous iterations of data fitting for a wide range of turbulent flows. A common set of values are as follows: $C_\mu=0.09$, $\sigma_k=1.00$, $\sigma_\varepsilon=1.30$, $C_{1\varepsilon}=1.44$, and $C_{2\varepsilon}=1.92$.

Practice Problem

The section introduces a benchmark problem for 2D transient computational fluid dynamics, which was described in Sharma (2012).

Problem Description

In a 1 m by 1 m square liquid container as shown in Fig. 24.7, the liquid inside the container moves as the lid on the top boundary continuously moves from the left to the right at a velocity of 1 m/s. Assuming the water on the top boundary moves at the same speed as the lid, please predict the distribution of v_x along $x=0.5$ m and

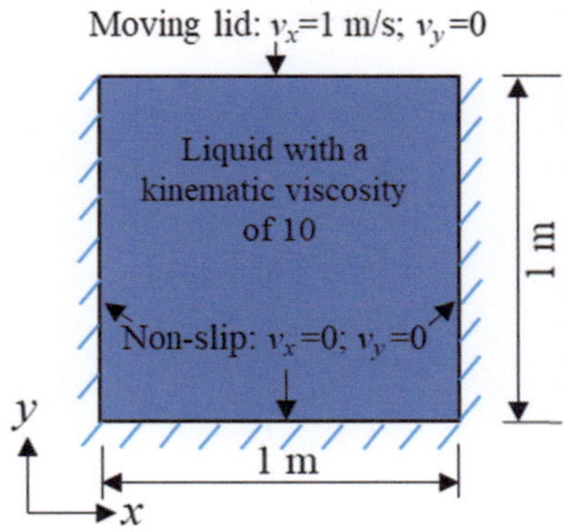

Fig. 24.7 Computational domain and boundary conditions

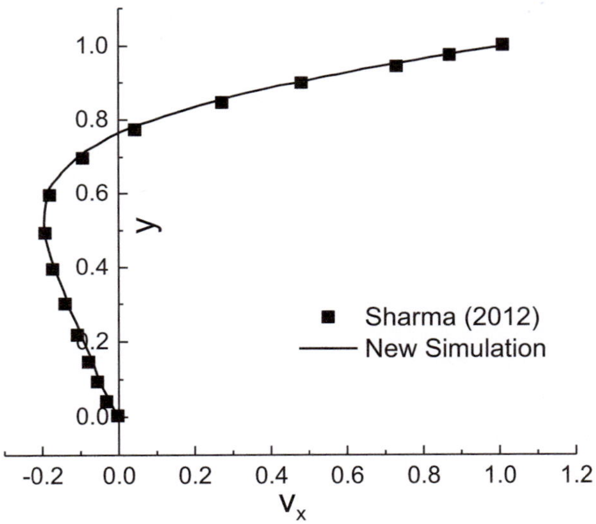

Fig. 24.8 x-velocity along the y-axis of the domain ($x = 0.5$ m)

v_y along $y = 0.5$ m, and compare them against the numerical results of Ghia et al. (1982). The kinematic viscosity is 10.

Results

The results obtained by the numerical simulation implemented with FlexPDE are compared against the results given in Sharma (2012). As shown in Figs. 24.8 and 24.9, a simple CFD implementation with a very coarse mesh and the finite element method can also generate acceptable results. Better comparisons can be obtained by further tuning the numerical model and solution techniques.

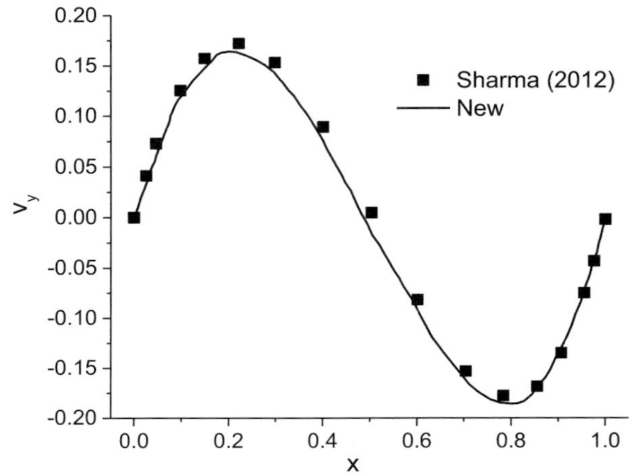

Fig. 24.9 y-velocity along the x-axis of the domain ($y = 0.5$ m)

Appendix: FlexPDE Script File

```
{ 2D CFD benchmark, Sharma, 2012, EPJ Web of Conferences 24}
TITLE 'Transient CFD-Incompressible with penalty'
COORDINATES CARTESIAN2
VARIABLES
  v(1e-8)=VECTOR(vx,vy)
  p(1e-8)
SELECT
  ERRLIM 0.01
DEFINITIONS
  rho=1
  nu=10
INITIAL VALUES
  p=0 v=VECTOR(0,0)
EQUATIONS
  p: DIV(v)=-10e-8*DT(p)
  v: -nu*DIV(GRAD(v))+DOT(v,GRAD(v))+GRAD(p)+DT(v)=0
BOUNDARIES
  REGION 1
    START(0,0)
      VALUE(vx)=0
      VALUE(vy)=0
    LINE TO (1,0)
      VALUE(vx)=0
      VALUE(vy)=0
    LINE TO (1,1)
      VALUE(vx)=1
      VALUE(vy)=0
    LINE TO (0,1)
      VALUE(vx)=0
      VALUE(vy)=0
    LINE TO CLOSE
```

```
TIME 0 TO 1
MONITORS
  FOR CYCLE=1
    VECTOR(v)
    CONTOUR(p)
PLOTS
  FOR CYCLE=1
    VECTOR(v)
    CONTOUR(p)
    ELEVATION(vx) FROM (0.5,0) TO (0.5,1)
    ELEVATION(vy) FROM (0,0.5) TO (1,0.5)
END
```

Chapter 25
Hydrodynomechanics: Fluid-Structure Interaction

Introduction

The term hydrodynomechanics refers to any phenomenon involving the mechanical response of a solid and the mass and momentum transfer in a liquid. In a more general sense, the scope can be extended from liquids to fluids. In this case, hydrodynomechanics can refer to, or more accurately, be related to three application areas: dynamic poroelasticity, acoustic-structure interaction, and complete fluid-structure interaction. The relationship between hydrodynomechanics and these areas are very complicated because the relevant applications were conducted and named more due to practical and historical interests rather than from the perspective of multiphysics. The following figure gives out a very typical example of fluid-structure interaction in petroleum engineering. Similar examples are the wind turbines and their foundations in lakes (Fig. 25.1).

Dynamic poroelasticity is the hydrodynamics occurring on the level of REV. It is the dynamic extension to poromechanics. By contrast, the broad definition of the fluid-structure interaction, including the acoustic-structure interaction and the complete fluid-structure interaction, involves solid and liquid interaction on the level of regions within the computational domain. From the perspective of the physical field, it is very similar to the dynamic poromechanics as it also involves the elastodynamics in the solid and acoustics in the fluid. However, they have different natures and thus appear as different types of multiphysics. As illustrated in Fig. 25.2, the dynamic poromechanics treats the porous material as an ensemble of REVs, each of which is composed of a liquid fraction and a solid fraction. Therefore, the solid and liquid are coupled at the scale of REV, which should be much smaller than the scale of the computational domain. Accordingly, the two fields are coupled on the level of mathematical equations. By contrast, fluid-structure interaction allows for coupling between solid and fluid regions which are at the scale of the computational domain. Therefore, the solid and liquid are coupled on the boundary between the solid and liquid regions within the computational domain. Different types of

Z. Liu, *Multiphysics in Porous Materials*,
https://doi.org/10.1007/978-3-319-93028-2_25

Fig. 25.1 Offshore oil platforms as a typical FSI example

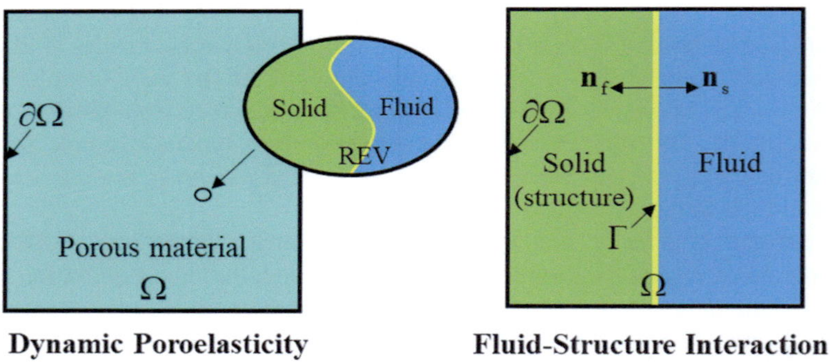

Fig. 25.2 Two types of hydrodynomechanics

equation systems will be used for different regions, depending on whether the region consists of a solid or a fluid. The coupling between different regions is achieved by ensuring the mechanical balance or continuity.

In this chapter, hydrodynomechanics will be introduced as three distinct areas: dynamic poroelasticity, acoustic-structure interaction, and complete fluid-structure interaction. The theory of dynamic poroelasticity has been given in poroelasticity, so this chapter only includes a short section to offer more details regarding its implementation and application. For the complete fluid-structure interaction, we will show the idea of region coupling and give detailed procedures for implementing such coupling. Next, the acoustic-structure interaction will be introduced in a similar way.

The differences and connections between the complete fluid-structure interaction and acoustic-structure interaction problems will be pointed out both in theory and in their applications. A simple example to show the implementation of the complete fluid-structure interaction will be presented in the final.

Dynamic Poroelasticity

The first application area of hydrodynomechanics is related to the dynamic poromechanics introduced in Chapter 20 for poroelasticity. However, from a different angle, it would be more accurate to state that dynamic poromechanics has a major overlap with hydrodynomechanics rather than to treat it as a special case of hydrodynomechanics. In fact, the complete equations in the dynamic poromechanics consist of momentum balance equations for both solid and liquid, which are mixed in any REV. The momentum balance equation for the solid is not merely the equation for the mechanical field (static) but also includes the inertial term. Therefore, it is the elastodynamics, or plastodynamics if plastic constitutive relationships are involved (which is rare). However, the momentum equation for the fluid is a wave propagation equation. We know that the wave equation in liquids can be obtained by degenerating the comprehensive Navier-Stokes equations. Therefore, the fluid part can still be described as hydrodynamics if acoustics in fluids can be viewed as a part of fluid mechanics in a broad sense. The governing equations for both the solid and liquid take the form of wave equations. These wave equations describe the mechanical waves in the solid and liquid. From this viewpoint, dynamic poroelasticity is the coupling between acoustic fields in two well-mixed phases. Despite the similar nature, the wave propagation equations in solids and liquids have similar nature but still differentiate in a few places. The major differences reside in the constitutive relationships: solids resist shear, while fluids do not. It is also worthwhile to mention that, in a broad sense, acoustics can also include wave propagation in solids.

The governing equations and constitutive relationships for dynamic poroelasticity have been introduced in Chap. 20. However, the practical treatment could be more complicated than the original equation system. The pressure fluctuations in the saturating fluid interact with different elastic waves (shear and longitudinal) in the porous material. As the pores are often small, the energy losses introduce damping to the system. Moreover, the porous domain can be coupled to the pure solid structures or pure fluid domains, introducing a multitude of multiphysics couplings.

There are several ways of describing the propagation of waves in porous materials. This first way is to directly solve the original equation system. This way accounts for the coupled propagation of elastic waves in the elastic porous matrix and pressure waves in the saturating pore fluid. This equation system can incorporate the damping effect of the pore fluid. A simpler approach is to treat the porous matrix and saturating pore fluid as a homogenized equivalent fluid. In this case, only the pressure waves are modeled with the consideration of losses. The loss models can range from pure empirical models based on measurements to analytical or semi-analytical phenomenological models; examples include the Delany-Bazley-Miki (Jeong 2011) and the Johnson-Champoux-Allard models (Doutres et al. 2010). However, if treated in this way, the simplified physical process is more like a monolithic physical process.

Complete Fluid-Structure Interaction

Mathematical Formulation

The above introduction indicates that the key component in the mathematical description of the fluid-structure interaction (FSI) is essentially not the governing equations. The governing equations for both elastodynamics and fluid dynamics/acoustics have been introduced in Part III of this book. For the solid region, or structure, the governing mechanisms can be formulated using the Navier's equation with the inertial term:

$$\rho_s \ddot{\mathbf{u}}_s = \nabla \cdot \boldsymbol{\sigma}_s + \mathbf{f}_s, \tag{25.1}$$

where $\boldsymbol{\sigma}_s$ is the stress tensor, $\mathbf{f}_s$ is the body force, ρ_s is the density of the solid, $\mathbf{u}_s$ the displacement vector, and $\ddot{\mathbf{u}}_s$ is the second temporal derivative of the displacement.

The governing equation for the fluid is the Navier-Stokes equations:

$$\frac{\partial \rho_f}{\partial t} + \nabla \cdot (\rho_f \mathbf{v}_f) = 0 \tag{25.2}$$

$$\rho \left(\frac{\partial \mathbf{v}_f}{\partial t} + \mathbf{v}_f \cdot \nabla \mathbf{v}_f \right) = \nabla \cdot \boldsymbol{\sigma}_f + \mathbf{f}_f. \tag{25.3}$$

The major physical mechanism to ensure the coupling of the fluid and solid regions is to satisfy the following conditions on the boundaries between the two regions (also frequently called interface or internal boundary), Γ:

$$\dot{\mathbf{u}}_s = \mathbf{v}_f, \tag{25.4}$$

and

$$\boldsymbol{\sigma}_s \cdot \mathbf{n}_s = -\boldsymbol{\sigma}_f \cdot \mathbf{n}_f. \tag{25.5}$$

The first equation ensures the mass conservation or no separation, while the second ensures the momentum balance or mechanical equilibrium. The first condition (Eq. 25.4) can be replaced by the following equation when the profile of the boundary between the fluid and solid regions is smooth in both time and space:

$$\mathbf{u}_s = \mathbf{u}_f. \tag{25.6}$$

According to the way to achieve the above couplings, the complete FSI techniques can be categorized into two classes: conforming mesh methods and nonconforming mesh methods. The conforming mesh methods track the location of the interface and enforce Eqs. 25.5 and 25.6 explicitly. Gradual updates on the mesh during the solution process are usually needed as the boundary between the two regions is displaced, and therefore, the mesh within each region needs to conform with these boundary displacements. This type of method can be implemented conveniently with existing FDM, FVM, or FEM programs. A simple approach is to first enforce the nonslip boundary condition for the fluid on the fluid-solid boundary:

$$\mathbf{u}_{\mathrm{f}} = 0. \tag{25.7}$$

On the solid side, a Neumann boundary condition can be prescribed so that the normal traction on the solid boundary is equal to the fluid pressure:

$$\boldsymbol{\sigma}_{\mathrm{s}} \cdot \mathbf{n}_{\mathrm{s}} = -\boldsymbol{\sigma}_{\mathrm{f}} \cdot \mathbf{n}_{\mathrm{f}} = -p\mathbf{n}_{\mathrm{s}}, \tag{25.8}$$

where p is always positive and the negative sign in front of p is used to ensure that the pressure acts as a compressive stress on the solid along its boundary with the fluid.

The location of the boundary between the two regions is tracked using a moving mesh algorithm, which is available in many numerical simulation programs these days. The moving boundary is induced by the fluid pressure acting on the deformable solid structure. Therefore, the mesh displacement on the boundary between the two regions is equal to the displacement of the solid,

$$\mathbf{u}_{\mathrm{mesh}} = \mathbf{u} \ \text{ on } \Gamma \tag{25.9}$$

The movement of the mesh within the fluid and solid regions can be obtained based on the boundary movement. A common way is to use the diffusion equation, which "diffuses" the boundary displacement into the interior of each region:

$$\nabla^2 \mathbf{u}_{\mathrm{mesh}} = 0. \tag{25.10}$$

On these boundaries where no boundary displacement occurs, a Neumann boundary condition can be used:

$$\mathbf{n} \cdot \nabla \mathbf{u}_{\mathrm{mesh}} = 0. \tag{25.11}$$

In addition, it is also common to use the velocity of the mesh movement as either the dependent variable or boundary conditions for this mesh diffusion field. There are still other ways to consider the moving mesh, which will not be detailed here. The equation systems coupled in the above way can be solved either implicitly or explicitly (partitioned method).

The second category is the nonconforming mesh methods, most notably the immersed boundary method (Peskin 1977, 2002). This type of method requires many treatments in addition to the existing discretization. These methods usually enforce the velocity consistency on the boundary between the two regions instead of the displacement consistency. More details can be found in Hou et al. (2012).

Application

The complete fluid-structure interaction (FSI) involves the fluid dynamics and the mechanical field. This type of coupled phenomenon encompasses any process involving a deformable or moving structure and a surrounding or internal fluid flow. Slightly different from the acoustic-structure interaction, the complete FSI theoretically involves any interaction between the fluid and solid structure, although the focus is more placed on low frequency and large-strain applications.

One typical example is the flow around a solid obstacle as shown in Fig. 25.3. As the fluid flow encounters a structure, the structure is deformed due to the pressure exerted by the moving fluid. If the magnitude and variation of the deformations of the structure are small, the fluid flow will not be greatly affected by the resistance force from the solid. Due to this reason, we can simplify this problem as a one-way problem in which the stresses result from the force exerted by the fluid on the solid is a major concern.

However, if this is a relatively soft plate instead of a rigid body, then a large deformation of the solid structure will be caused. This large deformation will in return significantly impact the flow condition. This is what is shown in the above figure, in which two-way fluid-structure interaction was simulated. In a turbulent flow, the force on the solid will vary, leading to changes in the deformation of the plates. Consequently, a very strongly-coupled two-way interaction can be produced.

It is also possible to have a condition in which the magnitude of the deformation of the solid is small, while the time variation is large, for example, when the major

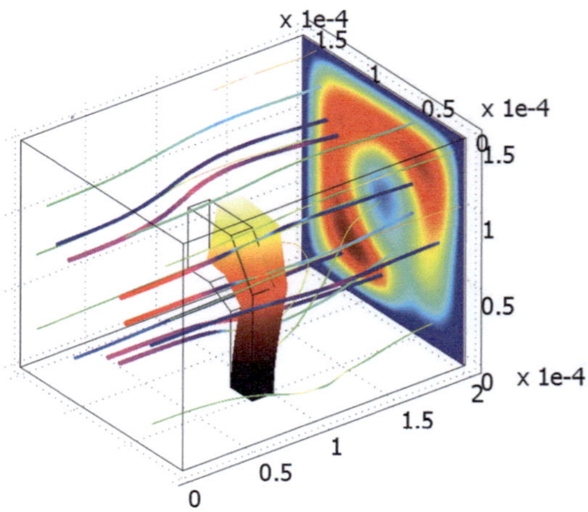

Fig. 25.3 A flexible obstacle in a fluid channel is bending under the viscous and pressure forces imposed by the fluid (flow streamlines are colored by flow velocity; deformed beam is colored by the deformation along the flow direction; contours are for flow velocity distribution) [COMSOL example library]

frequency is larger than several Hz. Under such a condition, these small structural deformations may still lead to pressure waves in the fluid. These pressure waves result in the radiation of sound from vibrating structures. Such problems can be treated as an acoustic-structure interaction using a wave equation rather than a complete fluid-structure interaction with the N-S equations.

Acoustic-Structure Interaction

In the acoustic-structure interaction (ASI), we have an acoustic field in a fluid, either a liquid or a gas, and a mechanical field in a solid which is in contact with the fluid. There are two major differences between the ASI and the complete FSI. First, the N-S equations in the fluid region degenerate into a wave equation. Second, the displacement of the solid structure is generally small due to the magnitude of the disturbance, while the frequency is higher than that in complete FSI. These two differences lead to the different treatments in the ASI.

The governing equation for the fluid is the mechanical wave equation. We here use the wave equation with the pressure as the dependent variable to facilitate the following formulation of coupling,

$$\nabla^2 p_{\mathrm{f}} - \frac{1}{c^2}\ddot{p}_{\mathrm{f}} = 0, \tag{25.12}$$

where c is the wave velocity.

The governing equation of the solid is still the Navier's equation with the inertial term:

$$\rho_{\mathrm{s}}\ddot{\mathbf{u}}_{\mathrm{s}} = \nabla \cdot \boldsymbol{\sigma}_{\mathrm{s}} + \mathbf{f}_{\mathrm{s}}. \tag{25.13}$$

As introduced in the previous section, the couplings for FSI require the consistency of the displacement (or velocity) and stress on the boundary between the fluid and solid. In acoustic-structure interaction problems, the first requirement is automatically met due to the negligible solid deformation. Therefore, we just need to ensure that the stresses on two sides of the fluid-solid boundary are balanced. Mathematically, this corresponds to the following two equations:

$$\mathbf{n}_{\mathrm{f}} \cdot \nabla p_{\mathrm{f}} = \rho\ddot{\mathbf{u}} \tag{25.14}$$

$$\mathbf{n}_{\mathrm{s}} \cdot \nabla \boldsymbol{\sigma}_{\mathrm{s}} = -p\mathbf{n}_{\mathrm{f}}. \tag{25.15}$$

The first equation represents a Neumann boundary condition for the fluid on the boundary, whereas the latter prescribes a Neumann boundary condition for the solid on the boundary. The two equations will generate a very strong coupling between the two fields by placing the dependent variable of one field into the matrix of the other in the process of discretization. Other ways to implement the above two equations, for example, treating them as Dirichlet boundary conditions by switching the terms on two sides of the equations, are also possible. These ways may require more bottom-level modifications to the discretization and solution procedures.

Application

The acoustic-structure interaction creates sound. In general, acoustics aim to understand the small and rapid changes in the acoustic pressure above and below the atmospheric pressure, which are usually described as pressure waves propagating in space and time. The ASI problems involve two physical fields: the acoustic field and mechanical field, if they are considered different fields. The coupling between the two fields can be both one-way and two-way, depending on the properties of the fluids and solids and the purpose of the analysis. Two-way couplings occur when both the acoustic pressure waves in the fluid and the vibrations of the solid are strong enough to affect each other significantly.

It is very common that a sound is generated as a moving structural object disturbs a gas or a liquid carrying the acoustic pressure (or sound) wave. The pressure wave in the fluid also produces vibrations in the solid. The interaction is by nature two-way. However, in many cases, the interaction in one direction is more dominant. The above one-way interaction can be observed in daily life. One example is a bass speaker, which moves its cone back and forth to compress and decompress the air to generate low-frequency sound. In the reserve direction, typical examples are the acoustic pressure waves that have been employed to produce vibrations in a solid for applications such as ultrasound imaging or nondestructive impedance testing.

ASI as a multiphysical process can further involve the thermal field. This is similar to the energy dissipation in which mechanical energy is converted into thermal energy. The difference is that here we focus on the energy associated with the waves. The energy can dissipate in two ways: via the viscosity in the fluids and via the attenuation associated with the damping in the solid. The viscous forces and damping can cause significant energy dissipation (or loss) near the walls of the structure in the acoustic boundary layer. In classic acoustics, the fluid is usually assumed to be inviscid and isentropic. Accordingly, slip boundary conditions are used on the boundaries between the solid and fluid, leading to solely coupling normal to the surface. By contrast, if the viscous and thermal losses are considered, both a mechanical and a thermal condition must be specified at the solid surface together with a nonslip condition to ensure continuity in the fluid velocity. That is, the movement of the solid is equal to the movement of the fluid. This leads to tightly coupled acoustic and mechanical fields.

Practice Problem

This section is about a very typical fluid-structure interaction problem in which a deformable solid obstacle is placed in a flow channel.

Problem Description

Figure 25.4 shows the cross-section of a flow channel in which water flows from the left to the right. The water flow velocity on the left boundary has a profile of $0.5y(\mathrm{H}-y)/\mathrm{H}^2$. The right boundary has a pressure of zero and no shear stress. The top and bottom sides are the walls of the flow channel, which can be approximated using the nonslip boundary condition. The flow channel has a length of 3 m (L) and a width (or height) of 1 m (H). An obstacle with a length of $H/4$ and a width of $L/40$ is fixed to the bottom wall at a location where the back face of the obstacle is $L/10$ from the left boundary. The deformation obstacle has the following mechanical properties: $\lambda = 10^7$ Pa and $G = 10^7$ Pa (Lamé constants). Please simulate the fluid flow and solid deformation in the process between the time when water starts moving into the channel and the time when the front of the flow passes the back face of the obstacle.

Governing Equation and Moving Mesh

In this example, the impact of the fluid movement on the solid deformation is achieved via the balance of the normal stresses (pressures) on two sides of inter-boundary: the solid and fluid regions. The influence of the solid deformation on the water flow is achieved by means of moving mesh. The solid and fluid regions are only analyzed using their own governing equations. Accordingly, in FlexPDE, the N-S equation is inactivated in the solid region using the command "INACTIVE,"

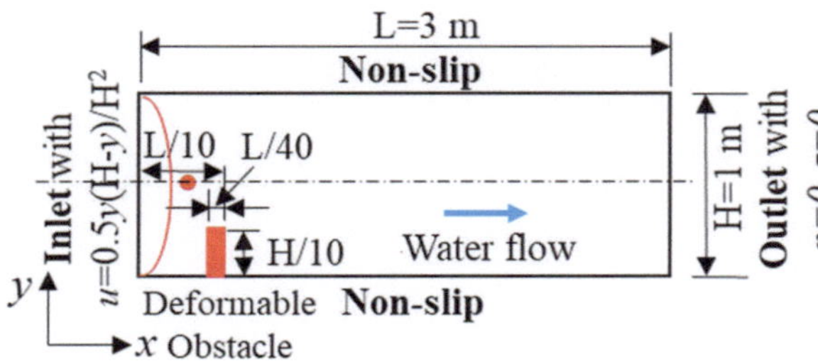

Fig. 25.4 Computational domain and boundary conditions

and likewise; the Navier's equation is inactivated in the fluid region. Four governing equations are needed as follows:

$$p\text{: rho}/K^*\text{DT}(p) + \text{DIV}(\text{rho}^* v) = 0$$
$$v\text{: DT}(\text{rho}^* v) + \text{DOT}(\text{rho}^* v, \text{GRAD}(v))$$
$$= -\text{GRAD}(p) + \text{mu}^*\text{DIV}(\text{GRAD}(v)) + 1/3^*\text{mu}^*\text{GRAD}(\text{DIV}(v))$$
$$\text{xm: DIV}(\text{GRAD}(\text{xm})) = 0$$
$$u\text{: DIV}(\text{sigma\_tensor}) = \text{VECTOR}(0,0)$$

The third equation is the governing equation for the moving mesh. This diffusion equation is used to predict the movement of the mesh within the domain based on the moving boundary.

Unfortunately, the above governing equation system contains six PDEs according to FlexPDE' rules (a vector in 2D is counted as two), which exceeds the PDE number limit in the student version. We can remove the displacement along the vertical direction, i.e., u_y, to enable the implementation of the model in the complimentary student version. Please simulate the above problem without u_y if you only have access to the student version. Here, the script file for the complete implementation of the above governing equation system is provided at the end of this chapter. The removal of u_y can be conveniently carried out, which is also a good practice for deepening the understanding of the mechanical field.

Results

Figures 25.5 and 25.6 illustrate the displacement of the obstacle and the velocity field at $t = 10^{-4}$ s. At this time, the front of the flow has not reached the front face of the obstacle yet. However, the increase in the pressure has already led to small displacements of the obstacle, which are not visible without magnification. Real deformation on the obstacle can be clearly observed at $t = 2 \times 10^{-4}$ s, when the front of the water flow is about to reach the back face of the obstacle, as shown in Fig. 25.7. The water pressure distribution explains the reason for this deformation: the pressure increases on the front face is much higher than that on the back face. It is noted that the accuracy of the simulation presented here may not be high due to the nonlinearity residing in the problem. Better discretization and solution techniques are desired for more satisfactory simulation results. We here just intend to use a very simple simulation to demonstrate the idea of fluid-structure interaction.

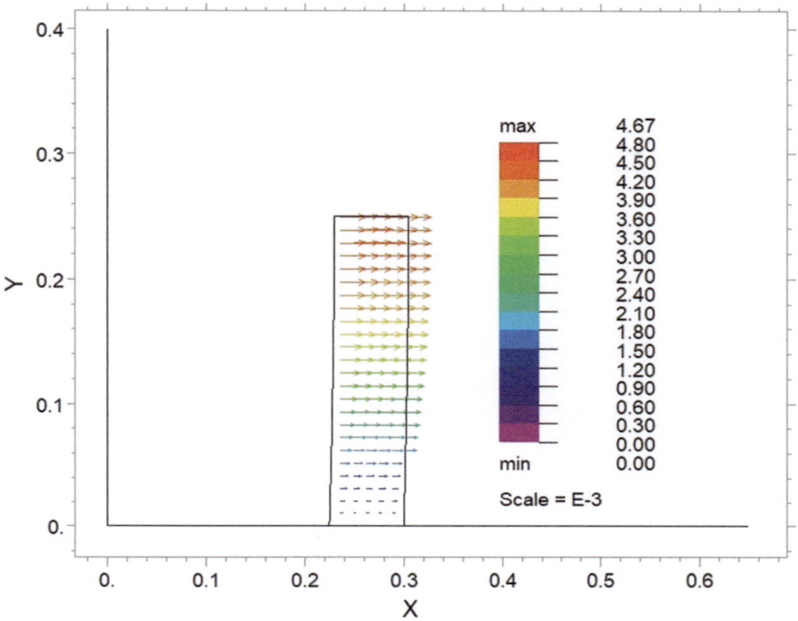

Fig. 25.5 Displacement field in the solid obstacle (arrows) at $t = 0.1$ ms

Fig. 25.6 Water velocity field (arrows) at $t = 0.1$ ms

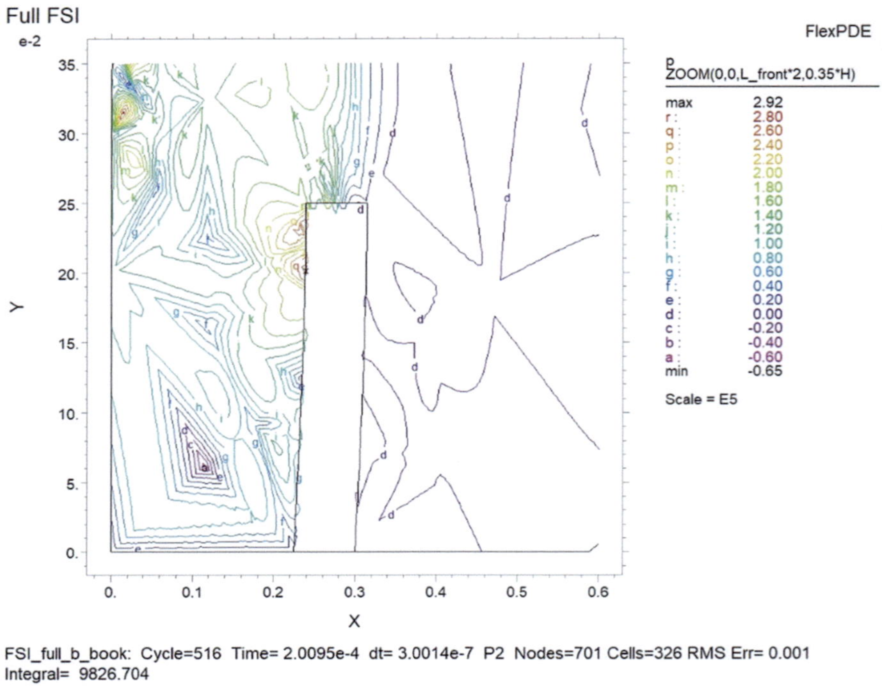

Fig. 25.7 Pressure distribution and deformed obstacle at $t = 0.2$ ms

Appendix: FlexPDE Script File

```
{ Full Fluid-Structure Interaction: Flow through a channel with a
flexible obstacle }
TITLE 'Full FSI' { the problem identification }
COORDINATES CARTESIAN2 { coordinate system, 1D,2D,3D, etc. }
SELECT
  ERRLIM 0.005
VARIABLES
  v(1e-8)=VECTOR(vx,vy)
  p(1e-8)
  xm=move(x)
  u(1e-8)=VECTOR(ux,uy)
!SELECT
  !NGRID =18
DEFINITIONS
  G=1e7
  lambda=1e7
  C=TENSOR((2*G+lambda,lambda,0),(lambda,2*G+lambda,0),(0,0,G))
! Stiffness matrix
  epsilon=VECTOR( dx(ux),dy(uy),0.5*(dx(uy)+dy(ux)) )
```

```
  sigma=dot( C,epsilon)
  { The above C and epsilon are defined using the Voigt notation;
sigma_tensor uses the tensor notation }
  sigma_tensor=TENSOR((XCOMP(sigma),ZCOMP(sigma)),(ZCOMP
(sigma),YCOMP(sigma)))
  H=1
  L=3
  L_front=L/10 ! x position of the front wall of the obstacle
  rho=1000
  mu=0.001
  K=1/4.4e-10
INITIAL VALUES
  p=0 ! only consider hydrodynamic pressure: gravity is excluded,
rho*9.81*(10-y)
  v=VECTOR(0,0)
EQUATIONS
  p: rho/K*DT(p)+DIV(rho*v)=0 !rho/K* !1e-3*0.001/0.01^2*div(grad
(p)) !VECTOR(0,0)
  v: DT(rho*v)+DOT(rho*v,GRAD(v))=-GRAD(p)+mu*DIV(GRAD(v))+1/
3*mu*GRAD(DIV(v))
  xm: DIV(GRAD(xm))=0
  u: DIV(sigma_tensor)=VECTOR(0,0) ! CONSTRAINTS { Integral
constraints }
BOUNDARIES { The domain definition }
  REGION 1 { For each material region }
    INACTIVE(u)
    START(0,0) { Walk the domain boundary }
      !NATURAL(p)=0
      VALUE(vx)=0.5*y*(H-y)/H^2
      VALUE(vy)=0
      VELOCITY(xm)=0
    LINE TO (0,H)
      !NATURAL(vy)=0
      VALUE(vx)=0
      VALUE(vy)=0
    LINE TO (L,H)
      VALUE(p)=0 { This is a key. Without it, the p distribution will be
wrong.
        The reason is unknown, possibly due to FlexPDE BC treatment. }
      NATURAL(vx)=0
    LINE TO (L,0)
      VALUE(vx)=0
      VALUE(vy)=0
    LINE TO CLOSE
  REGION 2
    INACTIVE(v,p)
    START (L_front,0)
      NATURAL(ux)=-p
      VALUE(uy)=0
      VALUE(vx)=0
      VALUE(vy)=0
      VALUE(xm)=L_front+ux
    LINE TO (L_front,H/4)
```

```
      NATURAL(ux)=0
      NATURAL(uy)=-p
      VALUE(vx)=0
      VALUE(vy)=0
      NATURAL(xm)=0
    LINE TO (L_front-L/40,H/4)
      NATURAL(ux)=p
      VALUE(uy)=0
      VALUE(vx)=0
      VALUE(vy)=0
      VALUE(xm)=L_front-L/40+ux
    LINE TO (L_front-L/40,0)
      VALUE(ux)=0
      VALUE(uy)=0
      NATURAL(xm)=0
    LINE TO CLOSE
TIME 0 TO 1
MONITORS
  FOR CYCLE=1
    VECTOR(v)
    CONTOUR(p) ZOOM(0,0,L_front*2,0.35*H)
    HISTORY(p) AT (0,H/2)
    ELEVATION(p) FROM (0,0) TO (0,H)
    ELEVATION(p) FROM (H/2,0) TO (H/2,H)
    VECTOR(u) ZOOM(0,0,L_front*2,0.35*H)
PLOTS
  FOR CYCLE=1
    VECTOR(v)
    CONTOUR(p)
END
```

Chapter 26
Thermoelectromagnetics

Introduction

Thermoelectromagnetics covers the multiphysics involving both the thermal field and the electromagnetic fields. This chapter introduces three major mechanisms in thermoelectromagnetics: Joule heating, dielectric heating, and induction heating. For each mechanism, the introduction will be carried out in the sequence of basic mechanisms, mathematical formulations, and applications. Additional knowledge that is needed for the implementation of thermoelectromagnetics involving these mechanisms, such as dielectric heating types, skin depth, and Poynting vector, will also be introduced. A practice problem will be presented in the last section for simulating microwave heating in a waveguide.

Joule Heating

Joule heating (also referred to as resistive or Ohmic heating) describes the process where the energy of an electric current is converted into heat as it flows through a resistive conductor. Joule heating is caused by the interactions between the moving particles constituting the current, which are, usually but not always, electrons, and the atomic ions constituting the body of the conductor. The moving particles accelerated by an electric field transfer some of their kinetic energy to the conductor ions via collisions. This energy transfer will lead to an increase in the kinetic energy of the conductor ions and consequently a temperature increase in the conductor. Via such a process, energy is transferred from the external electric field to the thermal field (Barrett and Cardello 2012). This is mostly true as collisions between the moving particles in the current and the conductor ions are inevitable, which exhibit as the fact that electric resistivity almost always exists at the macroscopic scale. Figure 26.1 clearly exhibits the heat generated by a circuit using a thermal image.

Z. Liu, *Multiphysics in Porous Materials*,
https://doi.org/10.1007/978-3-319-93028-2_26

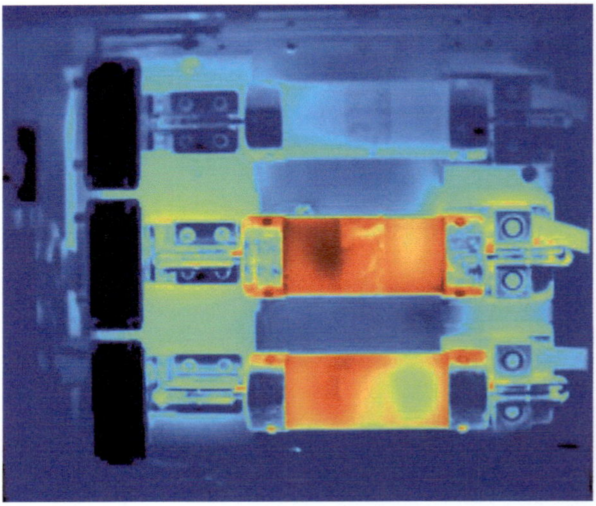

Fig. 26.1 Thermal image of fuses in a circuit (Fluke)

Mathematical Description

Most equations for a complete description of Joule heating have been introduced:

$$\nabla \times \mathbf{E} = -\frac{\partial}{\partial t}\mathbf{B} \tag{26.1}$$

$$\nabla \times \mathbf{H} = \mathbf{J}_{\mathrm{e}}^{\mathrm{f}} + \frac{\partial}{\partial t}\mathbf{D}. \tag{26.2}$$

$$\mathbf{D} = \varepsilon \mathbf{E}. \tag{26.3}$$

$$\mathbf{B} = \mu \mathbf{H} \tag{26.4}$$

$$\mathbf{J}_{\mathrm{e}}^{\mathrm{f}} = \sigma_{\mathrm{e}} \mathbf{E}. \tag{26.5}$$

An experimental observation underlying the formula for Joule heating is that energy dissipated per unit time is equal to the product of the energy dissipated per a unit charge passing through a resistor, i.e., a potential, and the amount of charge passing through resistor per unit time, i.e., a current: $P_{\mathrm{e}}^{\mathrm{f}} = IV$. When Ohm's law is also applicable, the formula can be written as

$$P_{\mathrm{e}}^{\mathrm{f}} = I^2 R = V^2/R. \tag{26.6}$$

The differential form of the Joule heating equation calculates the power per unit volume:

$$dP_e^f/dV = \mathbf{J}_e^f \cdot \mathbf{E}, \tag{26.7}$$

where $\mathbf{J}_e^f$ is the current density and $\mathbf{E}$ is the electric field. For a neutral plasma in a magnetic field and with a conductivity of σ, we have $\mathbf{J}_e^f = \sigma\mathbf{E}$; hence

$$dP_e^f/dV = \mathbf{J}_e^f \cdot \mathbf{E} = \mathbf{J}_e^f \cdot \mathbf{J}_e^f/\sigma = \rho|\mathbf{J}_e^f|^2 = \sigma|\mathbf{E}|^2, \tag{26.8}$$

where $\rho = 1/\sigma$ is the resistivity. This equation resembles the "I^2R" term in the macroscopic form of the Joule heating equation.

Applications and Adverse Effects

In some cases, Joule heating is utilized to design electric devices such as water heaters, while in others, it is an unwanted effect which could be detrimental to the safety and serviceability of electrical or electronic systems.

On one hand, the employment of Joule heating for various heating systems has a very long history. Applications utilizing Joule heating include direct use such as hot plates or pipes as shown in Fig. 26.2 and indirect use (via thermal expansion) such as microvalves for fluid control. Other heating mechanisms, such as induction heating to be introduced later, also boils down to the combination of Joule heating and dielectric heating to be introduced later.

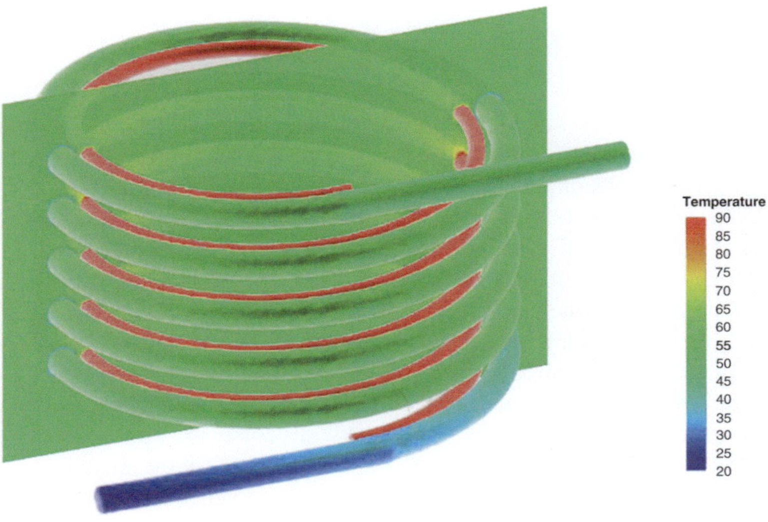

Fig. 26.2 Joule heating-water heating with Flow-3D [XC Engineering, 2009, online video]

On the other hand, efforts can be made to reduce the detrimental effects caused by Joule heating. Especially, Joule heating could be a concern in electrical system components such as conductors in electronics, electric heaters, power lines, and fuses. The generated heat degenerates or even melts the systems or components. Cooling designs are needed in many cases to prevent electric devices and their components from overheating. A voltage applied to a circuit can cause a current and consequently the Joule heating in the electrically conductive layer over the glass plate. Bending is then caused by the thermal expansion resulting from the differential displacements attributed to the material properties and temperature distributions.

Dielectric Heating

Dielectric heating is also known as electronic heating, RF heating, and high-frequency heating. It is the conversion of electromagnetic energy into thermal energy in dielectric materials caused by a high-frequency alternating electric field, radio wave, or microwave electromagnetic radiation. There are two slightly different heating types, i.e., RF heating and microwave heating. Both types of dielectric heating are induced by dipole rotation. In the RF heating, the applied electromagnetic radiation usually has a frequency between 3 and 300 MHz. Therefore, the heated objects are generally smaller than the wavelength. Microwave heating, distinct from RF heating, is associated with frequencies above 100 MHz. Due to the reason, an electromagnetic wave can be launched from a small dimension emitter and guided through space to the target to be heated.

Mechanism

Dielectric heating is a multiphysical phenomenon that involves electromagnetic waves and heat transfer. Theoretically, any common material that is exposed to electromagnetic radiation will be heated up. The rapidly varying electric and magnetic fields lead to four sources of heating. First, the electric field applied to a conductive material will cause electric currents. Second, a time-varying electric field will cause dipolar molecules, such as water, to oscillate and thus generates heat. Additionally, a time-varying magnetic field applied to a conductive material will also generate a current flow. Finally, there can also be hysteresis losses in certain types of magnetic materials.

Energy losses will be generated due to the current flow and the oscillation of the molecules in the material. These losses will dissipate in terms of heat and a subsequent rise in the temperature. As the temperature of the material changes, those properties characterizing the material's ability in generating heat, such as the electric conductivity and dielectric constants, will also change. In addition, the temperature change will also affect the material's ability in transferring heat such as the specific heat capacity and thermal conductivity. A comprehensive consideration of dielectric heating should allow for all of these couplings between the electromagnetic field and the thermal field. Despite the common omission of specific secondary couplings such as the effect on material properties in simplified analyses, the major heating mechanisms are always needed even in multiphysics simulations involving Joule heating.

Molecular rotation occurs in materials containing polar molecules, which have an electric dipole moment. These polar molecules will align themselves with an external electromagnetic field. In the case of an oscillating electromagnetic field, the polar molecules rotate continuously to align themselves with the fast-changing external field. This process is called dipole rotation or dipolar polarization. Heat is generated as the molecules gain kinetic energy. Moreover, collisions and other interactions will inevitably occur as the polar molecules rotate. This will further distribute the energy to the other molecules and atoms in the material. Dipole rotation is the major mechanism underlying dielectric heating. The mechanism can be clearly seen in microwave ovens, which operate most efficiently on liquid water. This is because water has a lot of polar molecules (water) rather than fats and sugars, which have less polar molecules.

Types of Dielectric Heating

By definition, dielectric heating refers to the heating of electrically insulating materials via dielectric loss resulting from an external alternating electromagnetic field. This changing field may be caused either by an electromagnetic wave propagating in free space, e.g., a guided wave in a microwave oven, or by a rapidly alternating electric field inside a capacitor. In the first condition, the freely propagating wave transfers the energy. In the latter case, although the heating is accomplished by changing the electric field inside the capacitive cavity at radio-frequency (RF) frequencies, no actual radio waves are generated or absorbed. Accordingly, the changing electric field may be viewed as an analogy to the electric component of an antenna's near field. In other words, this effect is the direct electrical analogy of magnetic induction heating, which is also near-field effect without classical radio waves.

Frequencies of the alternating field can determine the dielectric heating mechanism depending on the material type. This is attributed to the relationship between the frequencies and the size of the "particles" that they can best vibrate and the availability of different types of "particles" in the material. From a theoretical perspective, this physical correlation turns into the relationship between the frequency and the imaginary part of the permittivity as shown in Fig. 26.3, which controls the energy dissipation of the electromagnetic wave. Although high frequencies are generally better, lower-frequency dielectric heating, i.e., less than 10 MHz, can also generate significant heating effects in some materials especially liquids via some less common mechanisms. For example, a low-frequency alternating field can

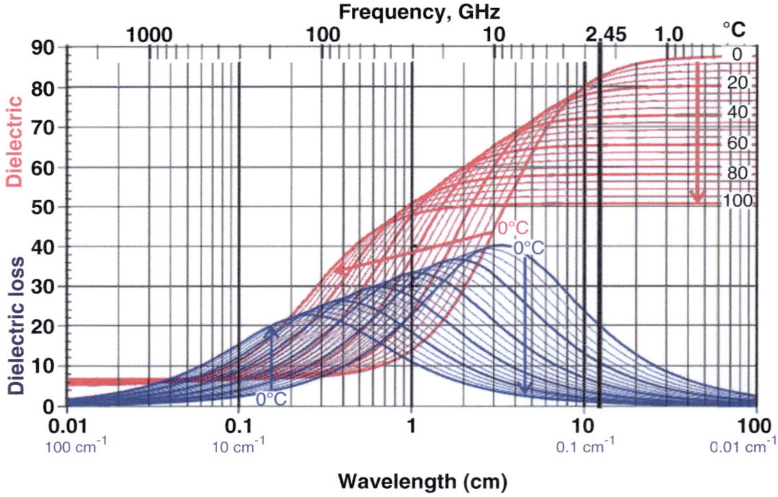

Fig. 26.3 Spectrum of the imaginary component of the dielectric constant. (Franks 1982; Buchner et al. 1999; Water Structure and Science site)

drag charged ions and make them move back and forth slowly in a liquid. This kinetic energy of these ions will be transferred to the liquid molecules via collisions, leading to an increase in the temperature of the solution.

RF heating covers the frequencies from 10 MHz to 100–100 GHz. Excluding special mechanisms such as the aforementioned ion-drag, frequencies larger than 10 MHz are usually required to cause efficient dielectric heating. In this frequency range or higher, dielectric heating is caused by molecular dipole rotation. RF dielectric heating with frequencies between 10 MHz and 100 MHz performs even better compared to microwave heating for quickly and uniformly heating certain food items and for killing parasites and pests in certain harvested crops. However, RF heating involves a contact process or near-contact process, though a physical contact is not necessary, which requires the distance between the electromagnetic radiator and the absorber is less than about 1/6th of a wavelength. Therefore, it is common to sandwich the material to be heated with metal plates, which generate an external field in which the sandwiched dielectric material works as a capacitor. The RF electric fields penetrate nonconductive materials far more deeply than higher-frequency microwaves; RF heating can be used to rapidly heat and prepare many non-electrically conductive food and agricultural items.

Higher frequencies, i.e., greater than 300 MHz, will enter the microwave range. In this case, the wavelength of the external electromagnetic field becomes shorter than the dimension of the heating space or the heating items. This is exactly what happens inside a microwave oven. In a microwave oven, the capacitor utilized in the above RF heating will no longer work as a heating space, but rather an antenna which generates and transmits far-field electromagnetic waves. The waves will be guided into an insulated space, i.e., microwave heating space, and eventually absorbed by the heating items to cause heating. The dipole-rotation mechanism in the microwave heating remains the same as that in the RF heating. Microwave penetrates to a distance defined by the skin effect. Microwave ovens for heating food are usually not set to the frequency for optimal absorption by water, i.e., 10 GHz, but, instead, to 2.5 GHz to allow a relatively deep penetration into the heated item.

Mathematical Formulation

In dielectric heating, the density of generated heat per a unit volume is

$$P_{\mathrm{e}}^{\mathrm{d}} = \omega \varepsilon_{\mathrm{r}}^{'} \varepsilon_0 |\mathbf{E}|^2, \tag{26.9}$$

where ω is the angular frequency of the external field, $\varepsilon_{\mathrm{r}}^{'}$ is the imaginary part of the complex relative permittivity of the heated material, ε_0 is the permittivity of vacuum, and $|\mathbf{E}|$ is the electric field strength. The imaginary part of the relative permittivity, which is also frequency-dependent, is a measure of the ability of a dielectric material to convert electromagnetic energy into heat. As we know, σ determines Joule heating. But in the case of a low σ and a high ω value, i.e., $\sigma \ll \omega \varepsilon_{\mathrm{r}}^{'} \varepsilon_0$, dielectric heating is the dominant mechanism for energy conversion from the electromagnetic field to the thermal field.

Application

The RF heating was proposed in the 1930s for heating dielectric materials. For example, US Patent US2147689A filed by the Bell Telephone Laboratories in 1939 proposed RF heating at 10–20 MHz. Such wavelengths were much longer than the dimensions of the heating space. Therefore, the near-field effects instead of electromagnetic waves are the primary heating approaches. In agriculture, RF dielectric heating has been accepted as an effective way to kill pests in certain food crops after harvest such as walnuts still in the shell. The RF heating owes the advantage that it can heat foods more quickly and uniformly. In medicine, the RF heating of body tissues has been used as a routine practice in diathermia and diathermy.

The most common type of microwave heating is the microwave oven. As shown in Fig. 26.4, a microwave generated by the waveguide feeder is transmitted into a heating space and exerted on a heated item. The distribution of the energy is determined by the geometry and the material of the oven, which in return determine the efficiency of the heating process. In addition, the efficiency of microwave heating also depends upon the material properties. Usually, it depends on the content of water, which has a lot of polar molecules accessible to dielectric heating. Besides, the position and shape of the food also affect the distribution of the electromagnetic

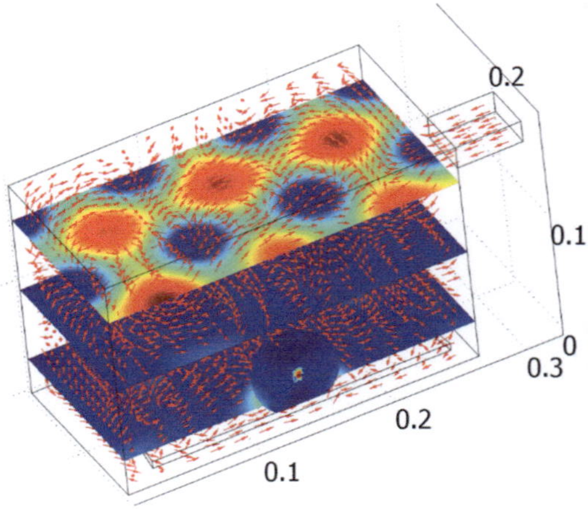

Fig. 26.4 Total energy flux (arrows) and energy intensity (contours) for a potato in a microwave (the rectangular block on the right represents a waveguide feed) [COMSOL example library]

field within the oven and consequently the heating result. That is why most microwave ovens rotate the heated items to improve the heating result. Microwave heating has also been used in the cancer treatment, in particular, hyperthermic oncology. This treatment involves localized heating of tumor tissues without damaging the healthy tissues around them. A thin microwave antenna is inserted into the tumor to heat the tumor. A coagulated region generated by the microwave heating could kill the cancerous cells.

Induction Heating

Mechanism

Strictly speaking, induction heating originates from both Joule heating and dielectric heating, which correspond to the thermal energy generated by the free electric current and bound electric current, respectively. The induction heating is triggered by a time-varying (alternating) current, usually at a high frequency. The alternating current will induce a time-varying magnetic field. The material to be heated is placed inside the magnetic field so that heat can be generated without touching the coil. In the heated item, or called workpiece here, both induced eddy electric currents and the relaxation in the electric dipoles will lead to energy dissipation, representing a conversion from the electromagnetic energy to thermal energy. Therefore, induction heating involves a multiphysics process including both electromagnetic fields and thermal field, but the key mechanisms for heat generation still boil down to Joule heating and dielectric heating.

The heating attributed to eddy currents is the major contributor. According to Rudnev et al. (2017), the dielectric heating effect does not exceed 7% of the heat generated by the eddy current in a great majority of induction heating applications. Possibly due to this reason, induction heating is explained as the heating process caused by the eddy currents in many places including this book. Following this understanding, induction heating is defined as the process of heating an electrically conducting object (usually a metal) by electromagnetic induction, through heat generated in the object by eddy currents.

In a narrow sense, induction heating is similar to the Joule heating effect, but with one important modification. The currents that heat the material are induced via electromagnetic induction. Therefore, it could be a non-contact heating process. Accordingly, only those with high electric conductivity, such as copper, gold, and aluminum, can be directly heated via induction. Food items are then heated by the metals.

Induction heating works better in specific metals due to additional heating mechanisms. For instance, induction can heat ferrous metals better than other materials. That is because the iron crystals in ferrous metals can be repeatedly magnetized and demagnetized by the alternating magnetic field. This will lead to hysteresis losses in addition to the normal induction heating.

Governing Equation: The Diffusion Equation

Based on the introduction in the Dielectric Heating section, we know that the electromagnetic field can also be described using potentials as follows:

$$\mathbf{B} = \nabla \times \mathbf{A}, \tag{26.10}$$

where **A** is the magnetic vector potential. From Maxwell's equations, it is known that

$$\mathbf{E} = -\frac{\partial \mathbf{A}}{\partial t} - \nabla \varphi. \tag{26.11}$$

Multiplying the electric field with σ, we have

$$\mathbf{J} = \sigma \mathbf{E} = -\sigma \frac{\partial \mathbf{A}}{\partial t} - \sigma \nabla \varphi = -\sigma \frac{\partial \mathbf{A}}{\partial t} + \mathbf{J}_{\mathrm{e}}^{\mathrm{s}}, \tag{26.12}$$

where $\mathbf{J}_{\mathrm{e}}^{\mathrm{s}}$ is the source current density in the induction coil. It follows from the Ampère's equation with a free charge current that

$$\nabla \times \frac{1}{\mu} \mathbf{B} = \mathbf{J}. \tag{26.13}$$

Substituting the definition of the magnetic flux density with the potential into the above equation, we obtain

$$\nabla \times \nabla \times \mathbf{A} = -\mu\sigma\frac{\partial \mathbf{A}}{\partial t} + \mu\mathbf{J}_{\mathrm{e}}^{\mathrm{s}} \tag{26.14}$$

for a homogeneous and isotropic medium independent of the field intensity. Expanding and imposing the Coulomb gauge condition $\nabla \cdot \mathbf{A} = 0$, we obtain

$$\sigma\frac{\partial \mathbf{A}}{\partial t} - \frac{1}{\mu}\nabla^2\mathbf{A} = \mathbf{J}_{\mathrm{e}}^{\mathrm{s}}, \tag{26.15}$$

which is called the diffusion equation. The equation can also be expressed in the same form but for different vectors instead of **A**, e.g., **J**, **E**, **H**, and **B**.

In the case of the magnetic vector potential, the above equation can be related for any physically observable electromagnetic induction phenomena such as eddy current, induced voltage, coil impedance, flaw impedance, and coil inductance. If the excitation current is assumed to be sinusoidal, then the eddy current will be sinusoidal as well. Consequently, a time-harmonic electromagnetic field can be introduced:

$$i\omega\sigma A - \frac{1}{\mu}\nabla^2\mathbf{A} = \mathbf{J}_{\mathrm{e}}^{\mathrm{s}}. \tag{26.16}$$

It should be noted that, in a ferromagnetic material, the harmonic approximation is not valid anymore. This is due to its time-dependent relative permeability, μ_{r}, which renders the eddy current nonsinusoidal.

Effective Volume: Skin Depth

To gain insights into the skin depth, Maxwell's equations are expressed as two second-order differential equations of **E** and **H**. For the purpose, let us assume that the material is linear, isotropic, homogeneous, and charge-free ($\rho = 0$). Then take curl on last two equations in Maxwell's equations

$$\nabla \times \nabla \times \mathbf{E} = -\mu\frac{\partial(\nabla \times \mathbf{H})}{\partial t} = -\mu\frac{\partial}{\partial t}\left(\sigma + \varepsilon\frac{\partial}{\partial t}\right)\mathbf{E}, \tag{26.17}$$

$$\nabla \times \nabla \times \mathbf{H} = \left(\sigma + \varepsilon\frac{\partial}{\partial t}\right)\nabla \times \mathbf{E} = -\mu\frac{\partial}{\partial t}\left(\sigma + \varepsilon\frac{\partial}{\partial t}\right)\mathbf{H}. \tag{26.18}$$

Applying the Jacobi identify to the two second-order homogeneous wave equations, we obtain

$$\nabla^2 \mathbf{E} - \gamma^2 \mathbf{E} = 0, \tag{26.19}$$

$$\nabla^2 \mathbf{H} - \gamma^2 \mathbf{H} = 0, \tag{26.20}$$

where $\gamma^2 = i\omega\mu(\sigma + i\omega\varepsilon)$ and γ is called the propagation constant of the medium. Frequently, the propagation constant is defined as

$$\gamma = \alpha + i\beta = i\omega\sqrt{\mu\varepsilon\left(1 - \frac{i\sigma}{\omega\varepsilon}\right)} \tag{26.21}$$

The solution to Eq. 26.19 for a uniform electric wave propagating in the positive z-direction is

$$E(z, t) = E_0 e^{-\alpha z} \cos(\omega t - \beta z). \tag{26.22}$$

Consequently, if an electric or a magnetic wave travels in a conductive medium, its amplitude starts to attenuate exponentially by a factor $e^{-\alpha z}$. The factor α is therefore called the attenuation constant, with the unit of Np/m (Neper/meter).

A good conductor satisfies $\sigma \gg \omega\varepsilon$. Under this condition, the propagation constant can be simplified to

$$\gamma = (1 + i)\sqrt{\pi f \mu\sigma}. \tag{26.23}$$

The skin depth is defined as the distance for which the amplitude of a plane wave decreases by a factor of $e^{-1} = 0.368$. Thus, the skin depth becomes

$$\delta = \frac{1}{\alpha} = \frac{1}{\sqrt{\pi f \mu\sigma}}. \tag{26.24}$$

When the electromagnetic waves reach the surface in a conductive medium, a current will start to flow within the surface and decrease exponentially toward the interior of the material.

Energy Conversion: The Poynting Vector

Poynting's theorem is a relation between the changing rate of the energy stored in the field and the energy flow. Multiplying Faraday's law and Ampère's law and ignoring the displacement current with the electric and magnetic field lead to

$$\oint_s (\mathbf{E} \times \mathbf{H}) \cdot \mathbf{n} dA = -\int_v \mathbf{H} \cdot \frac{\partial \mathbf{B}}{\partial t} dV - \int_v \sigma \mathbf{E} \cdot \mathbf{E} dV, \tag{26.25}$$

where the quantity $\mathbf{E} \times \mathbf{H}$ is known as the Poynting vector $\mathbf{P}_{\mathrm{EM}}$ in W/m^2. Poynting's theorem states that the sum of the Ohmic loss and the power absorbed by the magnetic field in the volume is equal to the total power input.

It is common to represent a harmonic variation in time as a complex function of coordinates, multiplied by the factor, $e^{i\omega t}$. The complex Poynting vector can, therefore, be written as

$$\mathbf{P}_{\mathrm{EM}} = \frac{1}{2}(\mathbf{E} \times \mathbf{H}), \tag{26.26}$$

and its integral is

$$\oint_s \mathbf{P}_{\mathrm{EM}} d\mathbf{S} = -\frac{1}{2} j\omega \int_v \mu |\mathbf{H}|^2 dV - \frac{1}{2} \int_v \sigma |\mathbf{E}|^2 dV. \tag{26.27}$$

The real part of the above equation determines the energy dissipated into heat (ohmic losses) in the volume V, and the imaginary part is equal to 2ω times of the mean value of the magnetic energy. The current from the real part is known as the eddy current and is the main heat source in induction heating.

Applications

Induction heating requires no physical contact between the heated item and the induction coil. This is especially suitable for heating applications in which a high degree of cleanliness is desired, such as in semiconductor manufacturing. The semiconductor industry has adopted induction heating for heating silicon. Another advantage is that the heat is generated inside the heated items as opposed to those from the surface in dielectric heating. As a result, a high heating efficiency can be achieved by avoiding heat losses from the surfaces.

Induction stoves are a great application of induction heating, which benefit from the above features. Figure 26.5 schematically illustrates the application of induction heating in a typical induction stove.

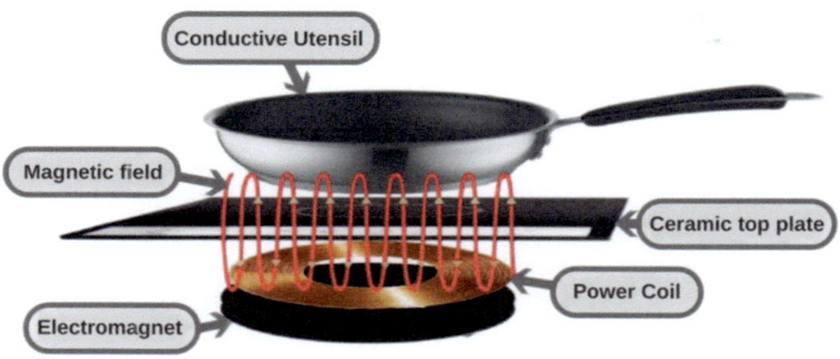

Fig. 26.5 Schematic of an induction stove (cooktop)

In such a stove, a coil is placed beneath the stove top. The coil will apply an electromagnetic field to a metal pot placed on the stove. This pot, which is usually made of highly electrically conductive materials, will be heated up quickly due to the induced electric currents in the pot. The pot then heats the food or other objects in it. Induction heating has also been applied to sealing, heat treatment, and welding.

Despite the benefits, induction heating can also be a detrimental effect. This is especially the case in transformers, in which induction heating leads to a waste of power. In the design and functioning of transformers, the eddy currents flowing inside the cores need to be eliminated to the minimum. These currents will cause heating in the transformer's magnetic core and other secondary problems such as damages to the structural integrity of relevant components.

Practice Problem

This section is developed to show the numerical implementation of dielectric heating.

Problem Description

As shown in Fig. 26.6, a long box with a cross section of 0.04 m by 0.02 m contains water. All the side surfaces and the two end surfaces (two ends of the longitudinal direction) are metals, which can be formulated as perfect electric conductors ($\nabla \times \mathbf{E} = \mathbf{0}$). One side surface is perfect magnetic conductors ($\nabla \times \mathbf{H} = \mathbf{0}$). The box contains water which has a relative permittivity of 78 + 12i. An electromagnetic wave with a frequency of 2.45 GHz was generated, whose wave number along the longitudinal direction is 0.4 of the total wave number. Please predict the temperature distribution on the path of $x = 0.02$ m at the end of 4 minutes.

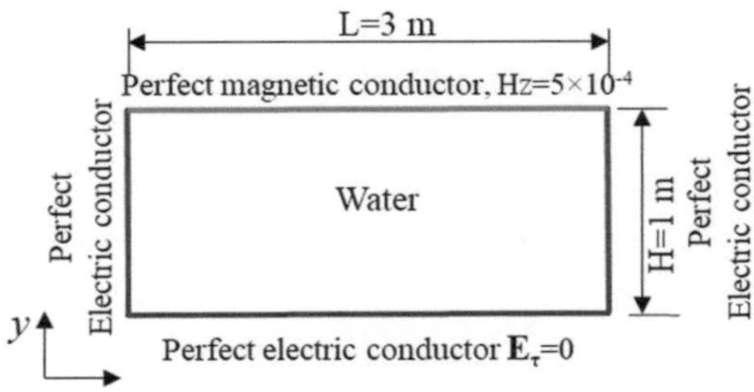

Fig. 26.6 Computation domain and boundary conditions

Governing Equation Natural Boundary Condition

The theoretical formulation of the problem involves both the thermal field and the waveguide theory introduced in Chapter 23 for electromagnetics. Governing equations for 2D need to be derived first following a procedure similar to the practice problem in Chapter 23. The following listed the governing equations written in the FlexPDE script. Please try to derive these equations first for practice.

Hz: DEL2(Hz) + k^2*Hz = 0.

Temp:DT(Temp) = 0.6/4.2e7*DEL2(Temp) + 0.5*omega*epsilon*epsilon_ri*(Ex^2 + Ey^2).

An additional constraint is needed:

INTEGRAL(Hz) = 0.

Results

The H field and E field are shown in Fig. 26.7 and Fig. 26.8, respectively. The temperature distributions are depicted in Fig. 26.9 and Fig. 26.10.

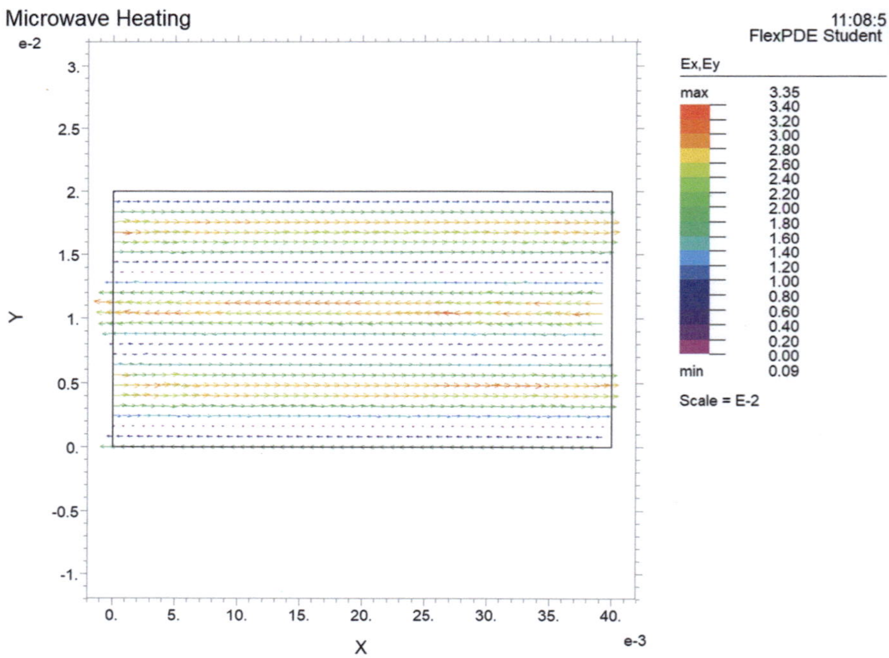

Fig. 26.7 Electric field

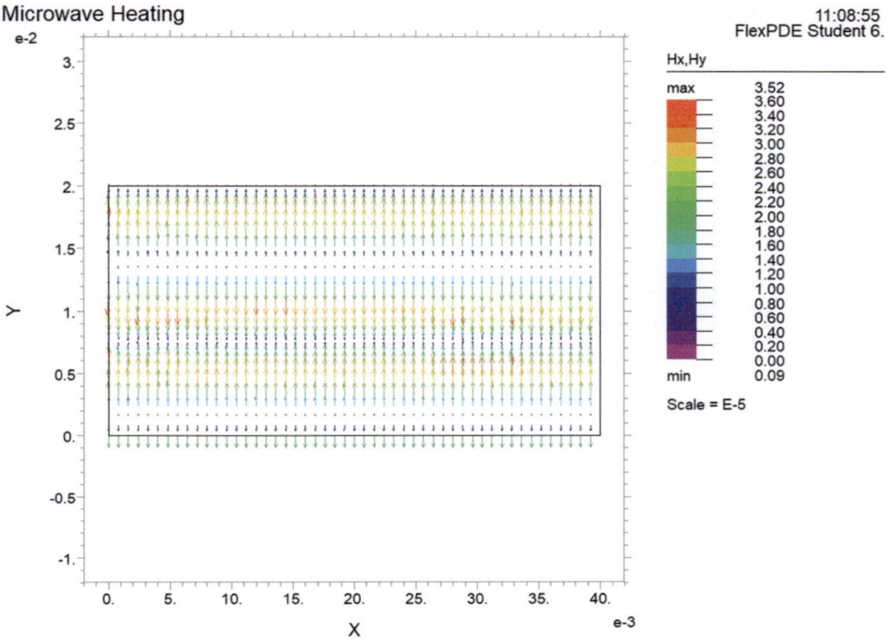

Fig. 26.8 Magnetic field

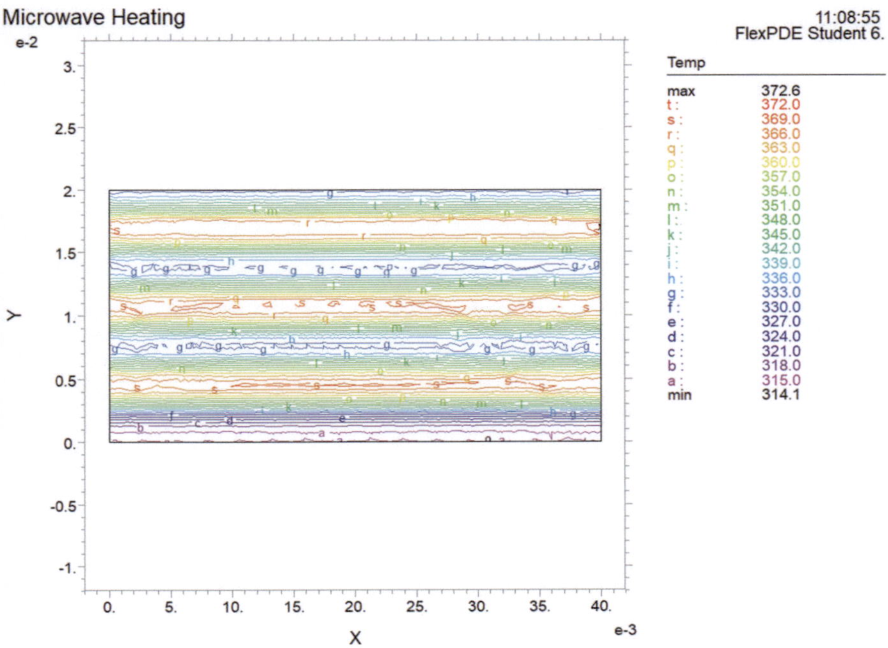

Fig. 26.9 Temperature distribution in the computational domain ($x = 0.02$ m) at $t = 240$ s

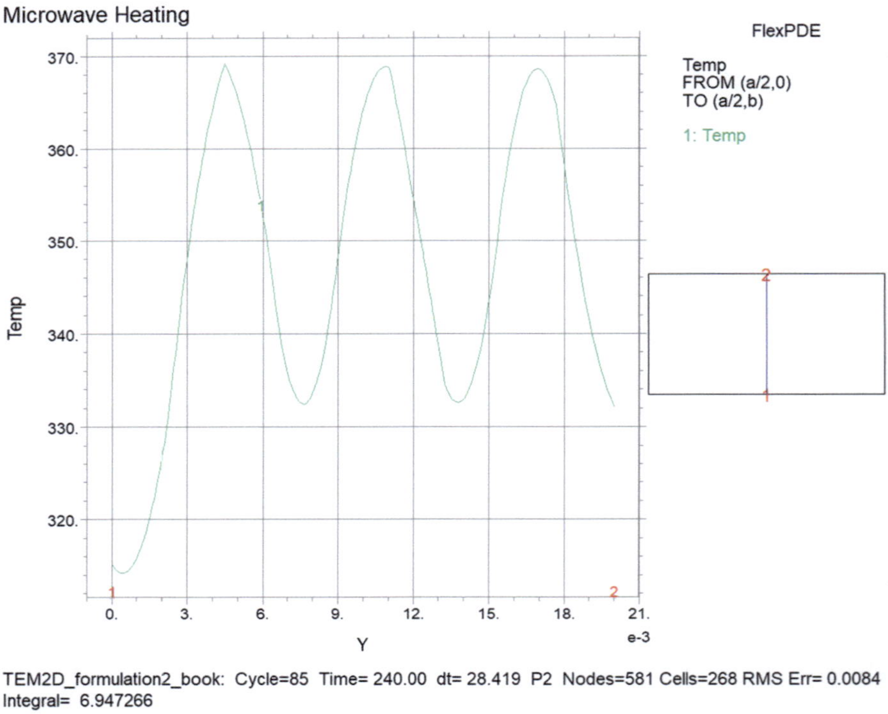

Fig. 26.10 Temperature distribution along y-axis of the computational domain ($x = 0.02$ m) at $t = 240$ s

Appendix: FlexPDE Script File

```
{ Microwave 3D with TE wave }
TITLE "Microwave Heating"
COORDINATES CARTESIAN2
SELECT
  ERRLIM 0.002
  NGRID 15 ! The student version with a course mesh may lead to
incorrect solutions
  REGRID off
VARIABLES
  Hz              ! z-component of the H field
  Temp            ! Temperature
INITIAL VALUES
  Temp=283.15
DEFINITIONS
  a = 0.04        ! Length of the wave guide
  b=0.02          ! Width of the wave guide
  mu=4*pi*1e-7
  epsilon=8.654e-12
  c=1/sqrt(mu*epsilon)
```

```
  f=2.45e10          ! Frequency of the TE microwave
  omega=f*2*PI       ! Radial frequency
  k=omega/c          ! Wave number
  kc=0.6*k           ! Wave number of the transverse wave components (x, y)
  kz=k-kc            ! Wave number of the longitudinal wave component (z)
  Hy=-kz/(kc^2)*dy(Hz)
  Hx=-kz/(kc^2)*dx(Hz)
  Ex=-omega*mu/(kc^2)*dy(Hz)
  Ey=omega*mu/(kc^2)*dx(Hz)
  epsilon_rr=78 ! Real part of the relative permittivity for water at
20 degC epsilon=epsilon_0*epsilon_rr
  epsilon_ri=12   ! Imaginary part of the relative permittivity
(heating source, 12 for water at 20 degC)
  z=1
EQUATIONS
  Hz: DEL2(Hz)+k^2*Hz = 0
  Temp: DT(Temp)=0.6/4.2e7*DEL2(Temp)
+0.5*omega*epsilon*epsilon_ri*(Ex^2+Ey^2)
CONSTRAINTS
  INTEGRAL(Hz) = 0 { since Hz has only natural boundary conditions, we
need an additional constraint to make the solution unique }
TIME 0 TO 240
BOUNDARIES { The domain definition }
  REGION 1
    START(0,0) VALUE(Hz) = 2e-5 NATURAL(Temp)=0 ! Perfect magnetic
conductor for one wall
  LINE TO (a,0) NATURAL(Hz) = 0 NATURAL(Temp)=0! Perfect electric
conductor for the other walls
  LINE TO (a,b) NATURAL(Hz) = 0
  LINE TO (0,b) NATURAL(Hz) = 0
  LINE TO CLOSE
MONITORS
  FOR Cycle=1
  CONTOUR(Hz)
PLOTS
  FOR Cycle=1
  CONTOUR(Hz)
  VECTOR(Ex,Ey)
  VECTOR(Hx,Hy)
  ELEVATION(Temp) FROM (a/2,0) TO (a/2,b)
  CONTOUR(Temp)
  CONTOUR(Temp)
END
```

Chapter 27
Electromagnetomechanics

Introduction

This chapter for electromagnetomechanics splits into magnetomechanics and electromechanics. Magnetomechanics deals with the steady-state force created by a magnetic field, which usually results from a permanent magnet or a constant current flowing through a conductor, and the mechanical responses caused by this magnetostatic (or magnetic) force. Accordingly, we will show the formulation of the force between a magnetic and a current using the Lorentz force law and that between magnets using both the Ampère model and the Gilbert model. For electromechanics, two major electromechanical mechanisms, i.e., piezoelectric effect and piezoresistive effect, will be discussed for their mathematical formulations and applications. In the final practice problem section, an example will be given for the electric potential generated by a lead zirconate titanate (piezoelectric) beam.

Magnetomechanics: Magnetostatic Force

Magnetomechanics deals with the steady-state force created by a magnetic field, which usually results from a permanent magnet or a constant current flowing through a conductor, and the mechanical responses caused by this magnetostatic (or magnetic) force. For example, a magnetic material such as iron suffers from a magnetostatic force when it is placed in a magnetic field. Figure 27.1 illustrates the deformation of a hollow nanocapsule made of sexithiophene (6 T) molecules in a strong magnetic field (20 Tesla) at different temperatures (Manyuhina et al. 2007). Some basics about magnetomechanics have been introduced in the electrostatics and magnetostatics. In this section, we will focus on the two types of magnetostatic forces.

Z. Liu, *Multiphysics in Porous Materials*,
https://doi.org/10.1007/978-3-319-93028-2_27

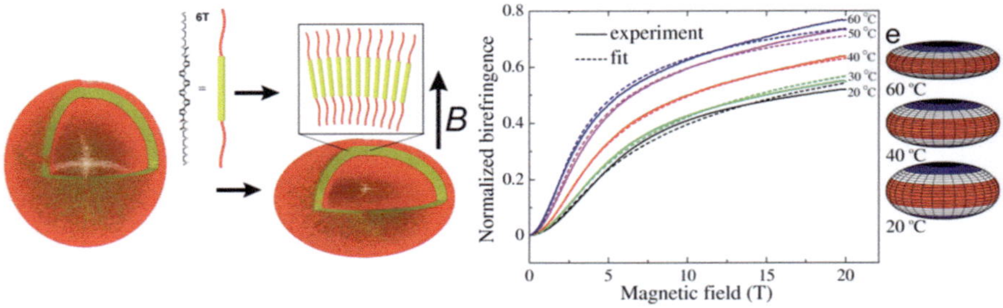

Fig. 27.1 Deformation of a hollow nanocapsule made of sexithiophene (6 T) molecules in a strong magnetic field (20 T) at different temperatures. (Manyuhina et al. 2007)

Force Between a Magnet and a Current

The force between a current and a magnet is the relatively easy one of the two types of magnetostatic forces from the perspective of numerical simulation. The force can be obtained by integrating (adding up) the Lorentz force created by all the moving charges within the current.

According to the Lorentz force law, the force that a magnetic field **B** exerts on a charge q moving at a velocity **v** is

$$\mathbf{F}_{\mathrm{m}} = q\mathbf{v} \times \mathbf{B}. \tag{27.1}$$

Like the electric field **E**, this formula was obtained with experiments, and it provides us with the definition of **B**.

In fact, the total force on a charged particle is

$$\mathbf{F} = Q(\mathbf{E} + \mathbf{v} \times \mathbf{B}). \tag{27.2}$$

It is noted that particles moving parallel to the magnetic field do not experience a magnetic force. Also, the magnetic force cannot do work, because it is always perpendicular to the motion of the particle. This force cannot change the speed of a particle but only alters its direction.

A line charge λ (unit: C/m) traveling down a thin wire produces a current **I**:

$$\mathbf{I} = \lambda\mathbf{v}. \tag{27.3}$$

For a thin wire, this direction is usually obvious; but we sometimes need to define a current as a vector for surface or volume currents.

The force on a wire due to an external magnetic field **B** is

$$\mathbf{F}_{\mathrm{m}} = \int \mathbf{v} \times \mathbf{B}dq = \int \mathbf{v} \times \mathbf{B}\lambda dl = \int \mathbf{I} \times \mathbf{B}dl. \tag{27.4}$$

Since the current and the line element of the wire are in the same direction, we can also write the above equation as

$$\mathbf{F}_{\mathrm{m}} = \int |\mathbf{I}| d\mathbf{l} \times \mathbf{B}. \tag{27.5}$$

Finally, if the current is constant along the wire, we have

$$\mathbf{F}_{\mathrm{m}} = |\mathbf{I}| \int d\mathbf{l} \times \mathbf{B}. \tag{27.6}$$

In 2D or 3D cases, we can define surface and volume current densities, respectively. Similar to Eq. 27.4, we can obtain the equation for the magnetic force as

$$\mathbf{F}_{\mathrm{m}} = \int (\mathbf{J} \times \mathbf{B}) d\Omega. \tag{27.7}$$

Force Between Magnets

Magnets exert forces and torques on each other due to the complex rules of electromagnetism, which make this type of magnetostatic force more complicated. The forces originate from the microscopic currents of electrically charged electrons orbiting nuclei and the intrinsic magnetism of fundamental particles that make up the material, such as electrons. Both of these can be well modeled as tiny loops of current, i.e., magnetic dipoles. These magnetic dipoles produce their own magnetic field and, in the meanwhile, are affected by external magnetic fields. Therefore, the most elementary force between magnets is the magnetic dipole-dipole interaction.

Two models are prevalent for calculating the forces between magnets: the Ampère and Gilbert models. The Ampère model is physically correct, while the Gilbert model is easier to use. The Ampère model is implemented directly using the mechanisms introduced above: all of the magnetic dipoles that make up the two magnets are determined, and then the net force on both magnets can be obtained by summing up all the interactions between the dipoles of the first magnet and that of the second. By contrast, the Gilbert model assumes that the forces between magnetic poles are caused by magnetic charges "smeared" on the poles. This model works well for predicting the forces between simple magnets where the distribution of the "magnetic charge" can be well modeled. However, it is worthwhile to mention that this model excludes some characteristics of magnetism, e.g., the relationship between the angular momentum and magnetic dipoles, and that the model is established based on the assumed magnetic charges.

Models for Magnetic Forces

Magnetic Dipole Moment

The magnetic field created by a magnet can be described using the field of the total magnetic dipole moment, **m**. The direction of the magnetic moment points from the south pole to the north pole of a magnet. Both the torque and force exerted on a magnet by an external magnetic field are proportional to that magnet's magnetic moment. In the chapter for electromagnetics, we have defined the magnetic dipole moment as the product of an infinitesimally small loop of current I with the area of the loop, **A**:

$$\mathbf{m} = I\mathbf{A}, \tag{27.8}$$

where the direction of **m** is normal to the area and the direction is determined by the current and the right-hand rule. This is, in fact, the definition of the magnetic dipole moment in the Ampère model.

In the Gilbert model, the magnetic dipole moment is assumed to be generated by two equal but opposite magnetic charges that are separated by a distance of d. The two separated magnetic charges form a tiny magnet. Then **m** can be defined in a way similar to that for the electric dipole moment **P** due to electrical charges:

$$\mathbf{m} = q_{\mathrm{m}}\mathbf{d}, \tag{27.9}$$

where q_{m} is the "magnetic charge." The direction of the magnetic dipole moment points from the negative south pole to the positive north pole of this tiny magnet.

Ampère Model

In the Ampère model, all magnetization results from the bound currents (or called Ampèrian currents) at the microscale. The total effect of these microscopic bound currents is to make the magnet behave as if there is a macroscopic electric current flowing in loops in the magnet, leading to a magnetic field normal to the loops. The Ampère model gives out the exact magnetic field both inside and outside the magnet. However, it is usually difficult to calculate the Ampèrian currents on the surface of a magnet. Every magnet produces a magnetic field which is strong near its poles.

In the Ampère model, the force on a magnetic dipole caused by a nonuniform magnetic field is essentially attributed to the Lorentz force on the "current loops" corresponding to the magnetic dipole. This force can be calculated as

$$\mathbf{F}_{\mathrm{m}} = \nabla(\mathbf{m} \cdot \mathbf{B}). \tag{27.10}$$

A permanent magnet is treated in the same way as an electromagnet: a magnetic contains small current loops throughout its volume, and its magnetic strength is proportional to the current and the surface of each loop and proportional to the density of current loops in the material.

It is often assumed that each small piece of a magnetic material has an equally strong magnetization, which is, however, not always true. This is because a magnet can change the magnetization of nearby magnets. While this effect is usually small between permanent magnets, it could be significant for electromagnetics. For example, if an electromagnet that consists of a wire wound around an iron core is brought close to a permanent magnet, the magnetization of that core can change drastically. Therefore, the Ampère model is suitable to compute the magnetic force field of permanent magnets, while for electromagnets, a magnetic circuit approach as introduced in the previous subsection may be a better option.

Gilbert Model

The Gilbert model assumes that the magnetic forces between magnets are due to the magnetic charges near the poles. This method is proposed via the analogy between the electric and magnetic fields. Since there is no magnetic current, the method is not physically based. However, this model provides better predictions in magnetic fields that are more dependent on the detailed shape and magnetization of the magnet rather than merely the magnetic dipole contribution.

It is necessary to mention that the Gilbert model is implemented based on the distributions of the "magnetic charges" over the magnetic poles. This fact renders this method only useful for simple configurations. Fortunately, many useful cases have simple configurations. For example, two poles that are small enough can be represented as single points, and then the force between these two infinitely small magnetic poles is given by

$$F_{\mathrm{m}} = \frac{\mu q_{\mathrm{m1}} q_{\mathrm{m2}}}{4\pi r^2}, \tag{27.11}$$

where q_{m1} and q_{m2} are the magnitudes of the magnetic charges on the poles, μ is the permeability of the intervening medium, and r is the distance between the poles.

Secondary Effect of Magnetostatic Force

The magnetostatic force can result in secondary effects on both the mechanical field and the magnetic field. As for the influence on the mechanical field, we have magnetostriction, which is the deformation of ferromagnetic materials as the result of an applied magnetic field. Ferromagnetic materials change their shape or dimensions during the process of magnetization due to their intrinsic properties. The magnetostrictive strain in such materials will change in response to the external magnetic field before the strain reaches the saturation value.

The intrinsic property responsible for magnetostriction is the unique structure of ferromagnetic materials: such a material consists of domains with uniform magnetic polarization in each domain. Upon applying a magnetic field, the boundaries between the domains shift and the domains rotate, producing a change in the dimensions of the material. It is common that crystals are usually nonisotropic due to the distinct structures along different directions. Therefore, it usually takes more energy to magnetize a crystalline material in one direction than another (magnetocrystalline). The material will tend to rearrange its structure to minimize the free energy of the system. This will induce a strain because different crystal directions have different lengths.

In the opposite way, the magnetostatic force can also change the property of the magnetic field. Especially, such a force can lead to magnetostriction and eventually changes the susceptibility of a material to an external magnetic field. This is the so-called Villari effect. Two other effects also exist in relation to magnetostriction. The Matteucci effect is the generation of a helical anisotropy of the susceptibility of a magnetostrictive material under the influence of a torque, whereas the Wiedemann effect is the twisting of the material when a helical magnetic field is applied.

Electromechanics

The piezoelectric effect and piezoresistive effect will be discussed in this section as the two major electromechanical mechanisms.

Piezoelectric

The piezoelectric effect states the relationship between the electric polarization and mechanical stresses/strains. In a broad sense, this implies a two-way process, via which the electric field and mechanical field can interact with each other. In a narrow sense, it only refers to the influence of the mechanical stresses/strains on the electric polarization. The piezoelectric effect relies on the structure and constituents of the material and can be characterized as a material property. This phenomenon is observable in many naturally occurring crystalline materials such as quartz, Rochelle salt, and human bone. A much more pronounced piezoelectric effect can be obtained via engineered materials, such as lithium niobate and lead (Pb) zirconate titanate (PZT). PZT sensors have been widely used in testing. Figure 27.2 shows two common bender element sensors made of PZT for soil testing.

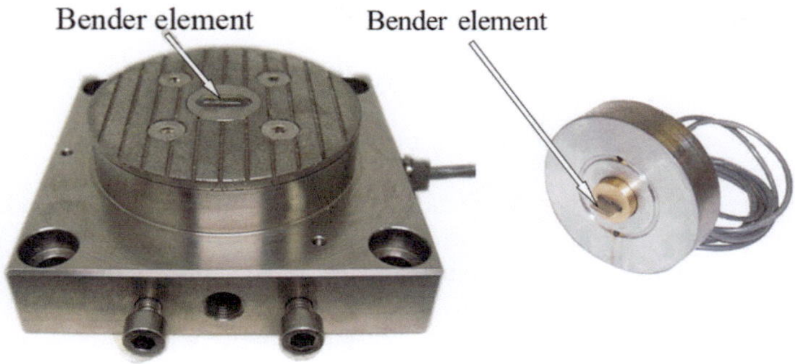

Fig. 27.2 PZT bender elements used in soil testing

Mechanism

The piezoelectric effect stems from the occurrence of electric dipole moments in solids. Dipoles tend to be aligned in regions called Weiss domains, while the alignments of different Weiss domains are usually random and can be much different. However, these random alignments can be unified by applying a strong external electric field at elevated temperatures. This process, called poling, is analogous to the magnetic poling. Not all the piezoelectric materials can be poled. As introduced before, the electric field in the material can be obtained via the density of electric dipole moments. In poled piezoelectric materials, the uniformly oriented dipoles will generate an electric field.

Under the influence of mechanical stresses/strains, the polarization **P** will change. This change could be attributable to either a reconfiguration of the dipole-inducing surrounding or a reorientation of molecular dipole moments. This change in polarization, either strength, direction, or both, may then manifest as piezoelectricity. The outcome depends on the orientation of polarization, the crystal's symmetry, and the applied mechanical stress. The polarization change appears as a variation of the surface charge density on the crystal surfaces. For example, a 1 cm^3 cube of quartz with an applied force of 2 kN can produce a voltage of 12,500 V. When an external electric field is applied, the process can also happen in the opposite way to generate stresses/strains, which is called the converse piezoelectric effect.

Mathematical Formulation

Piezoelectricity as a reversible process requires a fully coupled constitutive relationship that combines the constitutive relationships in both the electric field and mechanical field. For the electric field, we have the following relationship between the electric displacement field **D** and the electric field **E**:

$$\mathbf{D} = \boldsymbol{\varepsilon}\mathbf{E}. \tag{27.12}$$

For the mechanical field, we can recall Hooke's law for linearly elastic materials:

$$\mathbf{S} = \mathbf{s}\mathbf{T}, \tag{27.13}$$

where **S** is the strain, **s** is the compliance, and **T** is the stress. Here symbols for the mechanical field are different from the common ones, i.e., those in the other places of this book, due to the conflicts in the choice of symbols in the studies of the mechanical field and electric field.

To couple the two fields, one more material property, which is formulated as a matrix, needs to be added to describe the couplings between the two fields. This type of strong coupling, which is formulated using the following equations, is similar to that in poroelasticity:

$$\mathbf{S} = \mathbf{s}\mathbf{T} + \mathbf{d}^{\mathrm{T}}\mathbf{E} \tag{27.14}$$

$$\mathbf{D} = \mathbf{d}\mathbf{T} + \boldsymbol{\varepsilon}\mathbf{E}. \tag{27.15}$$

By employing the Voigt notation, we can obtain the following formulations for typical piezoelectric materials such as tetragonal PZT and $BaTiO_3$:

$$\begin{bmatrix} S_1 \\ S_2 \\ S_3 \\ S_4 \\ S_5 \\ S_6 \end{bmatrix} = \begin{bmatrix} s_{11} & s_{12} & s_{13} & 0 & 0 & 0 \\ s_{21} & s_{22} & s_{23} & 0 & 0 & 0 \\ s_{31} & s_{32} & s_{33} & 0 & 0 & 0 \\ 0 & 0 & 0 & s_{44} & 0 & 0 \\ 0 & 0 & 0 & 0 & s_{55} & 0 \\ 0 & 0 & 0 & 0 & 0 & s_{66} \end{bmatrix} \begin{bmatrix} T_1 \\ T_2 \\ T_3 \\ T_4 \\ T_5 \\ T_6 \end{bmatrix} + \begin{bmatrix} 0 & 0 & d_{31} \\ 0 & 0 & d_{32} \\ 0 & 0 & d_{33} \\ 0 & d_{24} & 0 \\ d_{15} & 0 & 0 \\ 0 & 0 & 0 \end{bmatrix} \begin{bmatrix} E_1 \\ E_2 \\ E_3 \end{bmatrix} \tag{27.16}$$

$$\begin{bmatrix} D_1 \\ D_2 \\ D_3 \end{bmatrix} = \begin{bmatrix} 0 & 0 & 0 & 0 & d_{15} & 0 \\ 0 & 0 & 0 & d_{24} & 0 & 0 \\ d_{31} & d_{32} & d_{33} & 0 & 0 & 0 \end{bmatrix} \begin{bmatrix} T_1 \\ T_2 \\ T_3 \\ T_4 \\ T_5 \\ T_6 \end{bmatrix} + \begin{bmatrix} \varepsilon_{11} & 0 & 0 \\ 0 & \varepsilon_{22} & 0 \\ 0 & 0 & \varepsilon_{33} \end{bmatrix} \begin{bmatrix} E_1 \\ E_2 \\ E_3 \end{bmatrix} \tag{27.17}$$

where the first equation represents the relationship for the converse piezoelectric effect (coupling from the electric field to the mechanical field) and the latter for the direct piezoelectric effect (coupling from the mechanical field to the electric field).

The piezoelectric coefficients, i.e., d_{ij}, can be determined as follows:

$$d_{ij} = \left(\frac{\partial D_i}{\partial T_j}\right)\bigg|_{E=0} = \left(\frac{\partial S_j}{\partial E_i}\right)\bigg|_{T=0}. \tag{27.18}$$

The mathematical formulations for other piezoelectric materials will be different, depending on the specific coupling mechanisms determined by the crystal structures.

Areas of Application

Piezoelectric materials are major players in the industries for sensors and actuators. The materials can be used for developing actuators, such as linear motors, rotary motors, and pumps, and sensors, such as load cells, pressure sensors, accelerometers, and gyroscopes. In soils, the commonly used geophones and bender elements (Fig. 27.2) are developed with piezoelectric materials. The materials have also been widely used in acoustics for the transducers for both generating and picking up wave signals.

However, piezoelectric materials have much wider use than sensors and actuators. For example, piezoelectric materials are also commonly utilized in microelectromechanical systems (MEMS). Piezoelectricity-based MEMS can be made for microscale chemical and biological sensors and high-precision mass measurement devices based on resonance. Piezoelectric materials are also used for energy harvesting by taking advantage of the direct piezoelectric effect. Kinetic energy can be converted to electricity via the process.

Piezoresistive

The piezoresistive effect is the change in the electrical resistivity of a semiconductor or a metal due to mechanical strains. In contrast to the piezoelectric effect, the piezoresistive effect causes a change in the electrical resistance (a material property) rather than the electric potential (a state variable).

Mathematical Formulation

In electrically conductive and semiconductive materials, strains can change the interatomic spacing and consequently the easiness for electrons to be raised into the conduction band. This leads to a change in the resistivity of the material. A linear relationship usually can be assumed between the resistivity and the strain within a certain range of strains. This proportionality is the piezoresistive coefficient and can be mathematically defined as

$$\rho_\sigma = \frac{\left(\frac{\partial\rho}{\rho}\right)}{\varepsilon}, \tag{27.19}$$

where ρ_σ is the piezoresistive coefficient, $\partial\rho$ is the change in the resistivity, $\rho$is the original resistivity, and ε is the strain.

Piezoresistive Coefficient

The change of the resistance in metals is usually due to the change of the geometry resulting from the applied mechanical stress. Accordingly, the resistivity can be easily formulated using Ohm's law. The piezoresistive effect in these cases is generally small but not negligible. Some metals owe piezoresistivity values that are much larger than what are predicted based only on the geometry. For example, the piezoresistivity of nickel is -13 times of the prediction using the geometry.

In semiconductors, the piezoresistive coefficient can be several orders of magnitudes higher than what is predicted using the geometrical effect. Due to the reason, it is common to use semiconductors to build strain gauges for the purpose of a very high coefficient of sensitivity. However, it is more difficult to handle semiconductors than metals for developing strain gauges. The reason is that semiconductors are generally more sensitive to environmental conditions. Common semiconductors with obvious piezoresistive effects include germanium, polycrystalline silicon, amorphous silicon, silicon carbide, and single-crystal silicon.

Application

The piezoresistive effect in some semiconductive materials, such as germanium, polycrystalline silicon, amorphous silicon, and single-crystal silicon, has been utilized for fabricating a wide range of pressure and acceleration sensors. Piezoresistors are the simplest form of piezoresistive devices. Because resistors made from piezoresistive materials, piezoresistors are usually used for mechanical stress measurements. Piezoresistors can be fabricated using a wide variety of piezoresistive materials. In addition, the piezoresistive effect has also been used in the fabrication of transducers, piezo-FETS, solid-state accelerometers, and bipolar transistors.

Practice Problem

The practice problem in this section requires simulating the electric potential generated by a simply supported plate made of a PZT material, which was described in Fernandes and Pouget (2003).

Problem Description

The plate is made of transversely isotropic PZT material. The transverse plane of the material (direction 1 and 2 or x and y) coincides with the plane of the plate. The constitutive relationships of the material are as follows:

$$\begin{bmatrix} S_1 \\ S_2 \\ S_3 \\ S_4 \\ S_5 \\ S_6 \end{bmatrix} = \begin{bmatrix} 139 & 77.8 & 74.3 & 0 & 0 & 0 \\ 77.8 & 139 & 74.3 & 0 & 0 & 0 \\ 74.3 & 74.3 & 115 & 0 & 0 & 0 \\ 0 & 0 & 0 & \frac{139-77.8}{2} & 0 & 0 \\ 0 & 0 & 0 & 0 & 25.6 & 0 \\ 0 & 0 & 0 & 0 & 0 & 25.6 \end{bmatrix} \text{GPa} \otimes \begin{bmatrix} T_1 \\ T_2 \\ T_3 \\ T_4 \\ T_5 \\ T_6 \end{bmatrix} + \begin{bmatrix} 0 & 0 & -5.2 \\ 0 & 0 & -5.2 \\ 0 & 0 & 15.1 \\ 0 & 12.7 & 0 \\ 12.7 & 0 & 0 \\ 0 & 0 & 0 \end{bmatrix} \frac{\text{C}}{\text{m}^2} \otimes \begin{bmatrix} E_1 \\ E_2 \\ E_3 \end{bmatrix} \tag{27.20}$$

$$\begin{bmatrix} D_1 \\ D_2 \\ D_3 \end{bmatrix} = \begin{bmatrix} 0 & 0 & 0 & 0 & 12.7 & 0 \\ 0 & 0 & 0 & 12.7 & 0 & 0 \\ -5.2 & -5.2 & 15.1 & 0 & 0 & 0 \end{bmatrix} \frac{\text{C}}{\text{m}^2} \otimes \begin{bmatrix} T_1 \\ T_2 \\ T_3 \\ T_4 \\ T_5 \\ T_6 \end{bmatrix} + \begin{bmatrix} 13.05 & 0 & 0 \\ 0 & 13.05 & 0 \\ 0 & 0 & 11.5 \end{bmatrix} \frac{\text{nF}}{\text{m}} \otimes \begin{bmatrix} E_1 \\ E_2 \\ E_3 \end{bmatrix} \tag{27.21}$$

As shown in Fig. 27.3, we here simplify the problem into a 2D problem, in which the plate is simplified into a simply supported beam with a length of 0.1 m and a thickness of 0.005 m. Please predict the distribution of the electric potential with

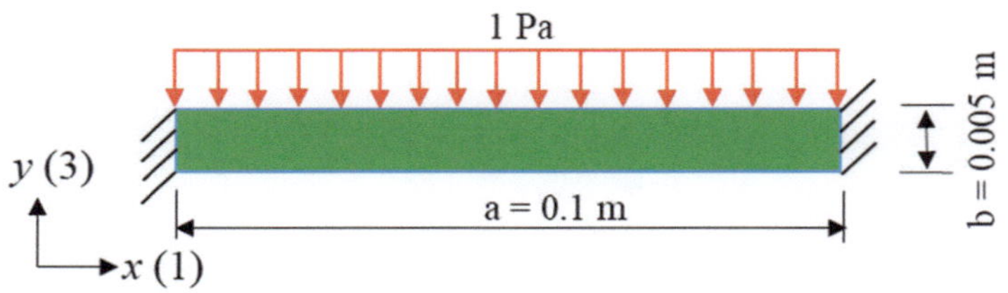

Fig. 27.3 Geometry and loading conditions

height (thickness) along the vertical axis of the beam (center) generated by a downward vertical distributed load of 1 Pa on the top of the beam.

Results

A 2D geometry was selected because a 3D description would require six dependent variables, which are not available in the student version of FlexPDE. In addition, the mechanical boundary condition of the simply supported beam cannot be easily implemented with given tools in FlexPDE. Due to this reason, we fixed the two ends of the beam for simplicity.

The vertical displacement of the beam is illustrated in Fig. 27.4, which can be validated using simple equations from the mechanics of materials. This is not done because it is not the focus. The distributions of the electric potential are shown in the following figures. The shapes of the curves match the normalized results obtained in Fernandes and Pouget (2003) (Figs. 27.5 and 27.6).

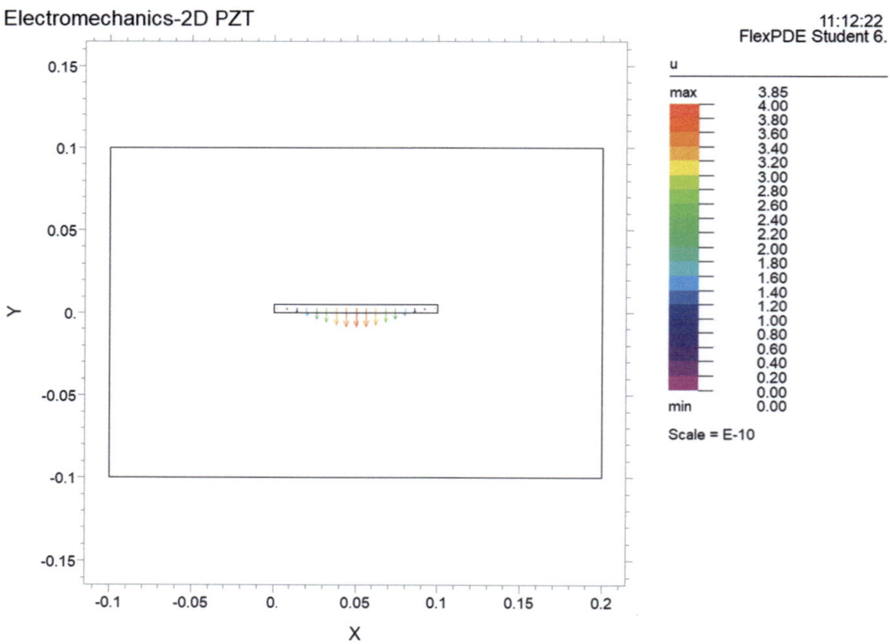

Fig. 27.4 Vertical displacement of the PZT plate

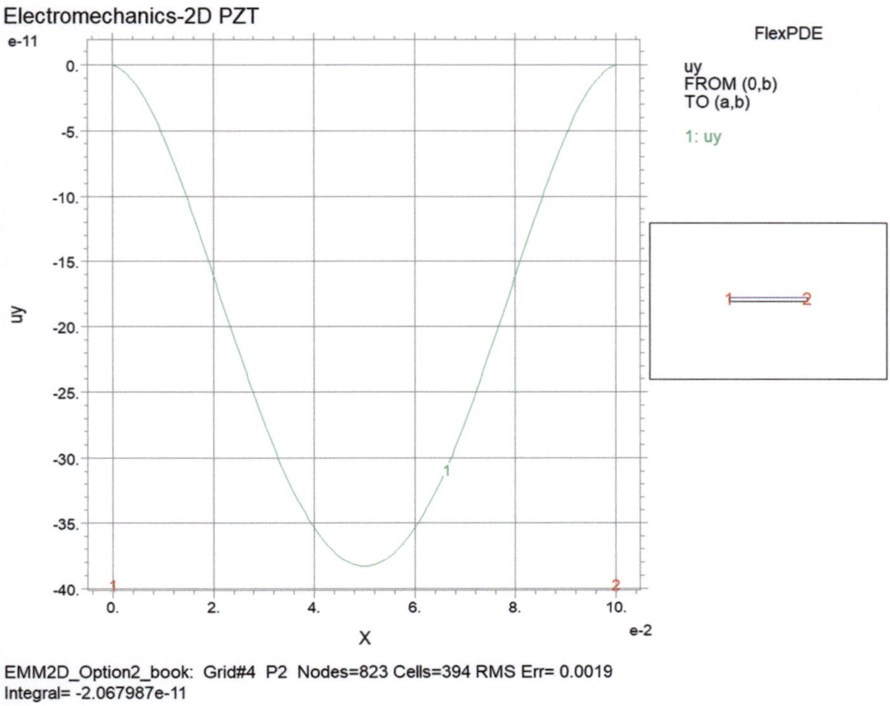

Fig. 27.5 Electric potential along the horizontal axis

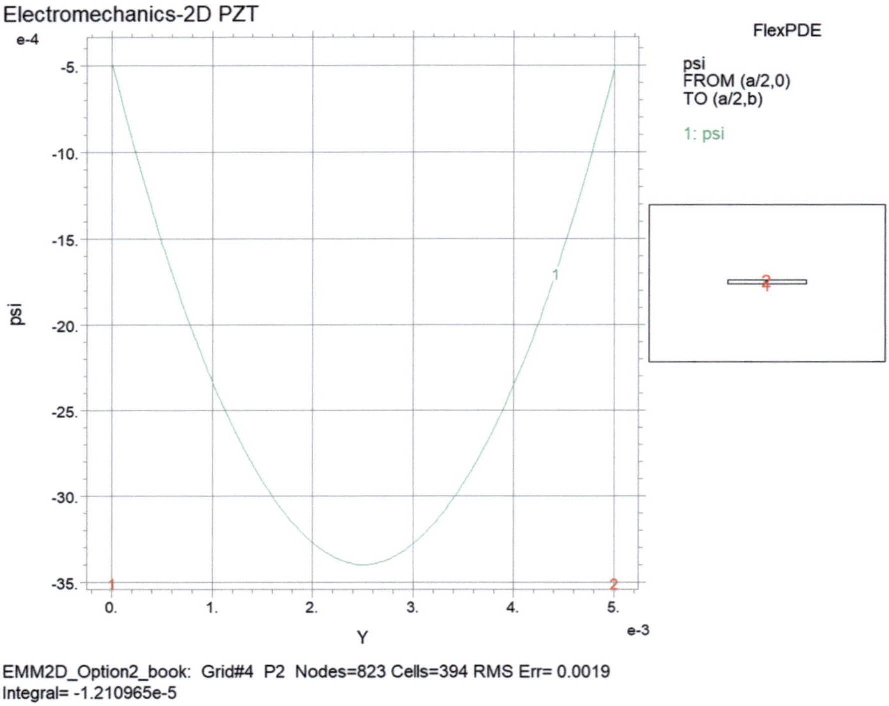

Fig. 27.6 Electric potential along the vertical axis

Appendix: FlexPDE Script File

```
{ 2D PZT Beam, Fernandes and Pouget, 2003, International Journal of
Solids and Structures, 40 }
TITLE 'Electromechanics-2D PZT'
VARIABLES
  u=VECTOR(ux,uy)
  psi
DEFINITIONS
  ! Dimensions of the beam and computational domains
  a=0.1     ! Length of the beam
  b=a/20   ! Thickness of the beam
  ! PZT constitutive relationship: sigma=C*epsilon+ee*e;
D=ee*epsilon+ep*E
  c11=139e9
  c12=77.8e9
  c33=115e9
  c13=74.3e9
  c44=25.6e9
  e31=-5.2
  e33=15.1
```

```
  e15=12.7
  ep11=13.05e-9
  ep33=11.5e-9
  E=-GRAD(psi)    { Electric field }
  C=TENSOR((C11,C13,0),(C13,C33,0),(0,0,C44)) ! Stiffness matrix
  ee=TENSOR((0,e31),(0,e33),(e15,0))
  ep=TENSOR((ep11,0),(0,ep33))
  epsilon=VECTOR( dx(ux),dy(uy),0.5*(dx(uy)+dy(ux)) )
  sigma=dot( C,epsilon) + dot(ee,E)
  D=dot(epsilon,ee) + dot(E,ep)
  { The above C and epsilon are defined using the Voigt notation;
sigma_tensor uses the    tensor notation }

  sigma_tensor=TENSOR((XCOMP(sigma),ZCOMP(sigma)),(ZCOMP
(sigma),YCOMP(sigma)))
EQUATIONS
  u: DIV(sigma_tensor)=VECTOR(0,0)
  psi: DIV(D)=0
BOUNDARIES
  REGION 1    { Air around the beam }
    INACTIVE(u)
    START(-a,-a)
      VALUE(psi)=0
    LINE TO (-a,a)
      VALUE(psi)=0
    LINE TO (2*a,a)
      VALUE(psi)=0
    LINE TO (2*a,-a)
      VALUE(psi)=0
    LINE TO CLOSE
  REGION 2    { PZT beam }
    START(0,0)
      VALUE(ux)=0
      VALUE(uy)=0
    LINE TO (0,b)
      NATURAL(uy)=-1
      NATURAL(ux)=0
    LINE TO (a,b)
      VALUE(ux)=0
      VALUE(uy)=0
    LINE TO (a,0)
      NATURAL(ux)=0
      NATURAL(uy)=0
    LINE TO CLOSE
MONITORS
PLOTS
  VECTOR(u)
  CONTOUR(XCOMP(sigma))
  ELEVATION(uy) FROM (0,0) TO (0,b)
  ELEVATION(uy) FROM (0,b) TO (a,b)
  ELEVATION(psi) FROM (a/2,0) TO (a/2,b)
END
```

Part V
Implementation Methods

Chapter 28
Finite Difference Method

Introduction: Numerical Analysis

Numerical analysis is about the use of numerical approximations as opposed to symbolic manipulations for the solution of mathematical problems. This approach originated from the hand calculations with large printed tables. After the birth of computers, automated calculations with computer programs have been promoting this type of analysis into a major approach in scientific studies and engineering applications. The overall goal of numerical analysis is to develop and apply algorithms to provide approximate but acceptable solutions to problems which are difficult to solve with analytical techniques.

Numerical analyses can be categorized into direct and iterative methods based on the control of calculation steps. Direct methods compute the solution to a problem within a finite number of steps, while iterative methods do not necessarily terminate in a finite number of steps. However, for practical purpose, analyses with iterative methods are usually terminated when the residual is smaller than a predefined allowable error.

The field of numerical analysis can be divided into multiple areas depending on the purpose of the analysis. The evaluation of the function value at a given point is one of the simplest applications of numerical analysis. Interpolation and extrapolation aim to find the value of an unknown function at a point with given values of that function in the ranges including and excluding that point, respectively. Regression is similar to interpolation; however, it requires finding out the mathematical formulation of a function instead of function values at given points. Another significant application is to compute the solution to a given equation. Optimization is implemented to search for the point at which a given function is maximized or minimized. Evaluating integrals numerically, which is called numerical integration or numerical quadrature, is the numerical analysis for calculating the value of a definite integral.

Z. Liu, *Multiphysics in Porous Materials*,
https://doi.org/10.1007/978-3-319-93028-2_28

Solving differential equation, especially partial differential equations, is another significant area of numerical analysis, as we know that many physical processes, including most of the monolithic and multiphysical processes introduced in the previous chapters of this book, can be described using partial differential equations. Therefore, solving these equations with initial and boundary conditions and material properties from real examples is like simulating or numerically reproducing the real physical processes. As a result, the area of numerical analysis is also termed numerical simulation. This is especially true when the solution process is enhanced with the computer-aided model establishment and graphical user interface-based postprocessing. These tools make numerical simulation appear as a virtual experiment carried out using computers. Notwithstanding the fancy and vague appearance, we need to bear in mind that those numerical experiments are implemented based on physical and mathematical laws underlying the partial differential equations. In fact, numerical simulation also contains pieces from other numerical analysis areas. For example, numerical integrals are usually needed for the finite element method.

This chapter will introduce one of the most straightforward numerical simulation methods: the finite difference method. We will show how to approximate derivatives using finite differences and discretize the equation and computational domain based on that. The discretization will be discussed for spatial and temporal derivatives sequentially. Then the treatment of the boundary and initial conditions will be explained with an example. A brief introduction to the error analysis in the finite difference method will also be provided. Most of the above introductions will be made in 2D. Therefore, the extension from 2D to 3D will be discussed to bridge the knowledge gap. In the final, MATLAB code for implementing the example in the boundary and initial condition section will be attached.

Following this chapter will be the introduction to the finite volume method, which is similar to the finite difference method in many aspects. The finite element method as a slightly more complicated and more popular type of numerical simulation method will be introduced finally.

Spatial Finite Difference and Derivative

One essential idea behind numerical simulation is discretization. For the purpose, we need to transform a continuous mathematical equation(s) into an algebraic equation. Accordingly, the computational domain will be discretized into a mesh or a grid which consists of multiple subdomains called cells or elements. The size of the matrices in the algebraic equation is determined by the mesh size, the dimensions of the computational domain, and the number of unknowns. Therefore, the discretization of the equations and that of the domains is correlated. Different ways or schemes can be used to approximate the same derivative, leading to different levels of accuracy. In addition, discretization of spatial derivatives and temporal derivatives is also different due to the difference between the spatial and temporal dimensions. For example, for a spatial dimension, we always know the conditions on the “start” end and need to march toward the “final” end, while for spatial dimensions, boundary conditions could be given on any ends (boundary segments) or even both. This leads to the different terms and ways used for these two types of discretization. The following two sections will be devoted to the finite differences for spatial and temporal dimensions sequentially.

Spatial Discretization

A finite difference is a sequence of the differences between function values as the independent variable changes with finite increments (Fig. 28.1). It usually has the mathematical expression of the form $f(x + h) - f(x)$. A really useful operation is to divide the finite difference by an increment in the independent variable, i.e., h, so that we can obtain the following quotient

$$D(u) = \frac{f(x+h) - f(x)}{h}, \tag{28.1}$$

where $D(u)$ is the finite difference approximation to the derivative.

The idea of the finite difference method is to use the above quotient, $D(u)$, to approach the gradient (derivative) at point x, point $x + h$, or any point between them. The geometric analogy of this approximation is to use a secant line to approximate a tangent line. When the interval gets infinitely small, the error of the above approximation will become infinitely small as well. Therefore, the derivative of a function f at a point x can be defined using the following limit:

$$f'(x) = \lim_{h \to 0} \frac{f(x+h) - f(x)}{h}. \tag{28.2}$$

If h has a non-zero value instead of approaching zero, then the above approximation can be written in a general way as

$$f'(x) = f'(x) + E(h^n), \tag{28.3}$$

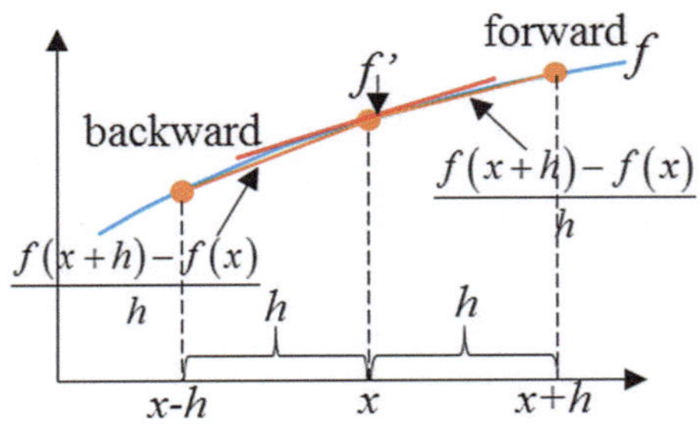

Fig. 28.1 Finite differences as approximations to derivatives

where E is the truncation error, which is a function of the nth power of the increment h. Usually, there are many ways to construct $D(u)$, and these different ways are called schemes in the context of the finite difference method. Equation 28.3 indicates that the scheme has an nth order accuracy. Schemes of different orders of accuracy can also be constructed for higher-order derivatives, which can be formulated in a way similar to Eq. 28.3.

For the first-order derivative, the most common schemes are forward, backward, and central difference schemes. A forward scheme has an expression of the form:

$$f'(x) = \frac{f(x+h) - f(x)}{h}. \tag{28.4}$$

It is called "forward" because we use the function value at a point in front of (along the direction of the corresponding axis) the point at which the derivative is calculated. Likewise, a backward scheme uses the function values at x and $x - h$ instead:

$$f'(x) = \frac{f(x) - f(x-h)}{h}. \tag{28.5}$$

We can also approximate the derivative using the function values at points on the two sides of the point of derivative as:

$$f'(x) = \frac{f(x+h/2) - f(x-h/2)}{h}. \tag{28.6}$$

A second-order derivative can be approximated using first-order derivatives as we can view the second-order derivative as a derivative of first-order derivatives. For example, in order to obtain the second-order derivative at the point x, we can use the derivatives at the points $x - h/2$ and $x + h/2$:

$$f''(x) = \frac{f'(x+h/2) - f'(x-h/2)}{h}. \tag{28.7}$$

The above approximation uses the central difference. We can also use central differences for the first derivative in the above equation, leading to:

$$f''(x) = \frac{\frac{f(x+h) - f(x)}{h} - \frac{f(x) - f(x-h)}{h}}{h} = \frac{f(x-h) - 2f(x) + f(x+h)}{h}. \tag{28.8}$$

For a derivative of any given order, we can construct different difference schemes by employing the function values at points around the point of interest using different weights. The general form of finite difference schemes can be formulated as

$$D(u) = \sum_{i=-m}^{n} a_i(x + ih), \tag{28.9}$$

where $m$and n are positive integers. These two integers are the numbers of points before and after the point where the derivative is calculated. The coefficients (weights) a_i and the corresponding values at the chosen points differentiate different schemes. The sum of the coefficients needs to be zero. The finite difference schemes introduced above are the most common ones and can be used to handle many common applications. However, in a few cases, we still want to employ more complicated schemes. This is especially necessary for problems with special features such as high nonlinearity or when there is a specific requirement for the solution such as accuracy. To construct more complicated schemes, we can resort to methods such as the method of undetermined coefficients and polynomial approximation to calculate the coefficients for any designed schemes. However, details for these methods will be excluded here considering that their use is less common in entry-level finite difference applications.

Temporal Finite Difference and Schemes

The finite difference in the time coordinate usually requires a different treatment from that in the spatial coordinates due to the different features of time and space. In the spatial finite difference context, forward and backward methods are usually adopted; by contrast, in the temporal context, we talk more about explicit and implicit methods. To differentiate the finite differences in space and time, subscripts will be used for spatial finite differences, while superscripts will be reserved for the temporal ones. An example is given in this subsection to introduce temporal finite differences where the spatial finite differences are also included for a complete introduction. Let us first reformulate the general transient partial differential equation in the following way

$$u_t = f(u), \tag{28.10}$$

where $f(u)$ in the equation contains all the terms except the temporal derivative term. First, for the temporal difference, let us use the following most common form as our purpose is to calculate the function value at the next time step $n + 1$ based on the known function value at the current time step n

$$\frac{u^{n+1} - u^n}{k} = f(u), \tag{28.11}$$

where k is the time step. An Euler forward scheme is used for the temporal derivative. The word "Euler" differentiates the temporal schemes from the spatial schemes.

The choice of u, i.e., u from which time step, determines the type of the temporal scheme. If all the unknown values are from the current step, then the scheme is explicit:

$$\frac{u^{n+1} - u^n}{k} = f(u^n). \tag{28.12}$$

The following solution can be readily obtained for this scheme:

$$u^{n+1} = u^n + k \cdot f(u^n). \tag{28.13}$$

We can see that an explicit form of the solution can be obtained conveniently. This provides convenience for the development of computer programs. Despite the advantage, the explicit scheme needs to satisfy conditions for the adopted time steps, i.e., k is small enough, to ensure stability. Otherwise, the numerical oscillation in the values of the dependent variables between steps will increase and eventually fail the solution process.

If $f(u)$ depends only on the dependent variables at the next step, then we will have the fully implicit scheme:

$$\frac{u^{n+1} - u^n}{k} = f\left(u^{n+1}\right). \tag{28.14}$$

An explicit mathematical function for the solution to the above equation is not always available. Even if one is available, more effort is usually required to derive the equation and to develop programs based on it. Despite the disadvantage, this fully implicit scheme is unconditionally stable – there is no requirement on the time step k for stability.

The above two types of schemes can also be mixed to construct other schemes

$$\frac{u^{n+1} - u^n}{k} = (1-\theta)f(u^n) + \theta f\left(u^{n+1}\right), \tag{28.15}$$

where θ is the ratio. θ values of 0 and 1 lead to explicit and fully implicit schemes, respectively, while those values between 0 and 1 produce partially implicit ones. Another common implicit scheme, the Newton-Nicolson scheme will be obtained when θ equals 0.5. Partially implicit schemes are also unconditionally stable.

To gain a better idea of these schemes, let us revisit a simple parabolic PDE which is common for time-dependent processes such as a transient heat transfer process:

$$\frac{\partial u}{\partial t} = \frac{\partial^2 u}{\partial x^2}. \tag{28.16}$$

There is only one spatial dimension and one temporal dimension. We discretize the space and time using a mesh, $x_1, \ldots, x_i, \ldots, x_I$, and a "time mesh," $t^0, \ldots, t^n, \ldots, t^N$. A uniform mesh size h and a constant time step k are adopted. Then the function value at any point is:

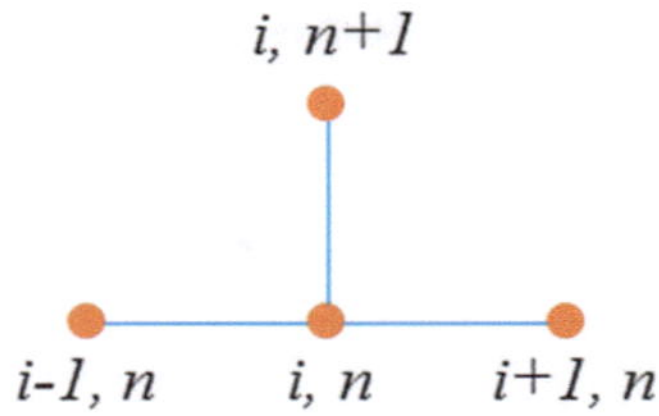

Fig. 28.2 Explicit method stencil

$$u(x_i, t^n) = u_i^n. \tag{28.17}$$

We can discretize the governing equation with the Euler forward scheme for time and the central difference scheme for space. The operation will lead to the following discretized form of the governing equation:

$$\frac{u_i^{n+1} - u_i^n}{k} = \frac{u_{i-1}^n - 2u_i^n + u_{i+1}^n}{h^2}. \tag{28.18}$$

The solution to the above equation is

$$u_i^{n+1} = u_i^n + r\left(u_{i-1}^n + u_i^n + u_{i+1}^n\right), \tag{28.19}$$

where $r = k/h^2$.

The stencils to illustrate this method are plotted in Fig. 28.2. As can be seen, the solution at one point at the new (next) step needs the information at this point and its two neighboring points from the current time step.

For the fully implicit method, the discretization will be as follows:

$$\frac{u_i^{n+1} - u_i^n}{k} = \frac{u_{i-1}^{n+1} - 2u_i^{n+1} + u_{i+1}^{n+1}}{h^2}. \tag{28.20}$$

The solution is:

$$-ru_{i-1}^{n+1} + (1 + 2r)u_i^{n+1} - ru_{i+1}^{n+1} = u_i^n. \tag{28.21}$$

The stencils for the full implicit method are shown in Fig. 28.3. Therefore, the function values at one point and its two neighboring points at the new step rely on the function value at the point from the current step.

Then a discretization of the above equation in the general form is

$$\frac{u_i^{n+1} - u_i^n}{k} = (1 - \theta)\frac{u_{i-1}^n - 2u_i^n + u_{i+1}^n}{h^2} + \theta\frac{u_{i-1}^{n+1} - 2u_i^{n+1} + u_{i+1}^{n+1}}{h^2}. \tag{28.22}$$

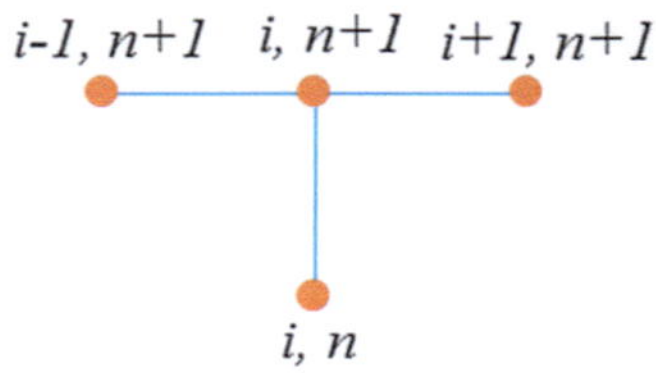

Fig. 28.3 Implicit method stencil

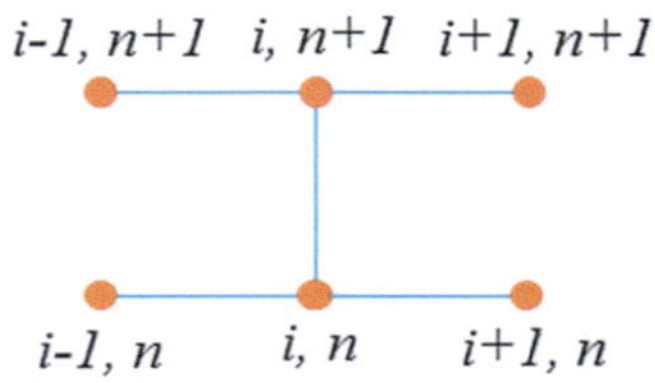

Fig. 28.4 Crank-Nicolson stencil

The solution to the above equation is

$$-\theta r u_{i-1}^{n+1} + (1 + 2\theta r)u_i^{n+1} - \theta r u_{i+1}^{n+1} = (1-\theta)ru_{i-1}^{n} + [1 - 2(1-\theta)r]u_i^n. \\ + (1-\theta)ru_{i+1}^{n} \tag{28.23}$$

Substituting $\theta = 1/2$ into the above equation will lead to the solution for the Newton-Nicolson method:

$$-\frac{1}{2}ru_{i-1}^{n+1} + (1+r)u_i^{n+1} - \frac{1}{2}ru_{i+1}^{n+1} = \frac{1}{2}ru_{i-1}^{n} + (1-r)u_i^n + \frac{1}{2}ru_{i+1}^{n}. \tag{28.24}$$

The relationship between the function values at different steps is shown in Fig. 28.4. The function values at one point and its two neighboring points at the next step rely on the function values at these three points at the current step.

The above introduction is made based on the application of the Euler forward scheme to the temporal derivative on the left-hand side of the equation. In fact, it is also common to use in the temporal derivatives one-step multistage schemes and multistep schemes, which will not be discussed in this book.

Boundary and Initial Conditions

The discretization will convert the mathematical description that is defined in a continuous geometric domain (and possibly time) into an algebraic equation defined at discrete mesh points (and possibly time points) in the domain (and time period). However, the spatial and temporal discretizations alone are not enough to generate such an algebraic equation. Shown in the following are the general forms of the algebraic equations for equilibrium problems, i.e., boundary value problem (BVP), and transient problems, i.e., initial-boundary value problem (IBVP), respectively

$$\text{BVP}: \mathbf{AU} = \mathbf{F}, \tag{28.25}$$

$$\text{IBVP}: \mathbf{A}^{n+1}\mathbf{U}^{n+1} = \mathbf{A}^{n}\mathbf{U}^{n} + \mathbf{F}, \tag{28.26}$$

where $\mathbf{A}$ is the stiffness matrix; $\mathbf{U}$is the array of unknowns, which are the values of the function at the grid points; $\mathbf{F}$ is the force matrix; the superscripts are the time steps: n is the current time step; and $n + 1$ is the next time step.

Boundary and initial conditions are needed to achieve the above algebraic equation. For boundary conditions, the Dirichlet boundary condition can be applied directly. For example, if the discretized domain starts with Point 1, then function values from neighboring points are needed to calculate the derivatives at this starting point. The Dirichlet boundary condition can then be assigned to "Point 0." "Point 0" is not necessarily explicitly considered in the algebraic equation, and thus its derivative does not need to be represented using finite differences. The application of the Dirichlet boundary condition $u = a$ at the left boundary just requires replacing all the function value at "Point 0," i.e., u_0, with a. The application of the Neumann boundary condition $u^{'} = \alpha$ on the left boundary (1D) involves the substitution of the following equation into the discretization equation of relevant points, e.g., Point 1, to cancel out u_0:

$$\frac{u_1 - u_0}{h} = \alpha. \tag{28.27}$$

The boundary conditions can be applied to other boundaries such as the right boundary in a 1D domain in a similar way.

The treatment of the initial condition is straightforward. We just need to find out the initial values at the grid points according to the initial condition,$\mathbf{U}^0$. Then the solution of the transient problem will involve an iterative process starting from $\mathbf{U}^0$. For example, the solution in the first step is obtained as

$$\mathbf{A}^1\mathbf{U}^1 = \mathbf{A}^0\mathbf{U}^0 + \mathbf{F}, \tag{28.28}$$

where $\mathbf{A}^0$ is calculated with $\mathbf{U}^0$.

To better demonstrate the discretization including boundary and initial conditions, let us recall the general example used in the previous section. A complete description of the problem is as follows:

$$\frac{\partial u}{\partial t} = \frac{\partial^2 u}{\partial x^2}, \tag{28.29}$$

$$u\big|_{x=0} = a, \tag{28.30}$$

$$\left.\frac{\partial u}{\partial x}\right|_{x=L} = \beta, \tag{28.31}$$

$$u|_{t=0} = u^0. \tag{28.32}$$

Let us discretize the domain with a length of L into I points, for which we have $I = \frac{L}{h} + 1$. The physical process to be simulated lasts a time period of Nk. Then the discretization starts with the temporal and spatial discretization at each point:

$$\text{Point}1 : \frac{u_1^{n+1} - u_1^n}{k} = \frac{u_0^n - 2u_1^n + u_2^n}{h^2}$$

$$\text{Point}2 : \frac{u_2^{n+1} - u_2^n}{k} = \frac{u_1^n - 2u_2^n + u_3^n}{h^2}$$

$$\cdots$$

$$\text{Point}\,i : \frac{u_i^{n+1} - u_i^n}{k} = \frac{u_{i-1}^n - 2u_i^n + u_{i+1}^n}{h^2}$$

$$\cdots$$

$$\text{Point}\,I : \frac{u_I^{n+1} - u_I^n}{k} = \frac{u_{I-1}^n - 2u_I^n + u_{I+1}^n}{h^2}$$

If we incorporate the above boundary conditions, we will get:

$$\text{Point}1 : \frac{u_1^{n+1} - u_1^n}{k} = \frac{a - 2u_1^n + u_2^n}{h^2}$$

$$\text{Point}2 : \frac{u_2^{n+1} - u_2^n}{k} = \frac{u_1^n - 2u_2^n + u_3^n}{h^2}$$

$$\cdots$$

$$\text{Point}\,i : \frac{u_i^{n+1} - u_i^n}{k} = \frac{u_{i-1}^n - 2u_i^n + u_{i+1}^n}{h^2}$$

$$\cdots$$

$$\text{Point}\,I : \frac{u_I^{n+1} - u_I^n}{k} = \frac{u_{I-1}^n - u_I^n + \beta h}{h^2}$$

The above series of equations can be represented using the following algebraic equation:

$$\begin{bmatrix} u_1^{n+1} \\ u_2^{n+1} \\ \vdots \\ u_i^{n+1} \\ \vdots \\ u_I^{n+1} \end{bmatrix} = \left(\begin{bmatrix} 1 & & & & \\ & 1 & & & \\ & & \ddots & & \\ & & & 1 & \\ & & & & \ddots \\ & & & & & 1 \end{bmatrix} + \frac{k}{h^2} \begin{bmatrix} -2 & 1 & & & & \\ 1 & -2 & 1 & & & \\ & & \ddots & & & \\ & & 1 & -2 & 1 & \\ & & & & \ddots & \\ & & & & 1 & -1 \end{bmatrix} \right) \begin{bmatrix} u_1^n \\ u_2^n \\ \vdots \\ u_i^n \\ \vdots \\ u_I^n \end{bmatrix}$$

$$+\frac{k}{h^2}\begin{bmatrix} a \\ 0 \\ 0 \\ \vdots \\ 0 \\ \beta h \end{bmatrix}. \tag{28.33}$$

The initial condition is applied in the first step of the iteration when $n = 1$. The discretized function value for the first step is:

$$U^0 = \begin{bmatrix} u^0 \\ u^0 \\ \vdots \\ u^0 \\ \vdots \\ u^0 \end{bmatrix}. \tag{28.34}$$

Error Analysis

Error analysis is an essential part of numerical simulation methods. But it is also worth pointing out that the error analysis is much less emphasized in engineering applications than in the studies of numerical simulation methods in mathematics. In many engineering applications, the error is examined by comparing the calculated results against analytical or experimental data instead of directly studying the error caused by the numerical schemes. This difference is especially obvious in the finite difference method, where the error associated with numerical schemes has been more extensively studied.

Because the algebraic equation is just an approximation, the numerical solution $\mathbf{U}$ at the spatial grid points and time steps is different from the true solution $\widehat{\mathbf{U}}$. Therefore, in general, $\widehat{\mathbf{U}}$ will not satisfy the above algebraic equation exactly. The discrepancy is the local (truncation) error:

$$\mathbf{A}\widehat{\mathbf{U}} = \mathbf{F} + \boldsymbol{\tau}, \tag{28.35}$$

where $\boldsymbol{\tau}$ is the local error. $\boldsymbol{\tau}$ is an array which has the same size as $\mathbf{U}$ (or $\widehat{\mathbf{U}}$). Therefore, $\boldsymbol{\tau}$ is the ensemble of the errors at each spatial and temporal discretization grid points. The true error is the global error, which is the difference between the numerical solution and the true solution

$$\mathbf{E} = \mathbf{U} - \widehat{\mathbf{U}}. \tag{28.36}$$

The local error and the global error are related via the following equation, provided that **A** is invertible:

$$\mathbf{E} = -\mathbf{A}^{-1}\boldsymbol{\tau}. \tag{28.37}$$

The above equation implies that

$$\|\mathbf{E}\| = \left\|-\mathbf{A}^{-1}\boldsymbol{\tau}\right\| \leq \left\|\mathbf{A}^{-1}\right\|\|\boldsymbol{\tau}\|. \tag{28.38}$$

A finite difference method is convergent if $\|\mathbf{E}\| \to 0$ as the mesh size or/and the step size approaches zero. The above three equations correspond to three types of criteria as the mesh size or/and the step size approaches zero:

Convergence: $\|\mathbf{E}\| \to 0$

Consistency: $\|\boldsymbol{\tau}\| \to 0$

Stability: $\|\mathbf{A}^{-1}\| <$ a finite constant

Therefore, the convergence can be ensured by two conditions: stability and consistency. The stability requires that $\|\mathbf{A}^{-1}\|$ is smaller than a constant as the mesh size or/and the step size approaches zero. The consistency requirement states that $\|\boldsymbol{\tau}\| \to 0$ as the mesh size or/and the step size approaches zero. The sign $\|\|$ means the norm of a matrix. For a given finite difference method, **A** is known. The stability can be assessed with the second-order norm of a matrix. The method will not be detailed here. Instead, we will investigate the consistency requirement, which is related to the local truncation error. Another reason to look into the local error is that the order of accuracy of finite difference schemes is determined by the local truncation error.

The local truncation error of finite difference schemes can be obtained via the following Taylor series

$$\begin{aligned} f(x+h) = f(x) + \frac{f'(x)}{1!}h + \frac{f^{(2)}(x)}{2!}h^2 + \ldots, \\ + \frac{f^{(n)}(x)}{n!}h^n + O\left(h^{n+1}\right) \end{aligned} \tag{28.39}$$

where $n!$ is the factorial of n and $O(h^{n+1})$ is the difference between the Taylor polynomial of degree n and the original function, which is a function of h^{n+1}. An approximation to the first derivative of the function "*f*" can be obtained by truncating the Taylor polynomial:

$$f(x+h) = f(x) + \frac{f'(x)}{1!}h + \ldots. \tag{28.40}$$

Similarly, we can obtain an approximation to the function at any given point around the point of interest:

$$f(x+h) = f(x) + \frac{f'(x)}{1!}h + O\left(h^2\right). \tag{28.41}$$

Based on the above Taylor expansion, we can readily obtain the error for any finite difference schemes such as the forward, backward, and central difference schemes.

Forward:

$$\frac{f(x+h)-f(x)}{h}=f'(x)+\frac{O(h^2)}{h}=f'(x)+O(h) \tag{28.42}$$

Backward:

$$\frac{f(x)-f(x-h)}{h}=f'(x)+O(h) \tag{28.43}$$

Central:

$$\frac{f(x+h)-f(x-h)}{h}=f'(x)+O(h^2) \tag{28.44}$$

Therefore, the forward, backward, and central difference schemes have the first- and second-order accuracy, respectively. The order of accuracy of temporal difference schemes can be obtained in a similar way.

Extension to 2D and 3D Dimensions

The extension of the above discretization to 2D and 3D cases can be implemented in different ways. The major challenge is that it is hard to deal with arrays more than two dimensions. In a 2D domain, the spatial discretization will lead to function values in the form of 2D arrays corresponding to the rows and columns of the points in the discretized 2D domain. We thus need to convert this 2D array into a 1D array, i.e., a column array, so that the stiffness matrices can be simplified from a 4D array to a 2D one.

The above extension can be implemented via both element-wise operations and matrix operations. For element-wise operations, we just need to find out the relationship between the dependent variable and the elements in the stiffness matrix. When conducting the conversion, the stiffness matrix is converted accordingly. This will be shown in the next chapter for the finite volume method. In a typical matrix operation, the conversion can be finished for a 2D domain in the following way

$$\frac{\partial^2 u}{\partial x^2}:(\mathbf{I}\otimes\mathbf{L})\mathbf{U} \tag{28.45}$$

$$\frac{\partial^2 u}{\partial y^2}:(\mathbf{L}\otimes\mathbf{I})\mathbf{U}, \tag{28.46}$$

where $\mathbf{I}$ is the identity matrix and $\mathbf{L}$ is the matrix generated by the stiffness matrix excluding any boundary conditions:

$$\mathbf{L} = \begin{bmatrix} -2 & 1 & & & & \\ 1 & -2 & 1 & & & \\ & \vdots & & & & \\ & & 1 & -2 & 1 & \\ & & & & \vdots & \\ & & & & 1 & -1 \end{bmatrix}. \tag{28.47}$$

The boundary conditions can be applied either by modifying the above **L** matrix or the stiffness matrix generated by the above operations specified in Eqs. 28.45 and 28.46.

Practice Problem

Problem Description

Develop FDM code to solve the problem described in Chapter 11. The material constituting the square area (1 m by 1 m) is water. The initial temperature is 273.15 K. The top boundary is set to 293.15 K, while the other boundaries are thermally insulated. Please calculate the temperature distribution at the end of 1 day.

Results

Figure 28.5 illustrates the temperature distribution.

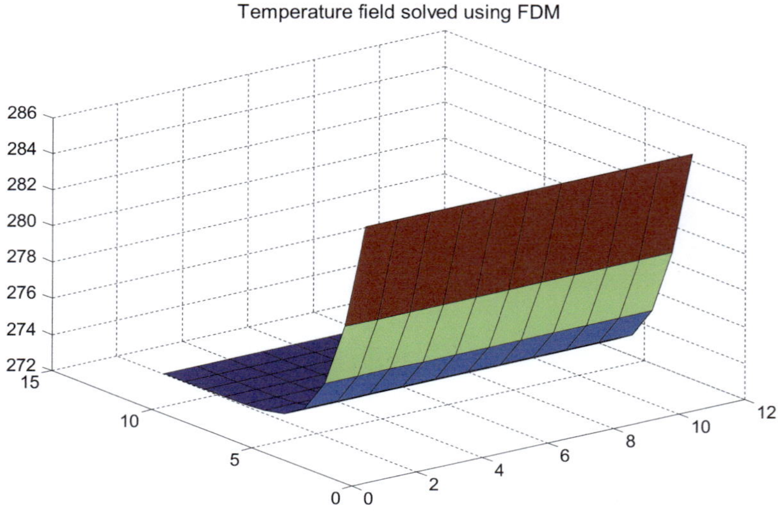

Fig. 28.5 Temperature distribution from the numerical simulation

Appendix: MATLAB Script File

```
clc;clear;
h=0.1;
k=60;
lambda=1;
rho=1000;
C=4200;
a=lambda/rho/C;
b=a*k/h^2;
m=(1/h+1);
U=273.15*ones(m^2,1);

figure(1)

Ky=[-2   1   0   0   0   0   0   0   0   0   0;...
     1  -2   1   0   0   0   0   0   0   0   0;...
     0   1  -2   1   0   0   0   0   0   0   0;...
     0   0   1  -2   1   0   0   0   0   0   0;...
     0   0   0   1  -2   1   0   0   0   0   0;...
     0   0   0   0   1  -2   1   0   0   0   0;...
     0   0   0   0   0   1  -2   1   0   0   0;...
     0   0   0   0   0   0   1  -2   1   0   0;...
     0   0   0   0   0   0   0   1  -2   1   0;...
     0   0   0   0   0   0   0   0   1  -2   1;...
     0   0   0   0   0   0   0   0   0   1  -1];

Kx=[-1   1   0   0   0   0   0   0   0   0   0;...
     1  -2   1   0   0   0   0   0   0   0   0;...
     0   1  -2   1   0   0   0   0   0   0   0;...
     0   0   1  -2   1   0   0   0   0   0   0;...
     0   0   0   1  -2   1   0   0   0   0   0;...
     0   0   0   0   1  -2   1   0   0   0   0;...
     0   0   0   0   0   1  -2   1   0   0   0;...
     0   0   0   0   0   0   1  -2   1   0   0;...
     0   0   0   0   0   0   0   1  -2   1   0;...
     0   0   0   0   0   0   0   0   1  -2   1;...
     0   0   0   0   0   0   0   0   0   1  -1];

F=[293.15;0;0;0;0;0;0;0;0;0;0]*ones(1,m);
%F=[293*ones(m,1);zeros(m^2-2*m,1);273*ones(m,1)]
F=F(:);

K_2D=kron(eye(m),Ky)+kron(Kx,eye(m));

for i=1:60*24
    t=i*k
    U=U+b*K_2D*U+b*F;
    U=reshape(U,m,m);
```

```
    surf(U)
    U=reshape(U,m^2,1);
    drawnow;
    title('Temperature field solved using FDM')
end
```

Chapter 29
Finite Volume Method

Introduction

The finite volume method (FVM) is another method for discretizing the continuous mechanics description of a physical process in terms of partial differential equations and other auxiliary equations into an algebraic equation(s) (LeVeque 2002; Toro 2013). Similar to the FEM, values are calculated at discrete grid points in a meshed geometry (and time steps). The "finite volume", or control volume, refers to the small volume surrounding each node point on a mesh.

One major difference between the FVM and the FDM is that the FVM is based on the integral form of the governing equations instead of the differential form. In the FVM, this discretization is conducted over each control volume, which can ensure that the flux entering the volume is identical to that leaving the volume. This endows FVM an advantage over other numerical simulation methods: the conservation of the conserved quantities can be ensured during the numerical process. This is very attractive for conservation problems such as fluid dynamics. This is possibly one of the major reasons why the FVM is predominant in computational fluid dynamics. The other major advantage of the FVM is that the method can be easily formulated to handle unstructured meshes, which provide higher efficiency and more flexibility for meshing.

In this chapter, we will first introduce the integral form of the governing equation and show how to formulate this equation into the form for the discretization in the FVM. Then we will discuss the discretization process using a 2D example by including the boundary and initial conditions. Extensions from 2D to multidimensional problems will be made to allow for structured 3D and even more complicated meshes. In the practice problem section, the problem solved in the FDM chapter will be solved again with the FVM.

Z. Liu, *Multiphysics in Porous Materials*,
https://doi.org/10.1007/978-3-319-93028-2_29

Integral Form of the Governing Equation

The following general differential form of conservation problems has been introduced at the beginning of the third part of this book:

$$\underbrace{\frac{\partial u}{\partial t}}_{\text{Accumulation}} \underbrace{+\nabla \cdot (u\mathbf{v})}_{\text{Advection}} = \underbrace{-\nabla \cdot (\alpha \nabla u)}_{\text{Diffusion (Conduction)}} + \underbrace{g}_{\text{Source}} . \tag{29.1}$$

The diffusion term in the equation also contains other mechanisms with similar mathematical formulations such as dispersion. The FVM requires that the above equation is satisfied over the control volume V_P around the centroid of the volume, leading to the following integral form of the governing equation:

$$\begin{aligned} &\int_t^{t+\Delta t} \left[\frac{\partial}{\partial t}\int_{V_P} udV + \int_{V_P} \nabla \cdot (u\mathbf{v})dV\right] dt, \\ &= \int_t^{t+\Delta t} \left[\int_{V_P} \nabla \cdot (\alpha \nabla u)dV + \int_{V_P} gdV\right] dt \end{aligned} \tag{29.2}$$

or we have

$$\frac{\partial}{\partial t}\int_{V_P} udV + \int_{V_P} \nabla \cdot (u\mathbf{v})dV = \int_{V_P} \nabla \cdot (\alpha \nabla u)dV + \int_{V_P} gdV. \tag{29.3}$$

Applying the general form of the Green's theorem, we reformat the above equation into

$$\frac{\partial}{\partial t}\int_{V_P} udV + \int_{\partial V} u\mathbf{v} \cdot d\mathbf{S} = \int_{\partial V} \alpha \nabla u \cdot d\mathbf{S} + \int_{V_P} gdV, \tag{29.4}$$

where $\mathbf{S}$ is the vectoral surface area of the control volume's boundary ∂V.

Let us first assume the variation of the dependent variable is linear over a control volume, which can ensure a second-order accuracy:

$$u(\mathbf{r}) = u_C + (\mathbf{r} - \mathbf{r}_C) \cdot (\nabla u)_C, \tag{29.5}$$

where the conserved quantity u is a function of the position vector $\mathbf{r}$ and the subscript C represents the centroid of the control volume under consideration.

In the above equation, the volume integral in the first term on the left-hand side of the equation can be reformulated as

$$\int_{V_P} udV = \int_{V_P} \left[u_C + (\mathbf{r} - \mathbf{r}_C) \cdot (\nabla u)_C\right] dV = u_C V_C. \tag{29.6}$$

The other terms can be transformed in a similar way, leading to the following equation:

$$u_C V_C = \int_t^{t+\Delta t} \left[\sum (\nabla u)_c \cdot \alpha_c \mathbf{S} + g_C V_C - \sum \mathbf{v} u_c \cdot \mathbf{S} \right] dt, \tag{29.7}$$

where the lowercase letter "c" in the subscripts indicates the centroid of the surface around the control volume. The summation is conducted over all the boundary segments, such as all the faces of a hexagon cell in an unstructured grid.

To numerically deal with the integration, the above equation can be further changed into

$$u_C{}^{t_2} = u_C{}^{t_1} + \frac{\delta t}{V_C} \left[\sum \alpha_c \mathbf{S} \cdot (\nabla u)_c + g_C V_C - \sum \alpha_c \mathbf{S} \cdot \mathbf{v} u_c{}^{t_1} \right], \tag{29.8}$$

where t_2 and t_1 are two consecutive time points for temporal integration, for which $t_2 - t_1 = \delta t$, and all the terms on the right-hand side of the equation are based on $u_C{}^{t_1}$ or the value(s) of u_C from even earlier time steps.

Boundary and Initial Conditions

This section introduces the discretization of the governing equation with the inclusion of the boundary and initial conditions. The treatments of the boundary and initial conditions are similar to those in the FDM. The boundary conditions can be included via the discretized equation in the control volumes on the boundary. The values of the dependent variable assigned via the Dirichlet boundary condition are included in the $u_C{}^{t_1}$ and $(\nabla u)_c$. The other terms can also include these boundary function values if they are functions of the dependent variable. The values of the derivatives on the boundary, which are given out by the Neumann boundary condition, can be used to replace the $(\nabla u)_c$ term directly.

A simple 1D example will be used to illustrate the discretization with the boundary and initial conditions in detail. However, it is worthwhile to mention that, in 1D problems or even many 2D problems with structured grids, the FVM can lead to the same algebraic equation as the FDM. The above integral equations use symbols that are common in the FVM. These symbols will be gradually changed to the symbols used in the FDM to compare and relate these two methods for a better understanding of discretization.

Let us first revisit the general parabolic equation with the typical boundary conditions as used in the previous chapter:

$$\frac{\partial u}{\partial t} = \frac{\partial^2 u}{\partial x^2}, \tag{29.9}$$

$$u|_{x=0} = a, \tag{29.10}$$

$$\left.\frac{\partial u}{\partial x}\right|_{x=L} = \beta, \tag{29.11}$$

$$u\big|_{t=0} = u^0. \tag{29.12}$$

In order to obtain an algebraic equation which is comparable to that in the FDM, we need to discretize the domain in the following way. The 1D domain with a length of L is divided into $I - 1$ portions with a uniform mesh size of h: there are two portions with a length of $h/2$ on the two ends (boundary) and $I - 3$ portions in the interior of the domain. Therefore, we have $I = \frac{L}{h} + 1$. The physical process to be simulated lasts a time of Nk. The general form of the integral equation including an accumulation and a diffusion term is as follows:

$$u_C{}^{t_2} = u_C{}^{t_1} + \frac{\Delta t}{V_C}\left[\sum \alpha_c \mathbf{S} \cdot (\nabla u)_c\right]. \tag{29.13}$$

This is a 1D problem, so there is no surface area between different control volumes (or cells). Let us use a unit area instead, thus $|\mathbf{S}| = 1$. $\mathbf{S}$ has a positive direction if it is on the right side of the cell and negative on the left. The centroid of the cell is the middle point of the cell. Accordingly, the above integral equation can be transformed into

$$u^{n+1} = u^n + \frac{k}{h}\left[\left(\frac{\partial u}{\partial x}\right)_{\text{right}} - \left(\frac{\partial u}{\partial x}\right)_{\text{left}}\right]. \tag{29.14}$$

The subscript "C" is removed as we use the function value in the middle of a cell u_C to represent the u of that cell. u^{n+1} and u^n are the values at the next and current time steps, respectively. $\left(\frac{\partial u}{\partial x}\right)_{\text{right}}$ and $\left(\frac{\partial u}{\partial x}\right)_{\text{left}}$ are the derivative on the right and left ends of the cell, respectively, whose signs in the equation are determined by the direction of $\mathbf{S}$ and $(\nabla u)_c$. The centroid of the boundary surface does not require special treatments as the boundary segments are two infinitesimally small points in this 1D domain.

Without considering the boundary conditions, the governing equation can be directly discretized as

$$\text{Point 1: } u_1^{n+1} = u_1^n + \frac{k}{h}\left(\frac{u_2^n - u_1^n}{h} - \frac{u_1^n - u_0^n}{h}\right).$$

$$\text{Point 2: } u_2^{n+1} = u_2^n + \frac{k}{h}\left(\frac{u_3^n - u_2^n}{h} - \frac{u_2^n - u_1^n}{h}\right).$$

$$\cdots$$

$$\text{Point } i\text{: } u_i^{n+1} = u_i^n + \frac{k}{h}\left(\frac{u_{i+1}^n - u_i^n}{h} - \frac{u_i^n - u_{i-1}^n}{h}\right).$$

$$\cdots$$

$$\text{Point } I\text{: } u_I^{n+1} = u_i^n + \frac{k}{h}\left(\frac{u_{I+1}^n - u_I^n}{h} - \frac{u_I^n - u_{I-1}^n}{h}\right).$$

The boundary conditions are implemented by changing the discretized equations in the boundary cells, i.e., Point 1 and Point I, with the following equations:

$u_0^n = a,$ at Point 1
$\left(\frac{\partial u}{\partial x}\right)\Big|_{\text{right}} = \beta,$ at Point I

The above set of equations with the boundary condition equations can be transformed into the following algebraic equation:

$$\begin{bmatrix} u_1^{n+1} \\ u_2^{n+1} \\ \vdots \\ u_i^{n+1} \\ \vdots \\ u_I^{n+1} \end{bmatrix} = \left(\begin{bmatrix} 1 & & & & & \\ & 1 & & & & \\ & & \ddots & & & \\ & & & 1 & & \\ & & & & \ddots & \\ & & & & & 1 \end{bmatrix} + \frac{k}{h^2} \begin{bmatrix} -2 & 1 & & & & \\ 1 & -2 & 1 & & & \\ & \vdots & & & & \\ & 1 & -2 & 1 & & \\ & & & \vdots & & \\ & & & 1 & -1 & \end{bmatrix} \right) \begin{bmatrix} u_1^n \\ u_2^n \\ \vdots \\ u_i^n \\ \vdots \\ u_I^n \end{bmatrix}$$
$$+ \frac{k}{h^2} \begin{bmatrix} a \\ 0 \\ 0 \\ \vdots \\ 0 \\ \beta h \end{bmatrix}. \tag{29.15}$$

The above equation is identical to the algebraic equation obtained using the FDM in the previous chapter. The initial condition is applied in the first step of the iteration with

$$U^0 = \begin{bmatrix} u^0 \\ u^0 \\ \vdots \\ u^0 \\ \vdots \\ u^0 \end{bmatrix}. \tag{29.16}$$

Another difference can be observed between the FVM and FDM: the FVM is discretized over the control volumes, while the FDM is discretized at the grid points. That is the reason why the meshing is done in the above specific way to obtain the same algebraic equation.

Multidimensional Problems

2D Structured Grids

The above example demonstrates the concept of the FVM in a 1D case. In 2D or 3D cases, the treatment is much more complicated, especially when complicated mesh structures and unstructured meshes get involved.

If we still use a structured grid such as the square mesh as below, we can still obtain the same discretization results as the FDM. However, for such a typical 2D grid, we now have more volumes (or cells) around the volume of interest. It is common to mark these neighboring volumes as "N" (North), "E" (East), "S" (South), and "W" (West) in FVM textbooks. The volume V_C and the vectoral area of its boundaries **S** will be the same. The $(\nabla u)_C$ at the centroid of a face or a control volume can be determined using the function values at the centroids of its two neighboring volumes. This is because the centroid of a boundary face is always the middle point of the line connecting the centroids of the two volumes sandwiching this face (Fig. 29.1).

In the above case, the integral form of the governing equation can be formulated as follows:

$$u_C{}^{n+1} = u_C{}^{n} + \frac{\alpha_c \Delta t}{V_C}\left[\sum \mathbf{S} \cdot (\nabla u)_c\right]. \tag{29.17}$$

The above integral governing equation can be discretized with the computational domain illustrated in the following figure (Fig. 29.2).

If we use an element-wise treatment, the above equation will lead to the following global algebraic equation.

$$\mathbf{U}^{n+1} = \mathbf{A}\mathbf{U}^{n}, \tag{29.18}$$

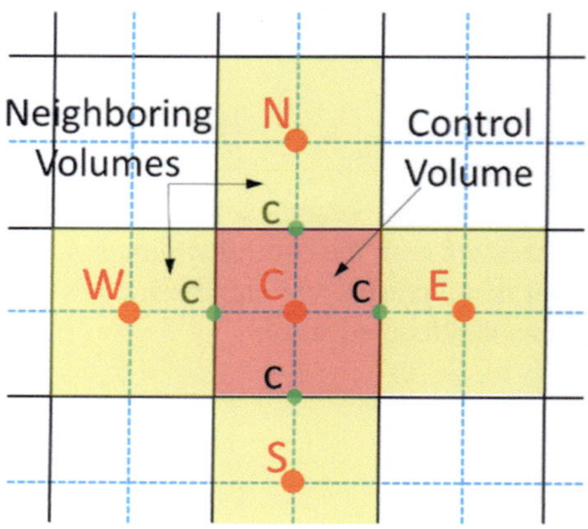

Fig. 29.1 Control volume and its neighboring volumes in a 2D structured grid

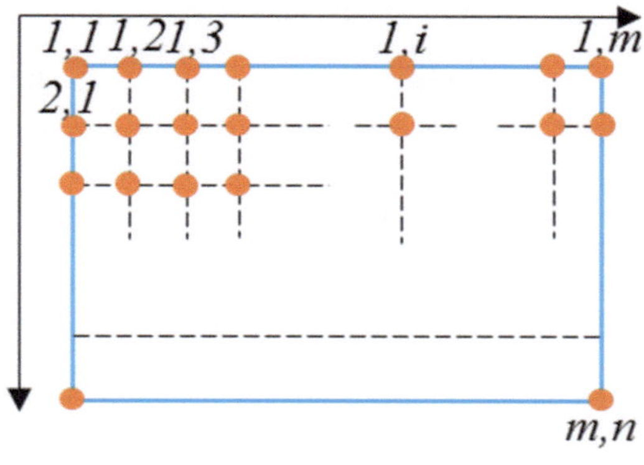

Fig. 29.2 Numbering of grid points

in which

$$\mathbf{U} = \begin{bmatrix} u_{1,1} \\ \vdots \\ u_{1,m} \\ u_{2,1} \\ \vdots \\ u_{m,n} \end{bmatrix}. \tag{29.19}$$

The value of the dependent variable at node (i,j) is located in the $(i+j)$th row of $\mathbf{U}$. Then the coefficients in the global stiffness matrix are a, b, d, and e corresponding to the north, south, east, and west neighboring volumes as shown below, respectively. The values of the coefficients are obtained from the integral form of the governing equation: $\frac{\alpha_c \Delta t}{dx \cdot dy}$, in which dx and dy are the mesh size along the $x$and y directions, respectively. The coefficients corresponding to the north and south control volumes are the $(i+j-m)$th and $(i+j+m)$th elements of the $(i+j)$th row.

$$\begin{array}{llllll} \text{Node} & i+j\text{-}m & i+j-1 & i+j & i+j+1 & i+j+m \end{array}$$

$$\mathbf{A} = \begin{bmatrix} \ddots & & & & & & & & \\ & & & & \vdots & & & & \\ & -a & \dots & -e & 1 & -d & \dots & -b & \\ & & & & \vdots & & & & \\ & & & & & & & & \ddots \end{bmatrix}. \tag{29.20}$$

More Complicated Grids

However, special treatments are needed for more complicated 2D/3D structured grids and unstructured grids. The first challenge is to determine the function values on the boundary faces of the volumes and those parameters depending on these values. Several terms defined at the centroid of the boundaries are generated when applying Green's theorem to generate the final integral form of the governing equation. Various convection differencing schemes have been proposed to determine the surface values. However, in arbitrarily unstructured meshes, it seems to be impractical to use any values other than the function values at the centroid of the control volumes to calculate these values. Therefore, the common convection differencing schemes are limited to the centroid function values of the control volume and its nearest (neighboring) volumes. One typical way is the central differencing, in which the surface value is determined based on the relative position between the two volumes on two sides of the surface, i.e., the control volume of interest and the other neighboring volume. Based on the linear variation assumed for the dependent variable in local volumes, the following general equation can be used to calculate the surface values.

$$u_{\mathrm{c}} = \frac{|\mathrm{cX}|}{|\mathrm{CX}|} u_{\mathrm{C}} + \frac{|\mathrm{cC}|}{|\mathrm{CX}|} u_{\mathrm{X}}, \tag{29.21}$$

where X represents the centroid of the neighboring volume and |CX| is the length of the line connecting C and X. Additionally, upwind differencing, in which the surface value is determined according to the direction of the flow, is also common. This method exhibits advantages in ensuring boundedness with a sacrifice of accuracy. Blended differencing as a mixture of the above two methods has also been proposed (Fig. 29.3).

The diffusion term also requires special attention. This is because in an unstructured grid, the direction of the outward normal, i.e., **S**, is not necessarily the same as the line connecting the centroids of the two volumes on two sides of the boundary face. The $\mathbf{S} \cdot (\nabla u)_{\mathrm{c}}$ splits into an orthogonal part and a non-orthogonal part as follows:

$$\mathbf{S} \cdot (\nabla u)_{\mathrm{c}} = \|\mathbf{S}\| \frac{u_{\mathrm{C}} - u_{\mathrm{X}}}{\|\mathbf{d}\|} + \mathbf{k} \cdot (\nabla u)_{\mathrm{c}}, \tag{29.22}$$

where $\|\mathbf{S}\|$ and $\|\mathbf{d}\|$ are the magnitudes of **S** and the vectoral length between the centroids of the control volume of interest and its neighboring volume (s) on the

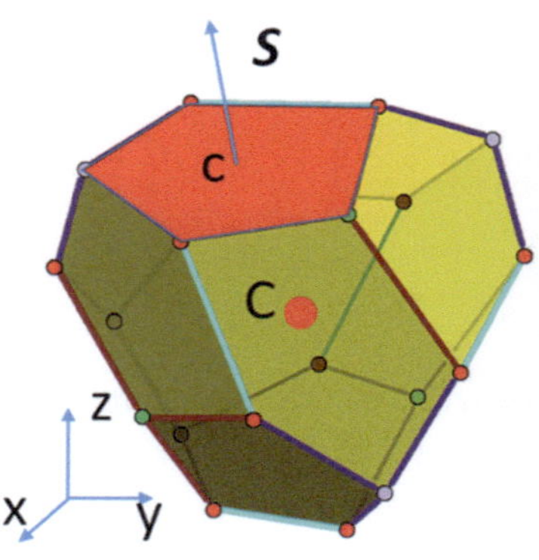

Fig. 29.3 Unstructured grid and treatment

other side of the boundary face **d**, respectively; u_X is the function value at the centroid of the neighboring volume; and **k** is a vector used to represent the non-orthogonal correction. Several decompositions are available to account for the non-orthogonal corrections such as the minimum correction approach, the orthogonal correction approach, and the over-relaxed approach (Jasak 1996).

Practice Problem

Problem Description

Develop FVM code to solve the problem described in Chap. 28. The material constituting the square area (1 m by 1 m) is water. The initial temperature is 273.15 K. The top boundary is set to 293.15 K, while the other boundaries are thermally insulated. Please calculate the temperature distribution at the end of one day.

Results

The following figure illustrates the temperature distribution (Fig. 29.4).

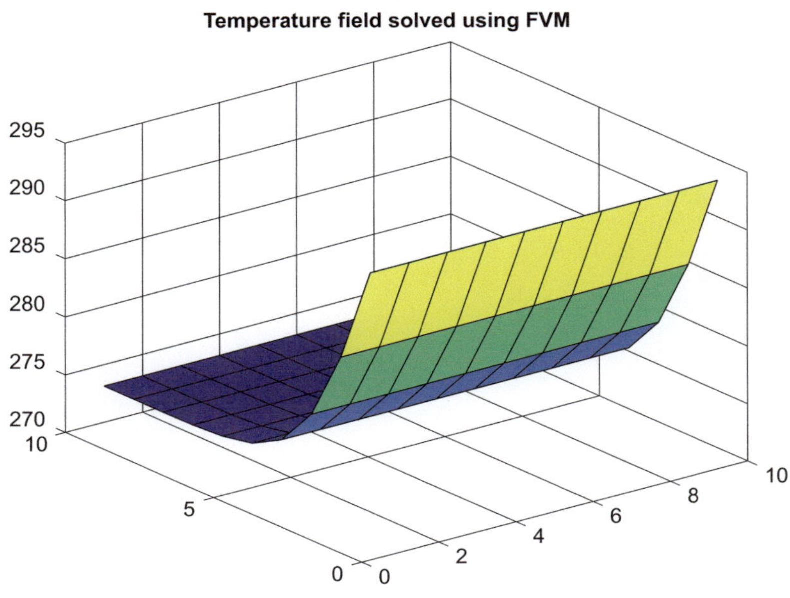

Fig. 29.4 Temperature distribution from the numerical simulation

Appendix: MATLAB Script File

```
% The initial temperature is 273.15.
% The top BC temperature is fixed at 293.15; while all the other BCs are
thermally insulated.
% A*U^n+1=B(U^n)
% Initiation
clear;
clc;

nx=10;
ny=10; %node numbers along x and y;
theta=0.5;
alpha=1/4.2e6; %thermal diffusivity;
Lx=1.;
Ly=1.; %dimensions of domain;
dx=Lx/(nx-1);
dy=Ly/(ny-1);
tTotal=3600*24;
dt=60; %total calculation of the physical time is 10;
Tinit=273.15; %initial temperature is 700;
Tboundary=293.15; %constant boundary temperature
Tmat(:,:)=Tinit*ones(nx,ny);

k=0;
F=alpha*dt/(dx*dy); %Coefficient associated with flux
a=F;b=F;d=F;e=F; %Coefficient in the A matrix associated with
neighboring cells
c=1+a+b+d+e;   %Coefficient in the A matrix associated with considered
cell
A=zeros(nx*ny,nx*ny);
B=ones(nx*ny,1)*Tinit;

for t=0:tTotal/dt
for i=1:nx
for j=1:ny
temp=(i-1)*ny+j;
%First row of cells (top, i=1) correspond to the first row of A and first
ny points of B
%Interior of the Domain
if(i~=1&i~=nx&j~=1&j~=ny)
   A(temp,temp)=c; %P
   A(temp,temp-ny)=-a; %N
   A(temp,temp+ny)=-b; %S
   A(temp,temp+1)=-d; %E
   A(temp,temp-1)=-e; %W

%Different Boundaries (BCs)
elseif(i==1)  %North/Top
   A(temp,:)=0;
   %A(:,temp)=0;
   A(temp,temp)=1;
   B(temp)=Tboundary;
```

```
elseif (i==nx&j~=1&j~=ny) %South/Bottom
   A(temp,temp)=c-b; %P
   A(temp,temp-ny)=-a; %N
   %A(temp,temp+ny)=-b; %S
   A(temp,temp+1)=-d;  %E
   A(temp,temp-1)=-e;  %W
elseif (j==1&i~=nx) %West/Left
   A(temp,temp)=c-e; %P
   A(temp,temp-ny)=-a; %N
   A(temp,temp+ny)=-b; %S
   A(temp,temp+1)=-d;  %E
   %A(temp,temp-1)=-e;  %W
elseif (j==ny&i~=nx) %East/Right
   A(temp,temp)=c-d; %P
   A(temp,temp-ny)=-a; %W
   A(temp,temp+ny)=-b; %E
   %A(temp,temp+1)=-d;  %S
   A(temp,temp-1)=-e;  %N
elseif (i==nx&j==1) %Left Lower Corner
   A(temp,temp)=c-b-e; %P
   A(temp,temp-ny)=-a; %N
   %A(temp,temp+ny)=-b; %S
   A(temp,temp+1)=-d;  %E
   %A(temp,temp-1)=-e;  %W
elseif (i==nx&j==ny) %Left Right Corner
   A(temp,temp)=c-b-d; %P
   A(temp,temp-ny)=-a; %N
   %A(temp,temp+ny)=-b; %S
   %A(temp,temp+1)=-d;  %E
   A(temp,temp-1)=-e;  %W
end
end
end
T=A\B;
B=T;
end

for i=1:nx,
for j=1:ny,
temp=(i-1)*nx+j;
Tmat(i,j)=T(temp);
end
end
%% plot Surface temperature on Z-axis
%X and Y axis as cordinates of the plate
[i,j]=meshgrid(1:1:nx,1:1:ny);
contourf(i,j,Tmat);
surf(Tmat);
title('Temperature field solved using FVM')
```

Chapter 30
Finite Element Method

Introduction

The finite element method (FEM) provides a much different way from the FDM and FVM for the discretization of the continuous mechanics description of a physical process. The discretization is made over a domain consisting of subdivisions called elements. These elements are essentially the same as those control volumes in the FVM from a geometric perspective. However, the FEM more emphasizes the grid points as the apexes of the element rather than the centroids, volume, and boundary faces that are emphasized in the FVM. The governing equation is discretized at all the grid points of the element to form a local algebraic equation. Since different elements share grid points, the local algebraic equation will be ensembled into the global governing equation to add up the coefficients from different elements for the same point.

Another major difference is that the FEM adopts the weak form of the governing equation instead of the differential form in the FDM or the integral form in the FVM. Usually, either the Galerkin method or variational methods from the calculus of variations are used to derive the weak form. The idea underlying the Galerkin and the variational methods is analogous to the principle of minimum energy in physics. In the FEM, we use polynomials of different orders, which also determine the order of the elements, to approach the true solution. The Galerkin method or variational methods will minimize the error of the approximation by fitting trial functions into the governing equation.

Due to the above characteristics, the FEM owes several attractive advantages. The FEM is a good choice for analyzing problems with complicated geometries. Second, since the FEM uses the weak form of the governing equation, the requirement on continuity is loosened. Therefore, it provides a better performance when handling discontinuities compared to the FDM and FVM. The FEM also shows good flexibility and good performance when dealing with moving mesh problems. In addition, the FEM is also suitable for problems with varying precision requirements across the computational domains. In virtue of these advantages, the FEM has been developed

Z. Liu, *Multiphysics in Porous Materials*,
https://doi.org/10.1007/978-3-319-93028-2_30

into a predominant numerical simulation technique in engineering. It is widely adopted in most of the multiphysics areas. One noticeable exception is the computational fluid mechanics, in which the FVM is predominant possibly due to its ability in ensuring conservation of water mass. The numerical simulation analysis with the FEM is usually termed the finite element analysis (FEA).

It is interesting to note that the FEM is studied in both mathematics and engineering. However, the FEM is usually introduced in significantly different ways in these two areas. In mathematics, the FEM is introduced as a special case of the Galerkin method. But in engineering, an introduction to the FEM is usually started from the variational methods or even the Ritz method. In addition, emphasis is placed on the performance of different types of elements on the overall performance of the FEM applications.

A typical procedure for implementing the FEM is slightly more complicated than that in the FDM and FVM. The first few steps are the establishment of the mathematical model including the geometrical domain and the discretization of this domain (meshing). This is not different from that in the FDM and FVM. The next step is to transform the governing equation into the weak form. Some or all the boundary conditions can be incorporated into the weak form of the governing equation(s). Then we will choose the order of the element. The shape and order of the element will determine the shape functions. The shape functions will be adopted to discretize the governing equation to obtain the local algebraic equation for a single element. The algebraic equations for all the elements within the computational domain are ensembled to obtain the global algebraic equation. The last step is to solve the global algebraic equation and to analyze and visualize the results.

In the following sections, at first, the focus will be placed on the Galerkin Method for obtaining the weak form of the governing equation. Then, using thermomechanics as an example, the discretization of the weak form of the governing equations with Neumann boundary conditions into a local algebraic equation will be explained. The discretization of the thermal field and mechanical field, including their couplings, will be introduced separately. Two common types of 2D elements, more specifically their shape functions, are given to help readers understand the discretization process. Other significant contents of the FEM, such as different element types, stiffness matrix ensembling for obtaining the global algebraic equation, and isoparametric elements, will not be detailed here, as they can be easily found in many reference books for the FEM. Finally, MATLAB FEM code is given for solving the problem that has been solved in the FDM and FVM chapters.

Galerkin Method and Weak Form

Suppose we have a linear differential operator D acting on a function u to produce a function, $p(x)$,

$$D(u(x)) = p(x). \tag{30.1}$$

In the minimal weight residual methods, which the FEM belongs to, we approximate u by a function, $\tilde{u}$, which is a linear combination of basic functions chosen from a linearly independent set:

$$u \cong \tilde{u} = \sum_{i=1}^{n} u_i \varphi_i. \tag{30.2}$$

Substituting the above approximation into the differential operator, D, in general, the result of the operation is not $p(x)$ anymore. Accordingly, an error or residual exists,

$$E(x) = R(x) = D(\tilde{u}(x)) - p(x) \neq 0. \tag{30.3}$$

The notion in the minimal weight methods is to force the residual to approximate zero in some average sense over the domain, that is,

$$\int_X R(x) W_i dx = 0, i = 1, 2, \ldots, n \tag{30.4}$$

where the number of weight functions W_i is exactly equal to the number of unknown constants u_i in $\tilde{u}$. The result is a set of algebraic equations (n) for the unknown constants, u_i. There are five common minimal weight residual methods, which are differentiated by the choices of the W_i's. These methods include the collocation method, sub-domain method, least squares method, Galerkin method, and method of moments.

In the Galerkin Method, the derivative of the approximating function is used as the weight function:

$$W_i = \frac{\partial \tilde{u}}{\partial u_i} = \varphi_i(x). \tag{30.5}$$

Then we have

$$\int_X [D(\tilde{u}(x)) - p(x)] \varphi_i(x) dx = 0. \tag{30.6}$$

Equation 30.6 is the weak formulation, or weak form, of the original differential equation, Eq. 30.1. $\varphi_i(x)$ is the shape function in the FEM. If the element type is fixed, then all the shape functions for each point will be known prior to the solution process. The shape functions for two simple types of elements will be given in a later section. The outcome of using the weak form is to convert the governing PDE, which requires the equality of the equation at every infinitesimal point within the domain and on the boundary, to the weak form of the equation, which requires that the above integral form of a function with the residual of the numerical approximation and the shape functions equals zero. The integral form will bring additional benefits: a relatively lower requirement on the continuity and differentiability of the original differential equation. This weak form is necessary in the FEM for obtaining the algebraic equation system. However, it is worthwhile to mention that the above way to obtain the weak form can be reinterpreted in a more straightforward way using test functions, which are random functions as follows:

$$v \cong \tilde{v} = \sum_{i=1}^{n} v_i \varphi_i. \tag{30.7}$$

The procedure is to multiply the original PDE with the above formation of the test function and then to integrate their product over the domain. As the test function is a random function and exists in every term of the integral form of the equation, v_i can be eliminated, and we will obtain the same weak formulation as Eq. 30.6. Detailed operations will be demonstrated in an example to be introduced in the following section.

Example of Discretization: Thermomechanics

Thermal Field

The governing equation of the thermal field, i.e., heat equation, is first recalled. When material properties are constants and advection is overlooked in solids, the heat equation can be reduced to

$$\rho C \frac{\partial (T)}{\partial t} = \nabla \cdot (\lambda \nabla T) + g, \tag{30.8}$$

where ρ is the density, C is the gravimetric heat capacity, T is temperature, λ is the thermal conductivity, $\mathbf{v}$ is the velocity of material particles, and g is the heat source. The above equation can be transformed into the weak form as follows:

$$\rho C \int_{\Omega} v \frac{\partial T}{\partial t} d\Omega = -\lambda \int_{\Omega} (\nabla v) \cdot (\nabla T) d\Omega + \int_{\partial\Omega} v \cdot (\lambda \nabla T) \cdot \mathbf{n} dS + \int_{\Omega} v g d\Omega, \tag{30.9}$$

where Ω is the computational domain, $\partial\Omega$ is the boundaries of the computational domain, v is the test function, $\mathbf{n}$ is the outward unit normal vector of the boundary, and $(\lambda \nabla \mathrm{T}) \cdot \mathbf{n}$ is the source on the boundaries.

In the FEM, the above equation in continuum mechanics will be discretized into an algebraic equation system. Taking the bilinear quadrilateral element (Quadrilateral 4 (Q4), see Fig. 30.1) element in 2D, for example, the values of the dependent variable T and test function v can be related to the shape functions N_{Ti} ($\mathbf{N}_T$) and test functions v_i ($\mathbf{v}$) of the element in the following way:

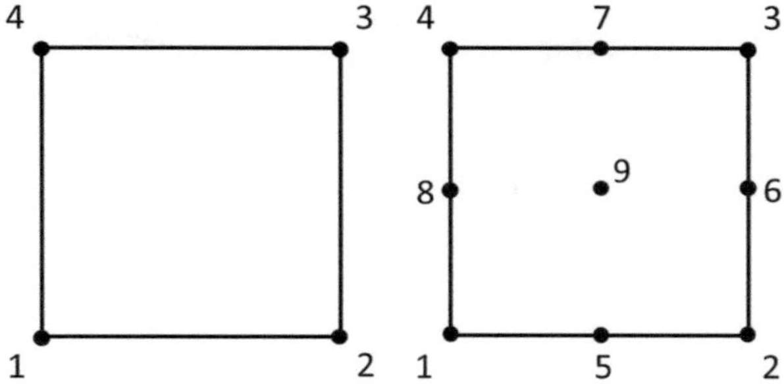

Fig. 30.1 Q4 and Q9 element

$$T = \sum_{i=1}^{4} N_{Ti} T_i = \mathbf{N}_T \cdot \mathbf{T},$$

$$v = \sum_{i=1}^{4} N_{Ti} v_i = \mathbf{N}_T \cdot \mathbf{v},$$

where,

$$\mathbf{N}_\mathrm{T} = [N_1 \;\; N_2 \;\; N_3 \;\; N_4],$$

$$\mathbf{T} = \begin{bmatrix} T_1 & T_2 & T_3 & T_4 \end{bmatrix}^T.$$

The discretized governing equation is formulated as

$$\rho C \int_\Omega (\mathbf{N}_\mathrm{T}\mathbf{v})^T \frac{\partial \mathbf{T}}{\partial t} d\Omega = -\lambda \int_\Omega (\mathbf{A}_\mathrm{T}\mathbf{N}_\mathrm{T}\mathbf{v})^T \cdot (\mathbf{A}_\mathrm{T}\mathbf{N}_\mathrm{T}\mathbf{T}) d\Omega + \int_{\partial\Omega} (\mathbf{N}_\mathrm{T}\mathbf{v})^T (-\mathbf{J} \cdot \mathbf{n}) dS + \int_\Omega (\mathbf{N}_\mathrm{T}\mathbf{v})^T g d\Omega, \tag{30.10}$$

where **N** is the shape function matrix and **A** is the matrix of the differential operator. The integration will be conducted using the Gauss-Legendre numerical integration. The time differential on the left side of the equation will be handled with an iterative process:

$$\left(\frac{\rho C}{\Delta t} \int_\Omega \mathbf{N}^T \mathbf{N} d\Omega \right) \left(\mathbf{T}^{n+1} - \mathbf{T}^n \right) = \left(-\lambda \int_\Omega \mathbf{B}^T \mathbf{B} d\Omega \right) \left[\theta \mathbf{T}^{n+1} + (1-\theta) \mathbf{T}^n \right] + \int_{\partial\Omega} \mathbf{N}^T (-\mathbf{J}^n \cdot \mathbf{n}) dS + \int_\Omega \mathbf{N}^T g^n d\Omega, \tag{30.11}$$

where $\mathbf{T}^{n+1}$ and $\mathbf{T}^n$ are the vectors of unknowns at the next and the current time steps, respectively. It is noted that the test function is eliminated due to its presence in all terms. Detailed expressions of the matrices in the above equation are as follows:

$\mathbf{A}_{\mathrm{T}} = \begin{bmatrix} \dfrac{\partial}{\partial x} \\ \dfrac{\partial}{\partial y} \end{bmatrix}$ (for equation with scalar dependent variables in 2D, independent of element type),

$$\mathbf{B}_{\mathrm{T}} = \begin{bmatrix} \dfrac{\partial N_1}{\partial x} & \dfrac{\partial N_2}{\partial x} & \dfrac{\partial N_3}{\partial x} & \dfrac{\partial N_4}{\partial x} \\ \dfrac{\partial N_1}{\partial y} & \dfrac{\partial N_2}{\partial y} & \dfrac{\partial N_3}{\partial y} & \dfrac{\partial N_4}{\partial y} \end{bmatrix},$$

where N_i's are the shape functions corresponding to the type of the adopted element. For example, the Q4 element has four functions N_1, N_2, N_3, and N_4.

There are two different ways to deal with the boundary source term specified by the Neumann boundary condition in the FEM. The first way is to calculate the term with shape functions associated with the nodes on the boundaries when discretizing the weak form of the governing equation, e.g., Eq. 30.9, while the other one is to establish the force (or source) vector matrix separately by multiplying the boundary source with the boundary length (area). For both of them, the boundary force needs to be formulated based on physical mechanisms such as convective heat transfer, which can be predicted with Newton's law of cooling:

$$-\mathbf{J} \cdot \mathbf{n} = h(T_{\mathrm{env}} - \mathbf{T}), \tag{30.12}$$

where T_{env} is the environmental temperature. The second method is employed in this code for the practice problem. When no heat source exists, the discretized governing equation can be expressed as

$$\mathbf{K}_{\mathrm{Ta}}\left(\mathbf{T}^{n+1} - \mathbf{T}^n\right) = \mathbf{K}_{\mathrm{Tc}}\left[\theta\mathbf{T}^{n+1} + (1-\theta)\mathbf{T}^n\right] + \mathbf{F}_{\mathrm{Ts}}{}^n + \mathbf{F}_{\mathrm{Tb}}{}^n, \tag{30.13}$$

where $\mathbf{K}_{\mathrm{Ta}}$ and $\mathbf{K}_{\mathrm{Tc}}$ are the stiffness matrices relevant to the accumulation (a) and conduction (c), respectively, and $\mathbf{F}_{\mathrm{Ts}}$ and $\mathbf{F}_{\mathrm{Tb}}$ are the force vectors that correspond to source on boundary surface (s) and body source (b, corresponding to g). The term "force" is adopted here to represent the generalized force, source. In the current study, no source (body force) is present, so $\mathbf{F}_{\mathrm{Tb}}$ is excluded. The stiffness matrices are calculated using the following equations:

$$\mathbf{K}_{\mathrm{Ta}} = \left(\frac{\rho C}{\Delta t}\int_{\Omega} \mathbf{N}_{\mathrm{T}}{}^T \mathbf{N}_{\mathrm{T}} d\Omega\right), \tag{30.14}$$

$$\mathbf{K}_{\mathrm{Tc}} = \left(-\lambda \int_{\Omega} \mathbf{B}_{\mathrm{T}}{}^{T} \mathbf{B}_{\mathrm{T}} dS \right). \tag{30.15}$$

When $\theta = 0$, the explicit scheme is used. The explicit scheme is believed to be conditionally stable depending on the values of time step and mesh size. When $\theta = 1/2$, the Crank-Nicolson scheme is produced. The full explicit scheme can be obtained by setting $\theta = 1$. Although θ may vary between 0 and 1, the most commonly used values are those for the fully implicit scheme and the Crank-Nicolson scheme.

If the explicit method is adopted, the solution at the current step can be obtained as

$$\mathbf{T}^{n+1} = \mathbf{C}_{\mathrm{T}} \mathbf{T}^{n} + \mathbf{M}_{\mathrm{T}} = \frac{\mathbf{K}_{\mathrm{Ta}} + (1-\theta)\mathbf{K}_{\mathrm{Tc}}}{\mathbf{K}_{\mathrm{Ta}} - \theta \mathbf{K}_{\mathrm{Tc}}} \mathbf{T}^{n} + \frac{\mathbf{F}_{\mathrm{Ts}}}{\mathbf{K}_{\mathrm{Ta}} - \theta \mathbf{K}_{\mathrm{Tc}}}. \tag{30.16}$$

Mechanical Field

In addition to the governing equation of the thermal field, we here also show the discretization of the mechanical field including the coupling from the thermal field. Accordingly, the influence of the deformation on the material properties, in both the thermal and mechanical fields, is neglected. The governing equation is the equation of equilibrium:

$$\nabla \cdot \boldsymbol{\sigma}_{\mathrm{e}} + \mathbf{g} = 0, \tag{30.17}$$

where $\boldsymbol{\sigma}_{\mathrm{e}}$ is the elastic stress (second-order tensor with Tensor notation) and $\mathbf{g}$ is the body force. The elastic stress is the difference between the total stress and the thermal stress as follows:

$$\boldsymbol{\sigma}_{\mathrm{e}} = \boldsymbol{\sigma}_{\mathrm{e}} - \boldsymbol{\sigma}_{\mathrm{t}} = \nabla \mathbf{u} - \alpha (T - T_{\mathrm{ref}}) \mathbf{I}, \tag{30.18}$$

where $\mathbf{I}$ is the identity matrix to allow for the fact that the thermal strain contributes to normal strains instead of deviatoric strains. It is seen from Eq. 30.18 that the element used for the mechanical field should be one order higher than that used for the thermal field to obtain smooth results for the strain. For example, if a linear element, Q4, is adopted for the thermal field, a quadratic Lagrangian element, such as Q9 (see Fig. 30.1), is needed for the mechanical field. $\mathbf{I}$ takes the following form:

$$\mathbf{I} = 1 \text{ in 1D},$$

$$\mathbf{I} = \begin{bmatrix} 1 & 0 \\ 0 & 1 \end{bmatrix} \text{ in 2D,}$$

$$\text{and } \mathbf{I} = \begin{bmatrix} 1 & 0 & 0 \\ 0 & 1 & 0 \\ 0 & 0 & 1 \end{bmatrix} \text{ in 3D.}$$

Then Navier's equation is obtained as the governing equation:

$$\nabla \cdot \{\mathbf{D}[\nabla \mathbf{u} - \alpha(T - T_{\mathrm{ref}})\mathbf{I}]\} + \mathbf{g} = 0, \tag{30.19}$$

where $\mathbf{u}$ is the displacement, $\mathbf{D}$ is the stiffness tensor, α is the coefficient of linear expansion, T_{ref} is the reference temperature for calculating thermal strain, and $\mathbf{g}$ is the body force. The above equation is transformed into the weak form as

$$\int_{\Omega} \nabla \mathbf{v} \cdot [\mathbf{D}\nabla \mathbf{u} - \alpha \mathbf{D}(T - T_{\mathrm{ref}})\mathbf{I}] d\Omega = \int_{\partial\Omega} \mathbf{v} \cdot [\mathbf{D}\nabla \mathbf{u} - \alpha \mathbf{D}(T - T_{\mathrm{ref}})\mathbf{I}] \cdot \mathbf{n} dS + \int_{\Omega} \mathbf{vg}. \tag{30.20}$$

The discretization of the above governing equation leads to

$$\begin{aligned} \int_{\Omega} (\mathbf{A}_{\mathrm{M}}\mathbf{N}_{\mathrm{M}})^T \cdot (\mathbf{D}\mathbf{A}_{\mathrm{M}}\mathbf{N}_{\mathrm{M}}) d\Omega \left[\theta \mathbf{U}^{n+1} + (1-\theta)\mathbf{U}^n\right] &= \int_{\partial\Omega} \mathbf{N}_{\mathrm{M}}{}^T \cdot (\boldsymbol{\sigma} \cdot \mathbf{n}) dS \\ &+ \int_{\Omega} \mathbf{N}_{\mathrm{M}}{}^T \mathbf{g} d\Omega - \alpha \int_{\Omega} (\mathbf{A}_{\mathrm{M}}\mathbf{N}_{\mathrm{M}})^T \mathbf{D}\mathbf{N}_{\mathrm{T}}(\mathbf{T}^n - T_{\mathrm{ref}})\mathbf{I} d\Omega. \end{aligned} \tag{30.21}$$

where $\boldsymbol{\sigma} \cdot \mathbf{n}$ is the boundary force vector. The above equation is simplified into

$$\begin{aligned} \int_{\Omega} \mathbf{B}_{\mathrm{M}}{}^T \mathbf{D}\mathbf{B}_{\mathrm{M}} d\Omega \left[\theta \mathbf{U}^{n+1} + (1-\theta)\mathbf{U}^n\right] &= \int_{\partial\Omega} \mathbf{N}_{\mathrm{M}}{}^T (\boldsymbol{\sigma} \cdot \mathbf{n}) dS + \int_{\Omega} \mathbf{N}_{\mathrm{M}}{}^T \mathbf{g} d\Omega \\ &- \alpha \int_{\Omega} \mathbf{B}_{\mathrm{M}}{}^T \mathbf{D}\mathbf{N}_{\mathrm{T}}(\mathbf{T}^n - T_{\mathrm{ref}})\mathbf{I} d\Omega. \end{aligned} \tag{30.22}$$

That is,

$$\mathbf{K}_{\mathrm{M}}\left[\theta \mathbf{U}^{n+1} + (1-\theta)\mathbf{U}^n\right] = \mathbf{F}_{\mathrm{M}} = \mathbf{F}_{\mathrm{s}} + \mathbf{F}_{\mathrm{b}} + \mathbf{F}_{\mathrm{T2M}}. \tag{30.23}$$

Then the solution at the current step can be obtained as follows in the explicit method,

$$\mathbf{U}^{n+1} = \mathbf{M}_{\mathrm{M}}\mathbf{U}^n + \mathbf{P}_{\mathrm{M}} = \frac{-(1-\theta)\mathbf{K}_{\mathrm{M}}}{\theta\mathbf{K}_{\mathrm{M}}}\mathbf{U}^n + \frac{\mathbf{F}_{\mathrm{s}} + \mathbf{F}_{\mathrm{b}} + \mathbf{F}_{\mathrm{T2M}}}{\theta\mathbf{K}_{\mathrm{M}}}. \tag{30.24}$$

The detailed matrices in the above equation for the calculation of the matrices of stiffness and force vectors are listed below:

$$\mathbf{A}_{\mathrm{M}} = \begin{bmatrix} \frac{\partial}{\partial x} & 0 \\ 0 & \frac{\partial}{\partial y} \\ \frac{\partial}{\partial y} & \frac{\partial}{\partial x} \end{bmatrix}$$ (for vector dependent variables in 2D, independent of the element type),

$$\mathbf{N}_{\mathrm{M}} = \begin{bmatrix} N_1 & 0 & N_2 & 0 & \cdots & N_9 & 0 \\ 0 & N_1 & 0 & N_2 & & 0 & N_9 \end{bmatrix}_{2\times 18},$$

$$\mathbf{B}_{\mathrm{M}} = \begin{bmatrix} \frac{\partial N_1}{\partial x} & & \frac{\partial N_2}{\partial y} & & & \frac{\partial N_9}{\partial x} & \\ & \frac{\partial N_1}{\partial y} & & \frac{\partial N_2}{\partial y} & \cdots & & \frac{\partial N_9}{\partial y} \\ \frac{\partial N_1}{\partial y} & \frac{\partial N_1}{\partial x} & \frac{\partial N_2}{\partial y} & \frac{\partial N_2}{\partial y} & & \frac{\partial N_9}{\partial y} & \frac{\partial N_9}{\partial x} \end{bmatrix}_{3\times 18}.$$

By solving the above coupled equation system, the numerical approximations to the unknowns, i.e., displacement and temperature, can be obtained for a typical thermomechanical problem. The stress (effective stress) will need to be calculated based on the unknowns with the following equation:

$$\begin{aligned} \boldsymbol{\sigma} &= \mathbf{D}\nabla\mathbf{u} - \alpha\mathbf{D}(\mathbf{T} - T_{\mathrm{ref}})\mathbf{I} \\ &= \mathbf{D}_{3\times 3}\mathbf{B}_{\mathrm{M}3\times 18}\mathbf{U}_{18\times 1} - \alpha\mathbf{D}_{3\times 3}\mathbf{N}_{\mathrm{T}1\times 4}(\mathbf{T} - T_{\mathrm{ref}})_{4\times 1}\mathbf{I}_{3\times 3}, \end{aligned} \tag{30.25}$$

where $\boldsymbol{\sigma}$ is the Cauchy stress tensor. But $\boldsymbol{\sigma}$ here it is represented by a vector due to the use of the Voigt notation.

Appendix: Typical Finite Elements (Q4 and Q9)

Q4

$$N_1 = \frac{1}{4}(1-\xi)(1-\eta);\ N_2 = \frac{1}{4}(1+\xi)(1-\eta);$$
$$N_3 = \frac{1}{4}(1+\xi)(1+\eta);\ N_4 = \frac{1}{4}(1-\xi)(1+\eta).$$
$$N_{1,x} = \frac{1}{2a}[-(1-\eta)];\quad N_{2,x} = \frac{1}{2a}(1-\eta);$$
$$N_{3,x} = \frac{1}{2a}(1+\eta);\quad N_{4,x} = \frac{1}{2a}[-(1+\eta)];$$

$$N_{1,y} = \frac{1}{2a}[-(1-\xi)];\ N_{2,y} = \frac{1}{2a}[-(1+\xi)];$$
$$N_{3,y} = \frac{1}{2a}(1+\xi); \qquad N_{4,y} = \frac{1}{2a}(1-\xi).$$

Q9

$$N_1 = \frac{1}{4}\xi(1-\xi)(1-\eta)\eta;\ N_2 = \frac{1}{4}\xi(1+\xi)(1-\eta)\eta;$$
$$N_3 = \frac{1}{4}\xi(1+\xi)(1+\eta)\eta;\ N_4 = \frac{1}{4}\xi(1-\xi)(1+\eta)\eta;$$
$$N_5 = -\frac{1}{2}(1-\xi)(1+\xi)(1-\eta)\eta;\ N_6 = \frac{1}{2}\xi(1+\xi)(1-\eta)(1+\eta)$$
$$N_7 = \frac{1}{2}(1-\xi)(1+\xi)(1+\eta)\eta;\ N_8 = -\frac{1}{2}\xi(1-\xi)(1-\eta)(1+\eta)$$
$$N_9 = (1-\xi)(1+\xi)(1-\eta)(1+\eta).$$
$$N_{1,x} = \frac{1}{2a}(1-2\xi)(1-\eta)\eta;\ N_{2,x} = \frac{1}{2a}(1+2\xi)(1-\eta)\eta$$
$$N_{3,x} = \frac{1}{2a}(1+2\xi)(1+\eta)\eta;\ N_{4,x} = \frac{1}{2a}(1-2\xi)(1+\eta)\eta$$
$$N_{5,x} = \frac{2}{a}\xi(1-\eta)\eta;\ N_{6,x} = \frac{1}{a}(1+2\xi)(1-\eta)(1+\eta)$$
$$N_{7,x} = -\frac{2}{a}\xi(1+\eta)\eta;\ N_{8,x} = -\frac{1}{a}(1-2\xi)(1-\eta)(1+\eta);$$
$$N_9 = -\frac{4}{a}\xi(1-\eta)(1+\eta).$$
$$N_{1,y} = \frac{1}{2b}\xi(1-\xi)(1-2\eta);\ N_{2,y} = \frac{1}{2b}\xi(1+\xi)(1-2\eta);$$
$$N_{3,y} = \frac{1}{2b}\xi(1+\xi)(1+2\eta);\ N_{4,y} = \frac{1}{2b}\xi(1-\xi)(1+2\eta);$$
$$N_{5,y} = -\frac{1}{b}(1-\xi)(1+\xi)(1-2\eta);\ N_{6,y} = -\frac{2}{b}\xi(1+\xi)\eta;$$
$$N_{7,y} = \frac{1}{b}(1-\xi)(1+\xi)(1+2\eta);\ N_{8,y} = \frac{2}{b}\xi(1-\xi)\eta;$$
$$N_9 = -\frac{4}{a}(1-\xi)(1+\xi)\eta.$$

where ξ and η are the local coordinates, while a and b are the sizes of the elements.

Practice Problem

Problem Description

Develop FEM code to solve the problem described in Chap. 28. The material constituting the square computational domain (1 m by 1 m) is water. The initial temperature is 273.15 K. The top boundary is set to 293.15 K, while the other boundaries are thermally insulated. Please calculate the temperature distribution at the end of 1 day.

Results

Figure 30.2 illustrates the temperature distribution.

Appendix: MATLAB Script File

The following MATLAB code for the FEM implementation of the thermal field problem was developed based on the MXFEM developed by Matthew Jon Pais (Pais 2011). A Q4 element was adopted. The quadratic Gauss integration was employed.

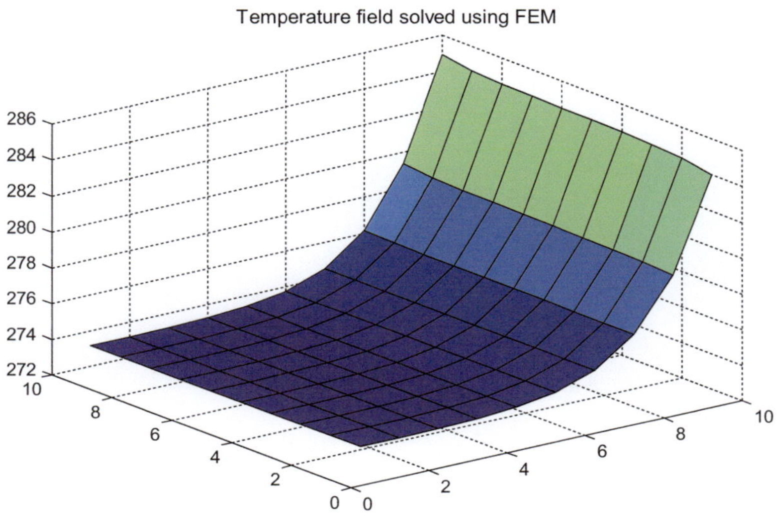

Fig. 30.2 Temperature distribution from the numerical simulation

main.m

```
% Main Program, Input, Calculation, and Plot

clear all; close all; clc; format compact; global DOMAIN PARA MAT SOL
IBC

%% Input
% PARA(1,2,3,4) = density(1), heat capacity(2), thermal conductivity
(3),
 % coefficient of convective heat transfer(4);
% SOL(1,2) = time step(1), time range of calculation(2);
% IBC = intial temperature(1),ambient temperature in K(2).

PARA  = [2700, 920, 0.75, 23];
SOL = [60,3600*24];
IBC = [273.15,293.15];

Height = 1;                % Domain Height
Width  = 1;                % Domain Width
LElem  = 1/9;              % Element Length

DOMAIN = [Width/LElem,Height/LElem,LElem,LElem];
MAT   = [15E9,0.3,20E6,0.3,2,1,Inf,30e-6];

connectivity;

%% Initialization
nXElem = DOMAIN(1);          % Number of elements in the x-direction
nYElem = DOMAIN(2);          % Number of elements in the y-direction
globalDOF_T=(nXElem+1)*(nYElem+1);
T0 = IBC(1);                 % Initial temperature
Tenv = IBC(2);              % Environment temperature
dt   =  SOL(1);
tTotal=  SOL(2);

T = sparse(1:globalDOF_T,ones(globalDOF_T,1),T0*ones
(globalDOF_T,1),globalDOF_T,tTotal/dt+1);            % Create the
solution matrix, T0 is assigned

tstep = 1;
[globalKa_T,globalK_T]=stiffnessMatrix_T;

%% Iteration
for t = dt:dt:tTotal
    tstep=tstep+1;
    globalF_T=forceVector_T(Tenv,T(:,tstep-1));
    T(:,tstep)=globalK_T*T(:,tstep-1)+globalKa_T\globalF_T;
end
```

```
%% Plot
figure(1)
T_final=reshape(T(:,end),(nXElem+1),(nYElem+1));
surf(T_final)
title('Temperature field solved using FEM')
```

connectivity.m

```
% This function defines connectivity information about the problem.
This includes defining
% node numbers for the problem, defining the XYZ coordinates of the node
numbers, creating
% a connectivity matrix, and % creating matrices storing the X and Y
coordinates for each element.

function connectivity
global CONNEC DOMAIN NODES XYZ

nXElem = DOMAIN(1);          % Number of elements in the x-direction
nYElem = DOMAIN(2);          % Number of elements in the y-direction
lXElem = DOMAIN(3);          % Elemental length in the x-direction
lYElem = DOMAIN(4);          % Elemental length in the y-direction
nNode  = (nXElem+1)*(nYElem+1);  % The number of nodes in the domain
NODES  = zeros(nNode,1);  % Initialize the matrix storing nodal
information
XYZ    = zeros(nNode,3);  % Initialize the matrix storing nodal
position

% Create global node numbering (NN), global xyz coordinates (XYZ), and
index of enriched nodes (NODES)
% XYZ = [NodeNumber,X-Coordinate,Y-Coordinate,Z-Coordinate;...]
% NODES = [NodeNumber]
nNode = 1;
NN = zeros(nXElem+1,nYElem+1);
for iYNode = 1:(nYElem+1)
    for iXNode = 1:(nXElem+1)
        NN(iXNode,iYNode) = nNode;
        XYZ(nNode,:) = [nNode (iXNode-1)*lXElem (iYNode-1)*lYElem];
        NODES(nNode,1) = nNode;
        nNode = nNode+1;
    end
end

% Create a global connectivity matrix (CONNEC) and elemental
coordinate matricies (Ex, Ey)
% CONNEC = [ElementNumber,LocalNode1,LocalNode2,LocalNode3,
LocalNode4,InclusionElementNum]
nElem = 1;
CONNEC = zeros(nXElem*nYElem,5);
```

```
for iYElem = 1:nYElem
    for iXElem = 1:nXElem
        N1 = NN(iXElem,iYElem);
        N2 = NN(iXElem+1,iYElem);
        N3 = NN(iXElem+1,iYElem+1);
        N4 = NN(iXElem,iYElem+1);
        CONNEC(nElem,1:5) = [nElem N1 N2 N3 N4];
        nElem = nElem+1;
    end
end
```

forcevector.m

```
% This function creates the global force vector from the user-supplied
input file.
% Tenv is the environment temperature, T is the solution at the
previous step

function [globalF]=forceVector_T(Tenv,T)
global DOMAIN MAT FORCE CONNEC PARA

nXElem  = DOMAIN(1);          % Number of elements in the x-direction
nYElem  = DOMAIN(2);          % Number of elements in the y-direction
le      = DOMAIN(3);          % Length of a side of an element
globalDOF_T=(nXElem+1)*(nYElem+1);
globalF = sparse(globalDOF_T,1);      % Create the global force vector

if isempty(FORCE) == 1, m = 1; else m = size(FORCE,1); end
if MAT(5) == 1, h = MAT(6); else h = 1; end

% Create the global force matrix from distributed loading

node = CONNEC((nYElem-1)*(nXElem)+1,5); % Initial node along top
edge
for i = 1:nXElem+1
    Force=PARA(4)*(Tenv-T(node));
    if (i == 1) || (i == nXElem+1)  % Convert distributed force to nodal
force
        globalF(node) = le*Force/2; % Elemental force in x-direction
for edge nodes
    else
        globalF(node) = le*Force;    % Elemental force in x-direction
    end
    node = node+1;              % Update node along top edge
end
end
```

stiffnessMatrix_T.m

```
% This function calculates the global stiffness matrix for the desired
% discontinuities defined by the user-supplied input.

function [globalKa,globalK] = stiffnessMatrix_T

global CONNEC DOMAIN NODES XYZ PARA SOL %MAT

nXElem  = DOMAIN(1);        % Number of elements in the x-direction
nYElem  = DOMAIN(2);        % Number of elements in the y-direction
lXElem  = DOMAIN(3);        % Length of elements in the x-direction
rho     =  PARA(1);
Cp      =  PARA(2);
Lambda=  PARA(3);
h       =  PARA(4);
dt      =  SOL(1);
tTotal=  SOL(2);
globalDOF_T=(nXElem+1)*(nYElem+1);
globalK   = sparse(globalDOF_T,globalDOF_T);

% Initialize the FE stiffness matrix
globalKa = sparse(globalDOF_T,globalDOF_T); % Define the global K of
the accumulation term
globalKc = sparse(globalDOF_T,globalDOF_T); % Define the global K of
the conduction term
globalKb = sparse(globalDOF_T,globalDOF_T); % Define the global K of
the boundary term

% Initialize the variables which will create the traditional sparse
matrix
nIndexT = 0;                              % Initialize index
uenrElem = nXElem*nYElem;                 % Number of elements
allRowT  = ones(uenrElem*16,1);           % Row indices
allColT  = ones(uenrElem*16,1);           % Column indices
allValTa  = zeros(uenrElem*16,1);         % Stiffness matrix values
allValTc  = zeros(uenrElem*16,1);
%allValTb  = zeros(uenrElem*16,1);

for iElem = 1:(nYElem*nXElem)
    N1  = CONNEC(iElem,2);                % Node 1 for current element
    N2  = CONNEC(iElem,3);                % Node 2 for current element
    N3  = CONNEC(iElem,4);                % Node 3 for current element
    N4  = CONNEC(iElem,5);                % Node 4 for current element
    NN  = NODES([N1 N2 N3 N4]',:);        % Nodal data for current element

    X1 = XYZ(N1,2); X2 = XYZ(N2,2); X3 = XYZ(N3,2); X4 = XYZ(N4,2); %
Nodal x-coordinates
```

```
    Y1 = XYZ(N1,3); Y2 = XYZ(N2,3); Y3 = XYZ(N3,3); Y4 = XYZ(N4,3); %
Nodal y-coordinates
    xyz = [X1 Y1;X2 Y2;X3 Y3;X4 Y4];                    % Nodal coordinate
matrix

    localKa = 0;            % Initialize stiffness for current element
    localKc = 0;            % Initialize stiffness for current element
    %localKb = 0;           % Initialize stiffness for current element
    local  = [N1 N2 N3 N4]; % Traditional index locations

    [gp,gw,J] = subDomain_T(3,xyz);
    xi        = gp(:,1);
    eta       = gp(:,2);

    for i = 1:length(gp)
        xi = gp(i,1); eta = gp(i,2);          % Gauss points
        W  = gw(i);                           % Gauss weights

        Ji  = [J(i,1) J(i,2);J(i,3) J(i,4)]; % Jacobian of subdomain
        detJ = det(Ji);                       % Determinant of the Jacobian

        N  = 1/4*[(1-xi).*(1-eta) (1+xi).*(1-eta) (1+xi).*(1+eta)
(1-xi).*(1+eta)];
        % Derivative of shape functions with respect to x
        Nx = 2/lXElem*1/4*[-(1-eta);1-eta;1+eta;-(1+eta)];
        % Derivative of shape functions with respect to y
        Ny = 2/lXElem*1/4*[-(1-xi);-(1+xi);1+xi;1-xi];

        Bu = [Nx(1)  Nx(2)  Nx(3)  Nx(4);...
              Ny(1)  Ny(2)  Ny(3)  Ny(4)];

        localKa = localKa + W*N'*(rho*Cp/dt)*N*detJ;
        localKc = localKc + W*Bu'*(-Lambda)*Bu*detJ;
        %localKb = localKb + zeros(8,8); % W*N'*(-h)*N*detJ; localKb
is not used here, but considered in forceVector.m instead

    end

    for row = 1:4
        for col = 1:4
            nIndexT = nIndexT+1;
            allRowT(nIndexT) = local(row);
            allColT(nIndexT) = local(col);
            allValTa(nIndexT) = localKa(row,col);
            allValTc(nIndexT) = localKc(row,col);
            %allValTb(nIndexT) = localKb(row,col);
        end
    end
end
```

```
globalKa = sparse(allRowT,allColT,allValTa,globalDOF_T,
globalDOF_T);
globalKc = sparse(allRowT,allColT,allValTc,globalDOF_T,
globalDOF_T);
%globalKb = sparse(allRowT,allColT,allValTb,globalDOF_T,
globalDOF_T);
%globalK  = eye(globalDOF_T,globalDOF_T)+globalKa\(globalKc
+globalKb);
globalK  = globalK+sparse(eye(globalDOF_T,globalDOF_T))+globalKa
\(globalKc);
```

subdomain_T.m

```
% This function subdivides enriched elements and determines the Gauss
points
% and weights to be used in the integration during the assembly of the
stiffness matrix.

function [gp,gw,J] = subDomain(npt,xyz)

corner = [1 2 3 4 1];
node   = [-1 -1;1 -1;1 1;-1 1];

% Find the triangles defining the subdomains
warning off MATLAB:delaunayn:DuplicateDataPoints
tri = delaunay(node(:,1),node(:,2)); % ,
{'Qt','Qbb','Qc','Qz','Q3'}

% Loop over subtriangles to get quadrature points and weights
[q,w] = gauss(npt,'QUAD');

pt = 1;
for e = 1:size(tri,1)
    coord = node(tri(e,:),:);
    xyzl  = xyz(tri(e,:),:);
    for i = 1:length(w)
        xi = q(i,1); eta = q(i,2);
        N  = [1-xi-eta;xi;eta];
        gp(pt,:) = N'*coord;
        gw(pt,1) = w(i)/2;
        J(pt,:)  = [-xyzl(1,1)+xyzl(2,1) -xyzl(1,2)+xyzl(2,2)...
                    -xyzl(1,1)+xyzl(3,1) -xyzl(1,2)+xyzl(3,2)];
        pt = pt+1;
    end
end
```

gauss.m

```
% This function contains various gauss integration points and weights
for
% quadrilateral and triangular elements.

function [Q,W] = gauss(npt,type)
if strcmp(type,'QUAD') == 1
    switch npt
        case 2
          np(1) = -0.577350269189626;
          np(2) =  0.577350269189626;

          nw(1) =  1.000000000000000;
          nw(2) =  1.000000000000000;
        case 3
          np(1) = -0.774596669241483;
          np(2) =  0.000000000000000;
          np(3) =  0.774596669241483;

          nw(1) =  0.555555555555556;
          nw(2) =  0.888888888888889;
          nw(3) =  0.555555555555556;
        case 4
          np(1) = -0.861136311594053;
          np(2) = -0.339981043584856;
          np(3) =  0.339981043584856;
          np(4) =  0.861136311594053;

          nw(1) =  0.347854845137454;
          nw(2) =  0.652145154862546;
          nw(3) =  0.652145154862546;
          nw(4) =  0.347854845137454;
        case 5
          np(1) = -0.906179845938664;
          np(2) = -0.538469310105683;
          np(3) =  0.000000000000000;
          np(4) =  0.538469310105683;
          np(5) =  0.906179845938664;

          nw(1) =  0.236926885056189;
          nw(2) =  0.478628670499366;
          nw(3) =  0.568888888888889;
          nw(4) =  0.478628670499366;
          nw(5) =  0.236926885056189;
        case 6
          np(1) = -0.932469514203152;
          np(2) = -0.661209386466265;
          np(3) = -0.238619186083197;
          np(4) =  0.238619186083197;
          np(5) =  0.661209386466265;
          np(6) =  0.932469514203152;
```

```
    nw(1) = 0.171324492379170;
    nw(2) = 0.360761573048139;
    nw(3) = 0.467913934572691;
    nw(4) = 0.467913934572691;
    nw(5) = 0.360761573048139;
    nw(6) = 0.171324492379170;
  case 7
    np(1) = -0.949107912342759;
    np(2) = -0.741531185599394;
    np(3) = -0.405845151377397;
    np(4) = 0.000000000000000;
    np(5) = 0.405845151377397;
    np(6) = 0.741531185599394;
    np(7) = 0.949109712342759;

    nw(1) = 0.129484966168870;
    nw(2) = 0.279705391489277;
    nw(3) = 0.381830050505119;
    nw(4) = 0.417959183673469;
    nw(5) = 0.381830050505119;
    nw(6) = 0.279705391489277;
    nw(7) = 0.129484966168870;
  case 8
    np(1) = -0.960289856497536;
    np(2) = -0.796666477413627;
    np(3) = -0.525532409916329;
    np(4) = -0.183434642495650;
    np(5) = 0.183434642495650;
    np(6) = 0.525532409916329;
    np(7) = 0.796666477413627;
    np(8) = 0.960289856497536;

    nw(1) = 0.101228536290376;
    nw(2) = 0.222381034453374;
    nw(3) = 0.313706645877887;
    nw(4) = 0.362683783378362;
    nw(5) = 0.362683783378362;
    nw(6) = 0.313706645877887;
    nw(7) = 0.222381034453374;
    nw(8) = 0.101228536290376;
  case 9
    np(1) = -0.968160239507626;
    np(2) = -0.836031107326636;
    np(3) = -0.613371432700590;
    np(4) = -0.324253423403809;
    np(5) = 0.000000000000000;
    np(6) = 0.324253423403809;
    np(7) = 0.613371432700590;
    np(8) = 0.836031107326636;
    np(9) = 0.968160239507626;
```

```
      nw(1) = 0.081274388361574;
      nw(2) = 0.180648160694857;
      nw(3) = 0.260610696402935;
      nw(4) = 0.312347077040003;
      nw(5) = 0.330239355001260;
      nw(6) = 0.312347077040003;
      nw(7) = 0.261610696402935;
      nw(8) = 0.180648160694857;
      nw(9) = 0.081274388361574;
    case 10
      np(1)  = -0.973906528517172;
      np(2)  = -0.865063366688985;
      np(3)  = -0.679409568299024;
      np(4)  = -0.433395394129247;
      np(5)  = -0.148874338981631;
      np(6)  =  0.148874338981631;
      np(7)  =  0.433395394129247;
      np(8)  =  0.679409568299024;
      np(9)  =  0.865063366688985;
      np(10) =  0.973906528517172;

      nw(1)  =  0.066671344308688;
      nw(2)  =  0.149451349150581;
      nw(3)  =  0.219086362515982;
      nw(4)  =  0.269266719309996;
      nw(5)  =  0.295524224714753;
      nw(6)  =  0.295524224714753;
      nw(7)  =  0.269266719309996;
      nw(8)  =  0.219086362515982;
      nw(9)  =  0.149451349150581;
      nw(10) =  0.066671344308688;
    case 12
      np(1)  = -0.981560634246719;
      np(2)  = -0.904117256370475;
      np(3)  = -0.769902674194305;
      np(4)  = -0.587317954286617;
      np(5)  = -0.367831498998180;
      np(6)  = -0.125233408511469;
      np(7)  =  0.125233408511469;
      np(8)  =  0.367831498998180;
      np(9)  =  0.587317954286617;
      np(10) =  0.769902674194305;
      np(11) =  0.904117256370475;
      np(12) =  0.981560634246719;

      nw(1)  =  0.047175336386512;
      nw(2)  =  0.106939325995318;
      nw(3)  =  0.160078328543346;
      nw(4)  =  0.203167426723066;
      nw(5)  =  0.233492536538355;
      nw(6)  =  0.249147045813403;
      nw(7)  =  0.249147045813403;
      nw(8)  =  0.233492536538355;
```

```
            nw(9)  = 0.203167426723066;
            nw(10) = 0.160078328543346;
            nw(11) = 0.106939325995318;
            nw(12) = 0.047175336386512;
    end

    % Adjust quadrature points based on dimension
    Q = NaN(npt*npt,2);
    W = NaN(npt*npt,1);
    n = 1;
    for i = 1:npt
        for j = 1:npt
            Q(n,:) = [np(i) np(j)];
            W(n)   = nw(i)*nw(j);
            n = n+1;
        end
    end
end
```

References

Abousleiman Y, Cheng A, Cui L, Detournay E, Roegiers J (1996) Mandel's problem revisited. Geotechnique 46(2):187–195

Acar YB, Alshawabkeh AN (1993) Principles of electrokinetic remediation. Environ Sci Technol 27(13):2638–2647

Allen MB III (1988) Basic mechanics of oil reservoir flows. In: Multiphase flow in porous media. Springer, New York, pp 1–86

Alshawabkeh AN, Yeung AT, Bricka MR (1999) Practical aspects of in-situ electrokinetic extraction. J Environ Eng 125(1):27–35

Anderson DM, Tice AR, McKim HL (1973) The unfrozen water and the apparent specific heat capacity of frozen soils. In: Second international conference on Permafrost, Yakutsk, USSR. North American Contribution, pp 289–295

Balanis CA (1999) Advanced engineering electromagnetics. Wiley, New York

Barrett AH, Cardello AV (2012) Military food engineering and ration technology. DEStech Publications, Inc., Lancaster

Biot MA (1941) General theory of three-dimensional consolidation. J Appl Phys 12(2):155–164

Biot MA (1956a) Theory of propagation of elastic waves in a fluid-saturated porous solid. I. Low-frequency range. J Acoust Soc Am 28(2):168–178

Biot MA (1956b) Theory of propagation of elastic waves in a fluid-saturated porous solid. II. Higher frequency range. J Acoust Soc Am 28(2):179–191

Biot M, Willis D (1957) The elastic coefficients of the theory of consolidation. J Appl Mech 24:594–601

Bjerrum L, Moum J, Eide O (1967) Application of electro-osmosis to a foundation problem in a Norwegian quick clay. Geotechnique 17(3):214–235

Buchner R, Barthel J, Stauber J (1999) The dielectric relaxation of water between 0 C and 35 C. Chem Phys Lett 306(1):57–63

Campbell GS (1974) A simple method for determining unsaturated conductivity from moisture retention data. Soil Sci 117(6):311–314

Campbell GS (1985) Soil physics with BASIC: transport models for soil-plant systems, vol 14. Elsevier, Amsterdam

Carcione JM, Morency C, Santos JE (2010) Computational poroelasticity-a review. Geophysics 75:75A229–75A243

Cary J (1965) Water flux in moist soil: thermal versus suction gradients. Soil Sci 100:168–175

Cary J (1966) Soil moisture transport due to thermal gradients: practical aspects. Soil Sci Soc Am J 30(4):428–433

Casagrande IL (1949) Electro-osmosis in soils. Geotechnique 1(3):159–177

Z. Liu, *Multiphysics in Porous Materials*,
https://doi.org/10.1007/978-3-319-93028-2

Casagrande L (1983) Stabilization of soils by means of electro-osmosis: state of the art. J Boston Soc Civil Eng 69(2):255–302
Celia MA, Binning P (1992) A mass conservative numerical solution for two-phase flow in porous media with application to unsaturated flow. Water Resour Res 28(10):2819–2828
Childs EC, Collis-George N (1950) The permeability of porous materials. In: Proceedings of the Royal Society of London A: Mathematical, Physical and Engineering Sciences, vol 1066. The Royal Society, pp 392–405
Cho J, Kim K, Chung K, Hyun S, Baek K (2009) Restoration of saline soil in cultivated land using electrokinetic process. Sep Sci Technol 44(10):2371–2384
Chung H, Kang B (1999) Lead removal from contaminated marine clay by electrokinetic soil decontamination. Eng Geol 53(2):139–150
COMSOL COMSOL example library. https://www.comsol.com/models/page/2
Corwin D, Plant R (2005) Applications of apparent soil electrical conductivity in precision agriculture. Comput Electron Agric 46(1):1–10
Courtney M (2015) Multiphysics brings the real world into simulations. E&T Magazine, vol 10
Coussy (2004) The atrium, Southern gate, Chichester, West Sussex PO19 8SQ. John Wiley & Sons Ltd, England
Cowin SC (1999) Bone poroelasticity. J Biomech 32(3):217–238
Cross M, Slone A (2005) FENet-multi-physics analysis (MPA) theme: a review of commercial MPA capability in 2005. In: Proceedings of the FENet Project Review Meeting, St. Julians, Malta, pp 119–130
Cyclopedia M. https://www.comsol.de/multiphysics
de Boer R (2000) Contemporary progress in porous media theory. Appl Mech Rev 53(12):323–370
De Vries DA (1963) Thermal properties of soils. In: Physics of plant environment. North-Holland Publishing Co., Amsterdam
Dede EM, Lee J, Nomura T (2014) Multiphysics simulation: electromechanical system applications and optimization. Springer-Verlag, London https://www.springer.com/us/book/9781447156390
Deen WM (1998) Analysis of transport phenomena. Oxford University Press, New York
Dennis MR, Glendinning P, Martin PA, Santosa F, Tanner J (2015) The Princeton companion to applied mathematics. Princeton University Press, Princeton
Derjaguin B, Landau L (1941) Theory of the stability of strongly charged lyophobic sols and of the adhesion of strongly charged particles in solutions of electrolytes. Acta Physicochim URSS 14 (6):633–662
Detournay E, Cheng A (1993) Fundamental of Poroelasticity, Chapter 5. In: Fairhurst C (ed) Comprehensive rock engineering, vol 2. Pergamon Press, Oxford
Doolittle JA, Brevik EC (2014) The use of electromagnetic induction techniques in soils studies. Geoderma 223:33–45
Doutres O, Salissou Y, Atalla N, Panneton R (2010) Evaluation of the acoustic and non-acoustic properties of sound absorbing materials using a three-microphone impedance tube. Appl Acoust 71(6):506–509
Energy2D web tool. http://energy.concord.org/energy2d/reynolds.html
Engineering X (2009) Simulation of a water heater with FLOW-3D
Esrig M, Gemeinhardt JP (1900) Electrokinetic stabilization of an illitic clay. J Soil Mech Found Div 92 (SM5, Proc Paper 490)
Fernandes A, Pouget J (2003) Analytical and numerical approaches to piezoelectric bimorph. Int J Solids Struct 40(17):4331–4352
Ferronato M, Castelletto N, Gambolati G (2010) A fully coupled 3-D mixed finite element model of Biot consolidation. J Comput Phys 229(12):4813–4830
Fischer HB (1966) Longitudinal dispersion in laboratory and natural streams.
Fisk M (2008) Simulation of induction heating in manufacturing. Thesis, Luleå Tekniska Universitet
Fluke Fluke Thermal Image. https://www.tequipment.net/fluke/thermal-imagers/

Franks F (1982) Water and Aqueous Solutions at Subzero Temperatures, Volume 7 in "Water, A Comprehensive Treatise", Springer. https://www.springer.com/series/10962
Fredlund DG, Rahardjo H (1993) Soil mechanics for unsaturated soils. Wiley, New York
Fredlund DG, Xing A (1994) Equations for the soil-water characteristic curve. Can Geotech J 31 (4):521–532
Fredlund D, Xing A, Huang S (1994) Predicting the permeability function for unsaturated soils using the soil-water characteristic curve. Can Geotech J 31(4):533–546
Frenkel J (2005) On the theory of seismic and seismoelectric phenomena in a moist soil. J Eng Mech 131(9):879–887
Gardner W (1958) Some steady-state solutions of the unsaturated moisture flow equation with application to evaporation from a water table. Soil Sci 85(4):228–232
Gassmann F (1951) Elastic waves through a packing of spheres. Geophysics 16(4):673–685
Geertsma J (1957) A remark on the analogy between thermoelasticity and the elasticity of saturated porous media. J Mech Phys Solids 6(1):13–16
Ghia U, Ghia KN, Shin C (1982) High-Re solutions for incompressible flow using the Navier-Stokes equations and a multigrid method. J Comput Phys 48(3):387–411
Gibbard J (2013) Open Hand Project Website. http://www.openhandproject.org/
Gouy M (1910) Sur la constitution de la charge électrique à la surface d'un électrolyte. Journal de Physique Théorique et Appliquée 9(1):457–468
Gray DH (1970) Electrochemical hardening of clay soils. Geotechnique 20(1):81–93
Groen D, Zasada SJ, Coveney PV (2014) Survey of multiscale and multiphysics applications and communities. Comput Sci Eng 16(2):34–43
Hackett RM (2016) Stress measures. In: Hyperelasticity primer. Springer, New York, pp 27–36
Hansson K, Šimůnek J, Mizoguchi M, Lundin L-C, Van Genuchten MT (2004) Water flow and heat transport in frozen soil. Vadose Zone J 3(2):693–704
Harlan R (1973) Analysis of coupled heat-fluid transport in partially frozen soil. Water Resour Res 9(5):1314–1323
Harlow FH, Nakayama PI (1968) Transport of turbulence energy decay rate. Report LA-3854, Los Alamos Scientific Laboratory, New Mexico
Heil M, Hazel AL, Boyle J (2008) Solvers for large-displacement fluid-structure interaction problems: segregated versus monolithic approaches. Comput Mech 43(1):91–101
Hou G, Wang J, Layton A (2012) Numerical methods for fluid-structure interaction-a review. Commun Comput Phys 12(2):337–377
Huang D, Xu Q, Cheng J, Lu X, Zhang H (2012) Electrokinetic remediation and its combined technologies for removal of organic pollutants from contaminated soils. Int J Electrochem Sci 7(5):4528–4544
Hunter RJ (2013) Zeta potential in colloid science: principles and applications, vol 2. Academic Press, London
Huweg AFS (2013) Modelling of electrokinetic phenomena in soils. University of Southern Queensland
Irwin N, Botz M, Greenkorn R (1996) Experimental investigation of characteristic length scale in periodic heterogeneous porous media. Transp Porous Media 25(2):235–246
Jame YW, Norum DI (1980) Heat and mass transfer in a freezing unsaturated porous medium. Water Resour Res 16(4):811–819
Jasak H (2006) Multi-physics simulations in continuum mechanics. In: Proceedings of 5th international conference of croatian society of mechanics, Trogir
Jasak H (1996) Error analysis and estimation for the finite volume method with applications to fluid flows. Ph.D. Thesis, Imperial College
Jeong C-H (2011) Guideline for adopting the local reaction assumption for porous absorbers in terms of random incidence absorption coefficients. Acta Acustica united with Acustica 97 (5):779–790

Kanit T, Forest S, Galliet I, Mounoury V, Jeulin D (2003) Determination of the size of the representative volume element for random composites: statistical and numerical approach. Int J Solids Struct 40(13):3647–3679

Kapoor R, Nemat-Nasser S (1998) Determination of temperature rise during high strain rate deformation. Mech Mater 27(1):1–12

Keyes DE, McInnes LC, Woodward C, Gropp W, Myra E, Pernice M, Bell J, Brown J, Clo A, Connors J (2013) Multiphysics simulations: challenges and opportunities. Int J High Perform Comput Appl 27(1):4–83

Kim S-O, Kim W-S, Kim K-W (2005) Evaluation of electrokinetic remediation of arsenic-contaminated soils. Environ Geochem Health 27(5):443–453

Kolmogorov AN (1962) A refinement of previous hypotheses concerning the local structure of turbulence in a viscous incompressible fluid at high Reynolds number. J Fluid Mech 13 (1):82–85

Koopmans RWR, Miller R (1966) Soil freezing and soil water characteristic curves. Soil Sci Soc Am J 30(6):680–685

Krishna R, Wesselingh J (1997) The Maxwell-Stefan approach to mass transfer. Chem Eng Sci 52 (6):861–911

Krzhizhanovskaya VV (2007) Sun S simulation of Multiphysics multiscale systems: introduction to the ICCS'2007 workshop. In: International conference on computational science. Springer, Berlin/Heidelberg, pp 755–761

Laboratories BT (1939) Method and apparatus for heating dielectric materials. Google Patents

Launder BE, Spalding DB (1974) The numerical computation of turbulent flows. Comput Methods Appl Mech Eng 3(2):269–289

Lautrup B (2005) Physics of continuous matter. Exotic and everyday phenomena in the macroscopic world. IOP, Bristol

LeVeque RJ (2002) Finite volume methods for hyperbolic problems, vol 31. Cambridge University Press, Cambridge, UK

Liu Z, Yu X (2011) Coupled thermo-hydro-mechanical model for porous materials under frost action: theory and implementation. Acta Geotech 6(2):51–65

Liu Z, Yu X (2013) Physically based equation for phase composition curve of frozen soils. http://www.liurg.org/documents/Physically%20Based%20Equation%20for%20Phase%20Composition%20Curve%20of%20Frozen%20Soils.pdf

Lo K, Ho K, Inculet I (1991) Field test of electroosmotic strengthening of soft sensitive clay. Can Geotech J 28(1):74–83

Logan BE (2012) Environmental transport processes. Wiley, New York

Luthin JN, Orhun A, Taylor GS (1975) Coupled saturated-unsaturated transient flow in porous media: experimental and numeric model. Water Resour Res 11(6):973–978

Mandel J (1953) Consolidation des sols (étude mathématique). Geotechnique 3(7):287–299

Manyuhina O, Shklyarevskiy I, Jonkheijm P, Christianen P, Fasolino A, Katsnelson M, Schenning A, Meijer E, Henze O, Kilbinger A (2007) Anharmonic magnetic deformation of self-assembled molecular nanocapsules. Phys Rev Lett 98(14):146101

Marc G, McMillan WG (1985) The virial theorem. In: Prigogine I, Rice SA (eds) Advances in chemical physics, vol 58. Wiley, Hoboken

Massabó M, Cianci R, Paladino O (2011) An analytical solution of the advection dispersion equation in a bounded domain and its application to laboratory experiments. J Appl Math 2011:493014

Mattson ED, Bowman RS, Lindgren ER (2002) Electrokinetic ion transport through unsaturated soil: 1. Theory, model development, and testing. J Contam Hydrol 54(1):99–120

McInnes K (1981) Thermal Conductivities of Soils from Dryland Wheat Regions of Eastern Washington. MS Thesis, Washington State University

Merxhani A (2016) An introduction to linear poroelasticity. arXiv preprint arXiv:160704274

Mitchell JK, Soga K (2005) Fundamentals of soil behavior. Wiley, New York

Moeendarbary E, Valon L, Fritzsche M, Harris AR, Moulding DA, Thrasher AJ, Stride E, Mahadevan L, Charras GT (2013) The cytoplasm of living cells behaves as a poroelastic material. Nat Mater 12(3):253–261
Moin P, Mahesh K (1998) Direct numerical simulation: a tool in turbulence research. Annu Rev Fluid Mech 30(1):539–578
Moser RD, Kim J, Mansour NN (1999) Direct numerical simulation of turbulent channel flow up to Re $\tau = 590$. Phys Fluids 11(4):943–945
Mualem Y (1976) A new model for predicting the hydraulic conductivity of unsaturated porous media. Water Resour Res 12(3):513–522
Mualem Y (1986) Hydraulic conductivity of unsaturated soils: prediction and formulas. In: Methods of soil analysis: Part 1-Physical and mineralogical methods. American Society of Agronomy, Soil Science Society of America, Madison, pp 799–823
Nepf H (1999) Drag, turbulence, and diffusion in flow through emergent vegetation. Water Resour Res 35(2):479–489
Newman GP, Wilson GW (1997) Heat and mass transfer in unsaturated soils during freezing. Can Geotech J 34(1):63–70
Noborio K, McInnes K, Heilman J (1996) Two-dimensional model for water, heat, and solute transport in furrow-irrigated soil: II. Field evaluation. Soil Sci Soc Am J 60(4):1010–1021
Noorishad J, Tsang C, Witherspoon P (1992) Theoretical and field studies of coupled hydromechanical behaviour of fractured rocks—1. Development and verification of a numerical simulator. Int J Rock Mech Mining Sci Geomech Abstracts 4. Elsevier:401–409
Ohm GS (1827) Die galvanische Kette, mathematisch bearbeitet. TH Riemann, Berlin
Overbeek JTG, Kruyt H (1952) Colloid science. In: Kruyt HR (ed) Kinetics of flocculation, vol 1. Elsevier, Amsterdam, pp 278–300
Pais M (2011), http://www.matthewpais.com/
Penta R, Ambrosi D, Shipley R (2014) Effective governing equations for poroelastic growing media. Q J Mech Appl Math 67(1):69–91
Perrin J (1904) Mécanisme de l'électrisation de contact et solutions colloidales. J Chim Phys 2:601–651
Perry RH, Chilton CH (1973) Chemical engineers' handbook. McGraw-Hill, New York
Peskin CS (1977) Numerical analysis of blood flow in the heart. J Comput Phys 25(3):220–252
Peskin CS (2002) The immersed boundary method. Acta Numerica 11:479–517
Philip J, De Vries D (1957) Moisture movement in porous materials under temperature gradients. EOS Trans Am Geophys Union 38(2):222–232
Polyanin (1997), http://www.navier-stokes-equations.com/lecturenotes;focus=CMTOI_de_dtag_hosting_hpcreator_widget_Download_14636605&path=download.action&frame=CMTOI_de_dtag_hosting_hpcreator_widget_Download_14636605?id=176396
Ponnamperuna FN, Tianco EM, Loy TA (1966) Ionic strengths of soil solutions composition variation with water content for desaturated soils. Soil Sci Soc Am J 35:436–442
Qi Y, Zhang TC (2015) Sorption and desorption of testosterone at environmentally relevant levels: effects of aquatic conditions and soil particle size fractions. J Environ Eng 142(1):04015045
Rempel A, Buffett B (1997) Formation and accumulation of gas hydrate in porous media. J Geophys Res Solid Earth 102(B5):10151–10164
Richards LA (1931) Capillary conduction of liquids through porous mediums. Physics 1 (5):318–333
Righetti PG, Van Oss CJ, Vanderhoff JW (1979) Electrokinetic separation methods. Elsevier/North-Holland Biomedical Press, Amsterdam
Roberts PJ, Webster DR (2002) Turbulent diffusion. ASCE Press, Reston
Rudnev V, Loveless D, Cook RL (2017) Handbook of induction heating. CRC Press, Boca Raton
Rudnick JW (2001) Linear Poroelasticity
Ruge JW, Stüben K (1987) Algebraic multigrid. Multigrid Methods 3(13):73–130

Rutqvist J, Börgesson L, Chijimatsu M, Kobayashi A, Jing L, Nguyen T, Noorishad J, Tsang C-F (2001) Thermohydromechanics of partially saturated geological media: governing equations and formulation of four finite element models. Int J Rock Mech Min Sci 38(1):105–127

Sagaut P (2006) Large eddy simulation for incompressible flows: an introduction. Springer Science & Business Media

Sahimi M (2011) Flow and transport in porous media and fractured rock: from classical methods to modern approaches. Wiley, Hoboken

Sawada S (1977) Temperature dependence of thermal conductivity of frozen soil. Memoirs of the Kitami Institute of Technology 9(1):111–122

Schanz M, Diebels S (2003) A comparative study of Biot's theory and the linear theory of porous media for wave propagation problems. Acta Mech 161(3):213–235

Schmitt FG (2007) About Boussinesq's turbulent viscosity hypothesis: historical remarks and a direct evaluation of its validity. Comptes Rendus Mécanique 335(9–10):617–627

Scott DM, Das DK, Subbaihaannadurai V (2006) A finite element computational method for gas hydrate. Part I: Theory. Petroleum Sci Technol 24(8):895–909

Sharma R (2012) Viscous incompressible flow simulation using penalty finite element method. In: EPJ web of conferences. EDP Sciences, p 01085

Spaans EJ, Baker JM (1996) The soil freezing characteristic: its measurement and similarity to the soil moisture characteristic. Soil Sci Soc Am J 60(1):13–19

Spalart PR (2009) Detached-eddy simulation. Annu Rev Fluid Mech 41:181–202

Stephanson O, Jing L, Tsang C-F (1997) Coupled thermo-hydro-mechanical processes of fractured media: mathematical and experimental studies, vol 79. Elsevier, Amsterdam https://www.elsevier.com/books/coupled-thermo-hydromechanical-processes-of-fractured-media/stephanson/978-0-444-82545-2

Stern O (1924) The theory of the electrolytic double-layer. Z Elektrochem 30(508):1014–1020

Subramaniyan AK, Sun C (2008) Continuum interpretation of virial stress in molecular simulations. Int J Solids Struct 45(14):4340–4346

Sullivan ME (2008) Remediation of arsenic and persistent organic contaminants using enhanced in-situ methods. Colorado State University, Fort Collins

Swenson RJ (1983) Comments on virial theorems for bounded systems. Am J Phys 51(10):940–942

Taber LA (2004) Nonlinear theory of elasticity: applications in biomechanics. World Scientific, Singapore

Tamagnini C, Jommi C, Cattaneo F (2010) A model for coupled electro-hydro-mechanical processes in fine grained soils accounting for gas generation and transport. An Acad Bras Cienc 82 (1):169–193

Terzaghi K (1925) Principles of soil mechanics, IV-settlement and consolidation of clay. Eng News-Record 95(3):874–878

Thilmany J (2010) Multiphysics: all at once. Mech Eng 132(2):39

Thomas H, He Y (1995) Analysis of coupled heat, moisture and air transfer in a deformable unsaturated soil. Geotechnique 45(4):677–689

Thomas H, He Y (1997) A coupled heat-moisture transfer theory for deformable unsaturated soil and its algorithmic implementation. Int J Numer Methods Eng 40(18):3421–3441

Thomas HR, King SD (1991) Coupled temperature/capillary potential variations in unsaturated soil. J Eng Mech 117(11):2475–2491

Thomas HR, Cleall PJ, Li Y, Harris C, Kern-Luetschg M (2009) Modelling of cryogenic processes in permafrost and seasonally frozen soils. Geotechnique 59(3):173–184

Toro EF (2013) Riemann solvers and numerical methods for fluid dynamics: a practical introduction. Springer Science & Business Media, Berlin/Heidelberg

Tsai D (1979) The virial theorem and stress calculation in molecular dynamics. J Chem Phys 70 (3):1375–1382

Van Brakel J, Heertjes P (1974) Analysis of diffusion in macroporous media in terms of a porosity, a tortuosity and a constrictivity factor. Int J Heat Mass Transf 17(9):1093–1103 van Dalen KN (2013) Multi-component acoustic characterization of porous media. Springer Science & Business Media

Van Genuchten MT (1980) A closed-form equation for predicting the hydraulic conductivity of unsaturated soils. Soil Sci Soc Am J 44(5):892–898

Verruijt A (1969) In: De Wiest RJM (ed) Elastic storage of aquifers in flow through porous media. Academic Press, New York

Verruijt A (2013) Theory and problems of poroelasticity. Delft University of Technology, Delft

Verwey E (1947) Theory of the stability of lyophobic colloids. J Phys Chem 51(3):631–636

Virkutyte J, Sillanpää M, Latostenmaa P (2002) Electrokinetic soil remediation-critical overview. Sci Total Environ 289(1):97–121

von Smoluchowski M (1903) Contribution to the theory of electro-osmosis and related phenomena. Bulletin International de l'Académie des Sciences de Cracovie 3:184–199

Wang HF (2000) Theory of linear poroelasticity. Princeton University Press, Princeton

Watanabe K, Flury M (2008) Capillary bundle model of hydraulic conductivity for frozen soil. Water Resour Res 44(12):1–9

Weinberg S (1977) The search for unity: notes for a history of quantum field theory. Daedalus 2:17–35

Wesseling P, Oosterlee CW (2001) Geometric multigrid with applications to computational fluid dynamics. J Comput Appl Math 128(1):311–334

Wikipedia Multiphysics. https://en.wikipedia.org/wiki/Multiphysics

Wilcox DC (1998) Turbulence modeling for CFD, vol 2. DCW Industries, La Canada

Yakhot V, Orszag SA (1986) Renormalization group analysis of turbulence. I. Basic theory. J Sci Comput 1(1):3–51

Yakhot V, Orszag S, Thangam S, Gatski T, Speziale C (1992) Development of turbulence models for shear flows by a double expansion technique. Phys Fluids A Fluid Dynamics 4 (7):1510–1520

Yuan J, Hicks M (2013) Large deformation elastic electro-osmosis consolidation of clays. Comput Geotech 54:60–68

Zhao C, Yang C (2012) Advances in electrokinetics and their applications in micro/nano fluidics. Microfluid Nanofluid 13(2):179–203

Index

Z. Liu, *Multiphysics in Porous Materials*,
https://doi.org/10.1007/978-3-319-93028-2

U

V

MIX
Papier aus verantwortungsvollen Quellen
Paper from responsible sources
FSC® C105338

If you have any concerns about our products,
you can contact us on
ProductSafety@springernature.com

In case Publisher is established outside the EU,
the EU authorized representative is:
Springer Nature Customer Service Center GmbH
Europaplatz 3, 69115 Heidelberg, Germany

Printed by Libri Plureos GmbH
in Hamburg, Germany